Global Ethics and Environment

As global capitalism expands and reaches ever-further corners of the world, practical problems continue to escalate and repercussions become increasingly serious and irreversible. These practical problems carry with them equally important ethical issues.

Global Ethics and Environment explores these ethical issues from a range of perspectives and using a wide range of case studies. Chapters focus on: the impact of development in new industrial regions; the ethical relationship between human and non-human nature; the application of ethics in different cultural and institutional contexts; environmental injustice in the location of hazardous materials and processes; the ethics of the impact of a single event (Chernobyl) on the global community; the ethics of transnational institutions.

This collection will both stimulate debate and provide an excellent resource for wide-ranging case study material and solid academic context.

Nicholas Low is an Associate Professor in the Faculty of Architecture, Building and Planning at the University of Melbourne. His book, *Justice, Society and Nature* (1998, Routledge), won the Harold and Margaret Sprout Award of the International Studies Association (USA) for the best book on ecological politics in 1998.

Royalties from this book will go to the Aboriginal Peoples of Australia.

Global Ethics and Environment

Edited by Nicholas Low

London and New York

First published 1999 by Routledge
2 Park Square, Milton Park, Abingdon, Oxon, OX14 4RN

Simultaneously published in the USA and Canada
by Routledge
270 Madison Ave, New York NY 10016

Routledge is an imprint of the Taylor & Francis Group

Transferred to Digital Printing 2008

© editorial and selection 1999 Nicholas Low, individual chapters
the contributors

Typeset in Galliard by
BC Typesetting, Bristol

British Library Cataloguing in Publication Data
A catalogue record for this book is available from the British Library

Library of Congress Cataloging in Publication Data
A catalogue record for this book has been requested

ISBN 0–415–19735–X (hbk)
ISBN 0–415–19736–8 (pbk)

Publisher's Note
The publisher has gone to great lengths to ensure the quality of this reprint
but points out that some imperfections in the original may be apparent.

This book is dedicated to Australia's Indigenous Peoples, in respect for their contribution to knowledge, in acknowledgement of grave injustice done to them in the course of European settlement, and in hope of future reconciliation.

Contents

Figures and tables

Figures

Tables

Contributors

Elmar Altvater is Professor of Political Economy at the Department of Political and Social Sciences of the Free University of Berlin, Germany. He is the author of several books, among them: *The Future of the Market, Grenzen der Globalisierung* (together with Birgit Mahnkopf) and *Kapital.doc*, a computerised comment on Marx's *Capital*.

Robert Bullard has written widely on and campaigned for environmental justice. He is Professor and Director of the Environmental Justice Resource Center, Clark Atlanta University, Atlanta, Georgia, USA. His books include *Dumping in Dixie: Race, Class, and Environmental Quality*, and *Unequal Protection: Environmental Justice and Communities of Color*.

John S. Dryzek is Professor of Political Science at the University of Melbourne. His books include *Discursive democracy: politics, policy, and political science, Rational ecology: environment and political economy, Democracy in Capitalist Times: Ideals, Limits, and Struggles* and *The Politics of the Earth: Environmental Discourses*.

Henrietta Fourmile is a direct descendant of the Yidindji Nations, traditional owners of the lands and waters of Cairns, Australia. She is Senior Research Fellow with the Centre for Indigenous History and the Arts at the University of Western Australia and recently completed two years with the United Nations Secretariat for the Convention on Biological Diversity.

Clive Hamilton is Executive Director of the Australia Institute, a public policy research centre in Canberra, Australia. He also teaches at the Australian National University. He has previously served as a senior public servant and an international environmental policy adviser.

David Harvey is Professor of Geography in The Johns Hopkins University, Baltimore, USA. He is author of *Social Justice and the City*, the recently reissued *Limits to Capital*, and the *Condition of Postmodernity*. His most recent book is *Justice, Nature and the Geography of Difference*, and a new work entitled *Spaces of Hope* is due out from Edinburgh University Press in the Spring of 2000.

Nicholas Low is Associate Professor in environmental planning at the University of Melbourne, Australia (Faculty of Architecture). His recent books (with Brendan Gleeson) include *Justice, Society and Nature*, *Consuming Cities*, and *Australian Planning*.

Arne Naess is a philosopher, environmentalist and mountaineer, winner of the Sonning Prize for contribution to European culture, the Mahatma Ghandi Prize, the Mountain Tradition Award, the Nordic Prize and the medal of the Presidency of the Italian Republic. He was founding editor of the journal *Inquiry*. His books include *Demoracy, Ideology and Objectivity*, *Communication and Argument*, *Ghandi and Group Conflict*, *Ecology, Community and Lifestyle*, and *Life Philosophy*.

Val Plumwood is an ecofeminist scholar who holds an Australian Research Fellowship at the University of Sydney. She is the author of *Feminism and the Mastery of Nature* and numerous papers in feminist, environmental and political philosophy. A forest activist, forest dweller and bushwalker.

Tom Regan is University Alumni Distinguished Professor at North Carolina State University, where he has taught philosophy for more than thirty years. Among his many books are *The Case for Animal Rights*, *Bloomsbury's Prophet: G.E. Moore and the Development of His Moral Philosophy*, *The Thee Generation: Reflections on the Coming Revolution*, and *Ivory Towers Should Not a Prison Make: Animal Rights, Activism, and The Academy*.

Deborah Bird Rose is a Senior Research Fellow at The Australian National University. She is the author of *Nourishing Terrains*, *Australian Aboriginal Views of Landscape and Wilderness*, *Dingo Makes Us Human* (winner of the 1992/3 Stanner Prize), and *Hidden Histories* (winner of the 1991 Jessie Litchfield Award). She has worked with Aboriginal claimants on land claims and in land disputes, and has worked with the Aboriginal Land Commissioner as his consulting anthropologist.

Vandana Shiva is a physicist and founder and Director of the Research Foundation for Science, Technology and Ecology. She is also a founding board member of the International Forum on Globalization. She has received many awards including the Alternative Nobel Prize in 1993. Her books include *The Seed Keepers*, *Biodiversity*, *Closer to Home: Women Reconnect Ecology, Health and Development Worldwide*, and *Ecofeminism*.

Kristin Shrader-Frechette is Alfred C. DeCrane Professor of Philosophy and Concurrent Professor of Biological Sciences at Notre Dame University, Indiana, USA. She has served on committees of the US National Academy of Sciences. She is Past-President of the Risk Analysis and Policy Association, and President of the International Society for Environmental Ethics.

Peter Singer is Ira W. DeCamp Professor of Bioethics in the University Center for Human Values at Princeton University. He was founding director of the

Centre for Human Bioethics at Monash University, Melbourne, Australia and foundation president of the International Association of Bioethics. His books include *Democracy and Disobedience, Animal Liberation, The Expanding Circle, Should the Baby Live?* (with Helga Kuhse) and *Ethics into Action.*

Karen J. Warren is Professor of Philosophy at Macalester College in St. Paul, Minnesota, USA. Her primary areas of scholarly interest are feminist philosophy, environmental philosophy, and critical thinking. She is the author of many refereed or invited articles, editor or co-editor of five books, including *Ecofeminism: Women, Culture, Nature,* and *Ecological Feminism,* and recipient of a first place award for the video *Thinking Out Loud: Teaching Critical Thinking Skills.*

Oran R. Young is Professor of Environmental Studies and Director of the Institute on International Environmental Governance, Dartmouth College, New Hampshire. His books include *Arctic Politics: Conflict and Cooperation in the Circumpolar North, Natural Resources and the State: The Political Economy of Resource Management,* and *Resource Regimes: Natural Resources and Social Institutions.*

1 Introduction

Towards global ethics

Nicholas Low

Introduction

'Global ecology' and 'global economy' are frameworks for describing what is happening in our world. But we have no guide to public action without 'global ethics'. Ethics tells us what we should do, how we should act. When we act together as a community or a society, then we need *political* ethics, as Aristotle explained (1976 edn). Since the 1970s our growing consciousness of the eco-logical crisis has confronted us with the need for global as well as local political action at the intersection between ecology and economy. This kind of action demands an ethic of the public sphere – the political, as well as the personal.[1]

The dominant ecopolitical ideas which emerged from environmentalism in the latter part of the twentieth century need to be expanded. The quest for 'ecologically sustainable development' has revealed a host of conflicting interests and demands whose resolution requires a conception of environmental justice – not least among them the conflict between human interests and those of the rest of nature. The slogan 'think globally, act locally' suggests that action can be mounted effectively *without* changing our global institutions, yet those institutions are already in transformation as a result of global ecology and global economy. They must be changed, and will be changed in the twenty-first century.

The Rio Declaration which came out of the United Nations Conference on Environment and Development in 1992 provided a broad set of political prin-ciples which are intended to guide the world towards ecological sustainability and social justice. The accompanying 'Agenda 21' identified, in general terms, policies which could implement those principles. However, to move beyond mere rhetoric means deciding between those industries and activities that are sustainable and those that are not. Such a decision cannot be left to individual transactions and personal morality. In a world of uneven consumption patterns, sustainable development raises major questions of international distribution. Deciding such matters inevitably raises the prospect of changing both the use and the allocation of social and ecological resources. Any fundamental change to resource allocation will have social distributional consequences, and the issue of justice therefore becomes a critical element of any formulation of sustainability. If ecological sustainability demands among other things the altering of human social development in order to secure its ecological footing, then this change will impinge upon the social distribution of environmental well-being both within and between nations.

There has been criticism of Western-style 'modernity' – of the social relations of the family, of the objectivity of science, of the emancipatory potential of the democratic state, of the idea of progress, and the universality of ethics (see Lyotard, 1984; Jänicke, 1990; Feyerabend, 1993; Jagger, 1980; Fraser, 1987; Bauman, 1993, 1995). Some critiques challenge not only existing institutions but also the basis for their social transformation. Yet not to intervene in a continuing process of transformation, to retreat from whatever is going on politically, is itself a political act. The existing system of competing national states and multinational corporations organized on the basis of market competition is patently inadequate to the purposes of conservation, sustainability and, ultimately, species survival.

A conference entitled 'Environmental Justice: Global Ethics for the 21st Century' was convened in 1997 at the University of Melbourne to explore these matters. The aim was to discuss the questions which arise from the idea of global ethics and to stimulate what Hannah Arendt (1977) called 'enlarged thought' on the subject. The agenda of the conference was to bring together people thinking about 'environmental ethics' – a vast field – with those thinking about political economy and society: global ecology and global economy; and also to attempt some integration of the urban and the non-urban dimensions of the ecological problem. We set out to attract scholars who were already making forays across these categorical boundaries. For all sorts of obvious and unavoidable reasons, given the limited capacity of a three-day event and the limited knowledge of its organizers, it has become unhappily obvious since the conference that many who could have made an important contribution did not attend. This book is published in the hope that further connections may be made and the debate advanced. The conference could not be, nor did it aim to be, definitive. Nevertheless, it marked a small step forward along what Arne Naess calls the 'long frontier' – the diversity of conceptions, knowledge and perspectives among those who seem to be heading in broadly the same direction.

This book is a record of the ideas offered by speakers at plenary sessions of the conference. The chapters which follow are based on their papers. The original papers have been revised and edited in the light of discussions which took place during the conference and subsequently. The order of presentation has been reshaped for an edited collection of essays for a wide readership, rather than being simply a conference 'proceedings' (more output from the conference can be found on the 'proceedings' website (http://www.arbld. unimelb.edu.au/envjust/papers/papers.html), in Low *et al.*, 1999, and in Gleeson and Low, forthcoming).

This book is structured as follows. Following this chapter is Arne Naess's introduction to the conference and overview of the field (Chapter 2). Nothing further needs to be said here by way of introduction about this essay from one of the world's leading philosophers of ecology, as ever not shrinking from the moral dilemmas and conflicts which lie ahead of us. Naess vividly confronts us with the problem of dealing fairly with our non-human neighbours on the planet. There follows a group of chapters which consider specific issues in which the idea of environmental justice is confronted: *prima facie* cases of environmental *injustice*. Next come a group of chapters which examine some of the underlying 'issues of principle' in environmental ethics. Finally there are four chapters reflecting on aspects of our global institutions and the ethical system they embody.

Environmental justice challenges

Part I of the book, 'Environmental Justice Challenges', contains two pairs of chapters, each with a distinctly different emphasis but each discussing actual events, issues and political struggles in which the question of justice arises through the experience of injustice. The first pair, the chapters of Bullard and Shiva respectively, is primarily about the (unjust) distribution of good and bad local environments. Both authors are activists and leaders in environmental struggles. The second pair, by Shrader-Frechette and Hamilton from a slightly less politically engaged vantage point, concerns justice issues of transnational and global scope.

Environmental justice, as such, was first articulated politically in the United States in both a creative leap and a challenge to the environment movement – which was also an invitation to make common cause. As Robert Bullard observes (Chapter 3), 'hazardous wastes and dirty industries follow the path of least resistance'. Such paths are constructed in societies by economic relations and the not-necessarily-intended and cumulative effects of political decisions. The path almost always leads to the neighbourhoods of the poor, the disenfranchised and excluded who, in the United States, are often also people of colour: Native Americans, African Americans and Latinos. The struggle for environmental justice developed out of the struggle for civil rights and against all forms of racism, and thus against *environmental* racism.

The idea of 'racism' gains meaning from the liberal ideal that every person is intrinsically of equal worth. This is a universal aspiration of the dominant political-economic culture, which also powers many different political movements, including socialism and feminism (as a movement) as well as the civil rights movement, in *opposition* to that culture. These movements insist that the dominant culture fails to live up to its own ideals. Here, then, we encounter in a concrete sense one of the major themes discussed in this book: the relationship between the universal and the particular in the ethics of the environment. Vandana Shiva (Chapter 4) argues that what is currently portrayed as the universal *interest* in the rhetoric of globalization is not: 'The global', she says, 'in this sense does not represent the universal human interest; it represents a particular *local* and *parochial* interest and culture, which has been globalized through its reach and control, its irresponsibility and lack of reciprocity' (p. 48). The global interest in the free-market economy and its institutional paraphernalia, the particular interpretation of freedom and equality is what Shiva targets here, not the basic idea of human freedom and equality.

The chapter by Shiva contains some interesting similarities to that of Bullard, as well as some important differences. Both describe 'situated' struggles – though arising within quite different political-cultural contexts: India and the United States. Both describe 'bottom-up' processes and local mobilization. Both look to local cultural and political traditions to find rhetorical ammunition for the fight. Both report successes through court action, and from that perhaps we may conclude that the idea of justice is still central to an independent judiciary. Both describe how action initiated at grassroots level has resulted in change at the institutional level, as it must in order to be effective. Both also invoke a range of broader ethical ideas connecting humans and the rest of nature, though these are not fully worked through in more abstract and rigorous terms.

As to the differences, the American 'environmental justice' movement contains an immanent critique of environmentalism as well as of American democracy as practised. The argument is that environmental laws have been used unfairly to create those 'paths of least resistance' and, implicitly I think, that the pluralist politics of the environment unfairly advantages the well off (environmentally as in other material ways). However, the system, which could be a fair one, is, according to Bullard, 'broken and needs to be fixed'. 'Human rights' is perhaps the dominant key in which the American ethical score is written.

Shiva, from a vantage point at the receiving end of environmental risks, is much less sanguine about the system. The system here is not just a national political economy, but a global one. And, for Shiva, this system needs to be radically changed, not just fixed, and changed in a way demanded by environmentalism. Perhaps the actual ethical aspirations to justice of a global economy centring on the United States are much less clear and explicit than those that Bullard supposes to hold within that nation. Indeed, this imbalance, or 'global apartheid', has been noted elsewhere (see, for example, Falk, 1995). Here it is precisely the lack of universality (of equal freedom) that is the problem. The ethical standards applied, albeit imperfectly, within North America do not apply to the rest of the world. On a more trivial level, Bullard describes a predominantly urban politics whereas Shiva spans the urban and rural, but this distinction is less important, I suspect, than their shared primary concern with 'the human', rather than with 'nature'. Nature is certainly invoked but as one with the exploited and oppressed. The possibility of the interests of humans and non-human nature diverging is not addressed.

The second pair of essays deal with a different kind of environmental problem, the sort of problem that does not create 'paths of least resistance' – or, if there are pathways, they are not directly affected by political-economic resistances. They are the environmental risks without frontiers that Ulrich Beck (1992) believed to be increasingly typical of our modern world. In fact, even these risks are not so undiscriminating. The primary impact of Chernobyl was felt by a nation afflicted by poverty. Global warming will impinge most drastically and soonest upon the poorest peoples of the Earth. But still, the impacts are indeed vastly widespread. Kristin Shrader-Frechette (Chapter 5) points out, with well-supported arguments, that half a million people – half of them outside the former Soviet Union – can be expected to die a premature death from cancer as a result of the single nuclear accident at Chernobyl. Clive Hamilton (Chapter 6) observes that some countries whose economies can best afford to adapt to a regime of reduced output of greenhouse gases are doing the least to induce adaptation. Of course, the small apparent adaptation (apparent because we do not know if even that will be complied with) currently agreed under the Climate Convention falls far short of what the vast consensus among disinterested climatologists insists is necessary to avert major climate change with all its terrible consequences.

These are injustices of a different order. It is not that these events have no distributional effects, nor even that they do not discriminate in the usual way between rich and poor, white and non-white, but that both the temporal and the spatial reach of the effects are massively greater. They bring into sharp focus the question of intergenerational justice: why should we be allowed to burden our successors with genetic damage, and with the all-encompassing environmental damage of climate change? This is a political question, a question of governance, on which the current dogma of rule by the *deregulated* market has little of moral value to contribute.

What is problematized in both Shrader-Frechette's and Hamilton's chapters is the distortion which occurs when utilitarianism is shorn of the ethical underpinnings of rights and needs. Shrader-Frechette points out that the 'justification principle' (adopted by the main international agency responsible for radiation protection) allows for damage to individuals by radiation to be offset by benefits to society in general. This appears to violate the right of the individual to environmental safety. It is not at all clear where the dividing line exists between the violation of a person's body by radiation 'in the public interest' and violation by imprisonment or torture, 'in the public interest' – or indeed whether there is or should be such a dividing line at all. Hamilton questions the adequacy of international scrutiny of arguments about the justice of distribution of costs of compliance with the global climate regime. Such little public scrutiny as there is appears to be subordinated to political considerations, as evidenced by the ease with which Australia was able to back its 'special case' at Kyoto with what appear to be spurious and tendentious arguments.

All that utilitarian microeconomics can say about such questions is that these are matters which have to be decided by political choice. But at the same time the neo-liberal ideology, which usually comes in the same package with micro-economics, expresses profound distrust of political choice and seeks to restrict it wherever possible, or to reduce it to an economic logic, and to shrink the public sphere and its ethics. Thus Hamilton notes the comments of the Australian government's chief economic adviser on climate change that it may be more efficient to evacuate small island states subject to inundation than to require industrialized countries like Australia to reduce their emissions. Evidently there are issues of global governance here whose resolution requires some kind of global ethic.

The chapters in the first part of the book raise the questions of environmental justice and global ethics in a concrete and practical sense. Conceptions of justice, however, are complex structures of ideas involving different ways of recognizing rights, deserts and needs, in which varying emphasis is given to each. On the face of it there seems to be a multiplicity of ways of conceiving of justice. The second part contains chapters which deepen the debates conceptually and explore this multiplicity.

Issues of principle

David Harvey (Chapter 7) captures something of the urgency and difficulty of our ethical predicament. On the one hand there is some evidence, he says, that the environmental justice movement in the United States is failing to realize its political potential;[2] on the other, there is more and more academic talk about the meaning of environmental justice. All the talk discloses a vast range of conceptual interpretations of environmental ethics and justice, but the words do not lead to more and better action: the need for action seems to require a focus around a single universally applicable interpretation. Thus Harvey says, 'Of course a universal environmental ethic is impossible – and of course it is desirable!' (p. 109).

To reinforce his first point, Harvey sets out some of the conceptual dimensions of environmental justice and their axes of difference: ecocentric versus anthropocentric, individualist versus communitarian, culturally embedded or universal, materialist or spiritual and many others. Harvey's list is by no means exhaustive but it illustrates the problem, namely that it is easier to find common ground in the shared experience of *injustice* than it is to move towards a shared understanding of *justice*. At the same time, the most reasonable way to approach the developing environmental crisis is, in Harvey's view, to regard ourselves as 'active agents caught within the web of life' and with the ability to change our political, social and (ultimately) ecological circumstances (at least those of our own making). Given that perspective, we need at least to find some kind of shared understanding of environmental justice as a guide to action.

Who, however, is 'we'? Here there seems to be a paradox. Environmental justice needs thinking out, otherwise public action will be arbitrary. But the more we think about 'justice', the more we may think ourselves apart. In a sense, perhaps, academics who are supposed to be professional thinkers are the extreme case here, so if academics can be induced to reach at least a common understanding, the means of doing so can perhaps serve as wider precedent. If we acknowledge that we each bring a particular conceptual paradigm to the analysis of our circumstances and what to do about them, we might be able to find common ground by a process of 'translation'. A conceptual paradigm is not unlike a language. Harvey draws on the work of James Boyd White (1990), for whom 'translation' 'is a word for a set of practices by which we learn to live with difference, with the fluidity of culture and with the instability of the self. It is not simply an operation of mind on material, but a way of being oneself in relation to another being' (quoted by Harvey, this volume, p. 118). This is an important idea which resonates with those of other contributors to this book.

Karen Warren (Chapter 8) also wants to reconcile universalism with contextualism. The issue is posed as one of 'abstract, impartial, absolute, universal perspectives versus concrete, local, historically specific, contextual perspectives' (p. 132). The issue is *between* these two world views rather than between perspectives *within* each; that is, between different absolutes or between different contextually situated perspectives. As with Harvey, Warren's question is 'What to do?' Like Harvey, Warren's ethic is oriented to praxis. She adopts a particular view of what it means to have or to abide by universal principles. She locates the groundwork of all ethics in 'care', and relates the application of ethical principles to their contribution to care: how they facilitate 'care practices'. Thus one may choose one's ethical principle (Kantian, utilitarian, feminist etc.), as 'fruits from a bowl' to suit a particular situation.

The idea of 'situated universalism', meaning that universals emerge as pragmatic guidelines (which therefore have to be consistently applied) for action from different cultural contexts, is not far from Harvey's position. Harvey might well also agree with Warren that 'caring' is an innate aspect of humanity, and that 'moral reasoning', or 'thinking with emotional intelligence', is by definition related to 'caring'. Caring, or morality, is nothing less than part of the biological foundation of our 'species being' (to translate into Marxian terminology). Isn't the principle of 'care' very similar to what Dworkin talks about as 'concern and respect' underlying his 'egalitarian plateau'?

> Government must treat those whom it governs with concern, that is, as human beings who are capable of suffering and frustration, and with respect, that is, as human beings who are capable of forming and acting on intelligent conceptions of how their lives should be lived.
>
> (Dworkin, 1977: 272)

According to Dworkin, on this 'plateau' 'all serious modern political theories find themselves' (Achterberg, 1997: 3). Warren, though, would not draw the line (if lines are to be drawn at all) just at *human* beings.

However, choosing moral principles as fruits from a bowl is rather a different matter. How do we know which fruit to choose? Warren's perspective recalls Walzer's 'spheres' of justice (Walzer, 1983; or perhaps Elster, 1992) in so far as different principles tend to emerge to embody 'care' in different departments of human activity. But for Warren the focus is rather more on the individual. Care practices are practices which 'maintain, promote, or enhance the well-being of relevant parties, or do not cause unnecessary harm to the well-being of those parties' (p. 139). Saying this may not help in cases of moral conflict but at least it eliminates the sort of repugnant moral reasoning in which the welfare of individuals is sacrificed to the principle, such as, for example, in the case of the harmless-sounding 'justification principle' for measuring 'allowable' radiation exposure (see above), or in the case of other utilitarian justifications of grave individual or social harms (e.g. inundation and evacuation of island nations in the name of efficiency). Warren argues that moral conflicts, which result from caring for different people whose interests (in well-being) conflict, cannot be avoided. The 'choice of fruits from a bowl' could suggest that moral choices are simply preferences (recall the 'apples and oranges' beloved of micro-economics!), but I understand Warren to mean (with reference to Benhabib, 1992) that a particular situation makes moral demands on us, including the need to negotiate with other people, which require us to balance a variety of moral principles, the right mixture being determined finally with reference to our emotional capacity for care and intellectual capacity for judgement.

The next four chapters, by Peter Singer, Tom Regan, Deborah Rose and Val Plumwood, take the question of difference into less familiar territory: away from the exclusively human and exclusively Euro-American. Singer and Plumwood define the core of the debate, while Regan and Rose elaborate different aspects of it.

Singer's position is well known but in a restatement of it he addresses his critics (Chapter 9). For Singer, as an ethical utilitarian (not to be confused with the subethical utilitarianism of microeconomics), the capacity to suffer pain is the key to moral considerability. He defends his position against both the traditional 'speciesist' ethic of the dominant Western tradition and against criticisms of the 'deep ecologists' that an 'animal liberation' perspective does not go far enough. In staking out his own ground he positions himself in relation to Regan, Plumwood and the humanist French philosopher Luc Ferry. The question is where one should draw a moral boundary in the natural world and on what grounds. Against Ferry (1995), Singer finds that a non-speciesist ethic is consistent with and is indeed to be found at the heart of enlightened humanism. Classical utilitarians such as John Stuart Mill and Jeremy Bentham entertained the idea of 'rights' but found it to be derivative of the capacity to suffer. Bentham explicitly included animals capable of suffering as within the same moral territory as humans. Singer denies that he wants to draw the line at 'the higher animals' (extending it to include, for example, fish, octopus and crustaceans – which 'should be given the benefit of the doubt') but insists on 'sentience' (experience) as the key. Thus trees and vegetable life are on the other side of the moral boundary because an ethic cannot be based on wrongs done to 'beings who are unable to experience in any way the wrong done to them, or any consequences of those wrongs' (p. 146).

The tendency to make such moral distinctions and to define moral boundaries is the principal subject of Plumwood's critique (Chapter 12). Moral argument in which distinctions appear is not neutral or innocent in creating power relations. When we make a moral boundary, Plumwood argues, we also advocate a power relationship in which those beyond the boundary are radically excluded from the kind of behaviour towards others considered right within it. Thus 'Others' are created which may be treated to exploitation (i.e. instrumentally) for the benefit of those within the pale. Singer refers to such others in earlier dualist conceptions: barbarians, slaves, women for example. Plumwood argues that, rather than extend the boundary, we should stop thinking in terms of boundaries.

There are both ethical and prudential reasons for such a step. Plumwood's ethical reasoning is grounded in a critique of the real structure of society and economy. The dualistic ethical structures of modern society (creating insiders and outsiders) help to channel the planet's ecological wealth to an ever smaller category of beings. Thus Plumwood argues that 'moral dualism organizes moral concepts so that they apply in an all-or-nothing way: for example, a being either has a full-blown "right" to equal treatment with humans, or it is not subject to any form of ethical consideration at all' (p. 191). But, says Plumwood, 'We have many opportunities to organize the ethical field differently' (*ibid.*). Thus 'care' 'can be applied to humans and also to non-human animals and nature more generally'. And 'ethically relevant qualities such as mind, communication, consciousness and sensitivity to others are organized in multiple and diverse ways across life forms that do not correspond to the all-or-nothing scenarios assumed by moral dualism' (*ibid.*).

Plumwood presents a powerful set of arguments explaining the ways in which moral dualism has served to entrench elite human interests by excluding and marginalizing subaltern groups and the non-human, and defining the benchmark against which others are measured (recalling the colonial paradigm). She then goes on to challenge first Singer's utilitarian position, then Regan's ethic of rights on the basis that they do not go nearly far enough in redirecting moral philosophy to give appropriate recognition to the non-human natural world. Plumwood's analysis introduces the fundamental question of power differentials and how power plays over moral landscapes – in a way which both recalls and extends the structural critiques of Karl Marx.

The chapters of Regan and Rose have a narrower, though no less compelling, focus. After positioning himself in relation to both Plumwood and Singer, Regan (Chapter 10) engages a particular opponent of 'animal rights', Carl Cohen (1986, 1997), whose Enlightenment philosophical tradition he,

Regan, shares. Cohen argues that the considerable harm done to animals in laboratory experiments is vastly (incalculably) outweighed by the good these experiments have done in improving human life. Here we have the utilitarian argument, which appears in the 'justification for radiation exposure' (as above), that social good may compensate for individual harm. Regan guides us through a refutation, showing precisely where and why Cohen's theory is inconsistent with Cohen's own underlying assumptions.

Regan defines an 'Enlightenment' tradition which includes himself, Singer and Cohen (and, by implication, Ferry), and excludes Plumwood. I find this surprising because, if we exclude Plumwood, we would also have to exclude Marx, and therefore Harvey, from the Enlightenment tradition. It may be instead that Singer and Regan represent one strand of the tradition and Plumwood and Harvey another; dissent from the structural make-up of modern society is also a feature of the Enlightenment. It does not seem to me true that a small departure from conventional thinking *necessarily* makes it easier to argue for the extension required for inclusion of the non-human world under the protection of ethics. Restructuring the way the argument is framed, as Plumwood does, may achieve the same result.

Deborah Rose (Chapter 11) offers her interpretation (from long observation) of Australian Aboriginal culture and describes the world this ethic evokes. It is a world where binary distinctions are not thought much of. The 'country' of the Aboriginal peoples is perceived as 'multidimensional' and the relationship of humans and non-humans as mutually dependent. Joining with Plumwood in attacking dualism, Rose observes that what are presented as *binary* constructs are in fact often *singularities*, since the 'other' is simply what the 'one' is not, the one pole positive, the other negative. This is really the core of the problem: 'A critical feature of the system is that the "other" never gets to talk back on its own terms'; thus, in the words of one government minister, 'self-determination' *is* 'economic independence'.

> The communication is all one way, and the pole of power refuses to receive the feedback that would cause it to change itself, or to open itself to dialogue. Power lies in the ability not to hear what is being said, not to experience the consequences of one's actions, but rather to go one's own self-centric and insulated way.
>
> (pp. 176–77)

This narcissistic way of dealing with Australia's Indigenous people has long supported the oppression and inclusion of Indigenous culture only when it can be made to fit European modes of behaviour (as, for example, Aboriginal artists into the European art market).

Rose draws on the work of Levinas (1989) and Fackenheim (1994) to generate a theoretical position in which dialogue with an entirely unfamiliar (and also alienated and marginalized) culture becomes possible. The universal ('abstract principles of justice') and the particular ('contextualized relationships of care') are not seen as mutually exclusive. Yet 'Environmental ethics conceived in dialogue must be both situated and open' (p. 182). Such dialogue must be based on recognition of the truth of colonial histories that have scarred and divided the social landscape, and requires the act of 'turning towards' one another – a passionate 'reaching out' because 'no singular self can mend the world' (Fackenheim cited by Rose).

The above six chapters mark out a considerable span of the 'long frontier' (see Naess, Chapter 2) of opposition to the dominant economistic paradigm of morality. As Harvey warned, it will be tempting for participants in the debates along the line to engage with one another instead of with that which they all confront. Somehow the construction of a counter-hegemony must encompass all these various positions while not arriving at a compromised consensus and always leaving open the possibility of debate. Perhaps a feminist stance which adopts 'inclusion' as a core idea may make such a construction easier to achieve, and the idea of 'translation' seems to have much to offer if 'reaching out' is to occur.

Global political justice and the world economy

The final part of the book pursues the theme of how broader conceptions of justice are being introduced into environmental policy debates and what scope there is for further expansion of these debates by global institutional means. I have argued elsewhere (Low and Gleeson, 1998) that such is the hegemonic power embedded in the governance of the global economy and its favoured institution 'the market' that only an expanded role for democracy at global level is likely to make room for the sorts of ideas canvassed above to have a major impact on public policy at *all* levels (for support of such a view see Held, 1995). Of course there are those who dispute the novelty or even reality of 'globalization' (see, for example, Hirst and Thompson, 1996). But there is no doubting the force of the neo-liberal belief system (or faith), which seems to be the less challenged the higher up one goes in the territorial hierarchy. Of course, also, there is much more to be said on these matters, but the four chapters here examine global institutions from different perspectives. Fourmile's chapter bridges between a Western style of argument and indigenous cultures on the question of biodiversity conservation. Young examines the role of justice as fairness in international regimes, and Dryzek invites us to think beyond the state in arguing for global democracy.

Where Deborah Rose (who is not an Indigenous Australian) provided an intimate glimpse of a particular culture, Henrietta Fourmile (who is) gives us a picture of the significance of indigenous cultures more generally as a central part of our planetary social heritage.[3] Fourmile (Chapter 13) explains that 'traditional ecological knowledge' (TEK) provides the essential link between biodiversity and cultural diversity. TEK is 'traditional' in the sense not of 'ancient' but of a culturally defined process of learning and *sharing* knowledge. In worrying about biological diversity we should also be aware that *cultural* diversity is reducing at an alarming rate. Globalization is as much a cultural as an economic phenomenon and brings the commodification of culture (subjecting cultural production to the demands of markets) and the elimination of languages. In particular, sharing of knowledge is threatened by the privatization of knowledge via Euro-American cultural institutions like patents and 'intellectual property rights' and products like the 'terminator' gene. Great care and respect is needed if the world is to share some of the content of indigenous knowledge to improve human life without at the same time condoning a rapacious exploitation which will end in the destruction of the cultures that are the seedbeds of such knowledge.

Oran Young (Chapter 14) shows how 'fairness' has come to play a significant role in the negotiation of international environmental regimes. A regime is usually the product of a treaty or agreement among a number of nations. Such a treaty is a contract, so perhaps it should not surprise us too much that the ideas of John Rawls (1971) are relevant to the conduct of the parties to such a contract. Rawls's 'veil of ignorance' with the aid of empirical observation of actual regimes becomes in Young's adaptation 'the veil of uncertainty'. Yet these regimes, Young points out, are also continuous exercises in 'problem-solving', so a mixture of factors come to influence behaviour. Thus, he writes:

> Fairness matters in the creation of international regimes but only as one element in a kind of causal soup that typically includes a variety of other factors operative at the same time. In addition, the role of fairness is a variable that assumes substantially different values across issue areas and even across regimes within the same issue area.

> (Chapter 14, p. 248)

Young seeks to show how an ethical norm such as 'fairness' is consistent with the assumptions of the dominant paradigm – the 'self-interestedness' of the governments of nation states in the case of ' realism'. Here there is a contrast and debate between such an intellectual position and those of a Shiva or a Plumwood (or Held, 1995; Athanasiou, 1998; Low and Gleeson, 1998; or Altvater, this volume, Chapter 16). In my view there is much to be said for keeping an open mind on this matter even while maintaining the debate and advocating this or that position. What happens in the climate change negotiations will perhaps provide a test of whether existing institutional norms are adequate to the ecological task or require more radical change. But whatever the outcome in this particular case, demonstrating the permeability of established positions to alternatives appears to be a necessary step in their evolution. There is something here of the 'translation' which Harvey considers so necessary. One might also say that a range of positions is bound to occur in the formation of any new hegemonic discourse which may evolve to replace the currently outworn one (of minimalist utilitarianism) – as the latter becomes ever more indefensible except in the terms of religious dogma, a stage, one might surmise, which usually precedes a sudden widespread rejection of 'the faith'.

In question, however, is whether this discourse will change in time to prevent large-scale, irreversible ecological damage. The means of such change, as Dryzek argues (Chapter 15), must involve a 'felicitous combination of democratic structure and ecological concern' – ecological democracy. But can we expect such a combination to occur where it seems to be most needed, at global level? Dryzek argues that, contrary to expectations, the prospects of ecological democracy at global level are good, but that we have to look beyond the state for signs of its development. Dryzek's point is that it is not just force and institutional rules that make things happen (and may be the subject of democratic influence), but also 'discourses'. A discourse is a shared set of assumptions and often unspoken understandings which 'enable its adherents to assemble bits of sensory information that come their way into coherent wholes' (Dryzek, this volume, p. 268). Minimalist utilitarianism is one such discourse. 'Discourses', says Dryzek, 'are intertwined with institutions; if formal rules constitute institutional hardware, then discourses constitute institutional software' (p. 269).

We have to look for signs of democracy, therefore, not only in the institutional hardware of global society, but also (and more promisingly) in the discursive software. Dryzek situates the main impetus for democracy in global civil society, which he defines as 'political association and public action not encompassed by the state on the one hand or the economy on the other' (p. 277). The 'software' of discourses spans both state and civil society. As I suggested above, even while the dominant economistic discourse is asserting itself ever more strongly, it is also retreating to the elite 'economics and accountancy' strongholds (to be found in state treasuries and global institutions of economic governance: IMF, OECD, the World Trade Organization, etc.) which are the most detached from real-world experience.

Dryzek argues that the form of the international governance system is not comparable to that of the nation state and it would be a mistake to try to create such a form at global level. Quite apart from the undesirable consequences of a global concentration of the power of force or violence, states, he says, are increasingly subject to economic constraints from the international capitalist system which restrict their capacity to respond to democratic demands. Thus he argues, perhaps with the example of the United States in mind, that a state analogue at global level would also be guided mainly by economic imperatives. Here I would mildly disagree, for the situation of a global state-like nexus of institutions (with appropriate separation of powers) would be profoundly different in relation to international capital from that of the nation state forced into competition with other such states (or regions or blocs) – though admittedly the example of the European Union is not particularly encouraging. Still, I do not read Falk's more recent essay (Falk, 1995) as an 'apology' (as Dryzek terms it) for his earlier advocacy of a nexus of state-like institutions operating under democratic norms at global level (Falk, 1975).

While change in discourse will certainly impinge upon institutional structures, it seems to me that institutional change at global level – through the establishment of powerful foci for democratic debate about ecological-ethical ends and means – might also help to establish new discourses.

Altvater's contribution (Chapter 16) is based on the first paper to be read, after Naess's, at the Melbourne conference. In this book it is placed at the end for a particular reason, namely to sound a severe warning – that the market system is inherently entropic, continuously using up resources and displacing costs in time and space. The world has not yet discovered the brakes or steering-wheel for this hypercharged engine of growth, and current world policy is to let it rip and hope for the best, while trying to patch up the economic damage caused by the overheated engine after it occurs. Already some voices within the economic system are warning that regulation of the market is a necessity for purely 'economic' reasons. The market system is the most successful the world has devised for generating a certain kind of wealth (of commodities) distributed in a certain (highly unequal) pattern. The particular kind of measurement of what produces this wealth (GDP) still guides public policy from the national level upwards. But this wealth, this pattern of distribution and this method of measurement are leading to an ecological catastrophe which will dwarf all previous human-made disasters – famines, wars and genocidal holocausts – and threaten the integrity of the planet itself. We should not become mesmerized by such an expectation of catastrophe, but unless we keep reminding ourselves of the reality of our present situation the world will continue on its present course; and this simply *has* to be changed.

Notes

1 This is not to deny that the 'personal' can also be political, as some feminists have argued. I would not, however, endorse the view that to speak of the 'public sphere' – or there being an important distinction between the public and the personal – necessarily involves gender bias.
2 Indeed, in some parts of the world (Australia is one example) environmentalism itself seems to be fading as a political force.
3 The two essays together surely exemplify an interweaving of cultures (cf. Tully, 1995) and mutual 'turning towards' of which Fackenheim (1994) writes.

References

Achterberg, W. (1997) 'Environmental Justice and Global Democracy', Paper given at the conference 'Environmental Justice: Global Ethics for the 21st Century', University of Melbourne, 1–3 October 1997 (to be published in B. J. Gleeson and N. P. Low (eds) (forthcoming) *Global Governance for the Environment*, London: Macmillan.
Arendt, H. (1977) *The Life of the Mind*, vol. 1: *Thinking*, New York and London: Harcourt Brace Jovanovich.
Aristotle (1976 edn) *Nicomachean Ethics*, Harmondsworth: Penguin.
Athanasiou, T. (1998) *Slow Reckoning: The Ecology of a Divided Planet*, London: Vintage Books.

Bauman, Z. (1993) *Postmodern Ethics*, Oxford: Blackwell.
—— (1995) *Life in Fragments: Essays in Postmodern Morality*, Oxford: Blackwell.
Beck, U. (1992) *Risk Society: Towards a New Modernity*, tr. M. Ritter, London: Sage.
Benhabib, S. (1992) *Situating the Self: Gender, Community and Postmodernism in Contemporary Ethics*, Cambridge: Polity Press.
Cohen, C. (1986) 'The Case for the Use of Animals in Biomedical Research', *New England Journal of Medicine* 315(14), 865–70.
—— (1997) 'Do Animals Have Rights?', *Ethics and Behavior* 7(2), 91–102.
Dworkin, R. (1977) *Taking Rights Seriously*, London: Duckworth.
Elster, J. (1992) *Local Justice*, New York: Russell Sage Foundation.
Fackenheim, E. (1994 [1982]) *To Mend the World: Foundations of Post-Holocaust Jewish Thought*, Bloomington: Indiana University Press.
Falk, R. A. (1975) *A Study of Future Worlds*, New York: Free Press.
—— (1995) *On Humane Governance: Towards a New Global Politics*, Cambridge: Polity Press.
Ferry, L. (1995 [1992]) *The New Ecological Order*, tr. C. Volk, Chicago: University of Chicago Press (originally published as *Le Nouvel ordre écologique: l'arbre, l'animal et l'homme*, Paris: Bernard Grasset).
Feyerabend, P. (1993) *Against Method: Outline of an Anarchistic Theory of Knowledge*, London: Verso.
Fraser, N. (1987) 'What's Critical about Critical Theory? The Case of Habermas and Gender', in S. Benhabib and D. Cornell (eds) *Feminism as Critique: Essays on the Politics of Gender in Late-Capitalist Societies*, Cambridge: Polity Press, pp. 31–56.
Gleeson, B. J. and Low, N. P. (eds) (forthcoming) *Global Governance for the Environment* London: Macmillan.
Held, D. (1995) *Democracy and the Global Order*, Stanford, CA: Stanford University Press.
Hirst, P. and Thompson, G. (1996) *Globalization in Question: The International Economy and the Possibilities of Governance*, Cambridge: Polity Press.
Jagger, A. (1980) *Feminist Politics and Human Nature*, Totowa, NJ: Rowman & Allanheld.
Jänicke, M. (1990) *State Failure: The Impotence of Politics in Industrial Society*, Cambridge: Polity Press.
Levinas, E. (1989) *The Levinas Reader*, ed. S. Hand, Oxford: Blackwell.
Low, N. P. and Gleeson, B. J. (1998) *Justice, Society and Nature: A New Exploration of Political Ecology*, London: Routledge.
Low, N. P., Gleeson, B. J., Elander, I. and Lidskog, R. (eds) (1999 forthcoming) *Consuming Cities: The Urban Environment in the Global Economy after Rio*, London: Routledge.
Lyotard, J.-F. (1984 [1979]) *The Postmodern Condition: A Report on Knowledge*, tr. G. Bennington and B. Massumi, Minneapolis: University of Minnesota Press.
Rawls, J. (1971) *A Theory of Justic*, Cambridge, MA: Harvard University Press.
Tully, J. (1995) *Strange Multiplicity, Constitutionalism in an Age of Diversity*, Cambridge: Cambridge University Press.
Walzer, M. (1983) *Spheres of Justice: A Defence of Pluralism and Equality*, New York: Basic Books.
White, J. B. (1990) *Justice as Translation: An Essay in Cultural and Legal Criticism*, Chicago: University of Chicago Press.

2 An outline of the problems ahead

Arne Naess

Introduction

After sixty years' participation in international conferences, most of which have been rather unsuccessful, it astonishes me that I feel that this one will make a difference. One reason why I feel this way is, I guess, the great areas of agreement among us. One of the remarkable agreements seems to be that it makes sense to speak of unfair or unjust policies in relation to non-humans. But there are many other areas where agreements or near-agreements were small twenty or thirty years ago.

We face an overwhelming danger of 'preaching to the converted'. It is essential to be clear about our disagreements. We shall then have a better chance to stand together in the ugly social and political conflicts ahead. We shall know better to what extent we can firmly rely on each other in those conflicts.

Development and environment

Some conflicts are called conflicts between development and environment. In the 1950s and 1960s the questions were asked: 'How can undeveloped countries change into developed ones?' 'Can and should the developed countries play a positive role in this process?' The 'underdeveloped' was defined in terms of unsatisfied needs of the vast majority of the population, especially material needs – the need, for instance, not to be harassed by the brutal state police of an authoritarian government. From the point of view of the people in both the so-called 'underdeveloped' and the 'developed' countries the needs were rightly considered real ones, not mere wants and desires. The use of the word 'needs' for the latter creates confusion. A minority in the so-called 'developed' countries considered it a question of justice to try to help the underdeveloped countries to become developed, but only in the sense of satisfying the obvious needs of the large majority. Aid, often misguided, was proposed and carried out on a minor scale. We learned that to reach the desperately poor was more difficult than expected, and that means other than humanitarian ones were necessary. Then a new maxim was created: 'Trade, not aid!' Articles were imported from poor countries, and shoppers found shelves marked 'products from . . .', where the name of the so-called 'underdeveloped' country could be seen. The volume of such goods was small compared with what was expected. Both strategies failed to have the hoped-for great impact.

Despite the fact that the Norwegian U-help programme (help to under-developed nations) reached 1.17 per cent of its GDP, a minor but well-established party agreed with me that the goal should be 2 or 3 per cent – and the projects should be more professional, involving extensive studies of the local cultural and social conditions, and the prolonged stay of experts: for years, not months. I succeeded in broadening the India projects. India ought to send people to us in Norway, with Norway of course paying the expenses, because *we* certainly also need help, though not of the material kind. Mutual help! But India sent only one person, a professor. Our social and spiritual needs were certainly not properly met.

The efforts to create vigorous trade did not succeed, and we got a new trend – or new 'discourses' as they are called today: those of the under-developed being assisted through huge loans to achieve industrialization. Their economic advance should be measured by economic growth and GDP. A terrifying vision (in my view) was created: that of a global search for ways of reaching the level of consumerism and waste characterizing the rich, industrial states, as if both environmental sustainability and life quality – the way you *feel* your life – can be neglected. Such things are clearly not proportional to consumption and waste, and neither is economic advance proportional to eco-nomic growth and GDP. As to the loans, justice requires them to be annulled.

If we wish to retain the general term 'development' we should class the rich industrial countries as *overdeveloped*. But I think it advisable only to use the complex term 'ecologically unsustainable development', here defining develop-ment as volume and direction of *change*. This makes development a kind of vector which, by definition, has the properties magnitude and direction.

A slogan used in the deep ecology movement is relevant here: 'Full richness and biodiversity of life on Earth'. The term 'richness', or 'abundance', as added to biodiversity, is essential because we should not often degenerate into eco-tourists. We need life all around us. The large Tysfjord in Norway is yearly invaded by about four hundred *Orca orca* whales – a bad name for them is 'killer whales'. Norwegians living along the fjord enjoy whale safaris in their neighbourhood, but not as mere tourists. All whale families of *Orca orca* and many old specimens have proper names. One hears cries of: 'Where is the so-and-so family?'; 'I have not seen Uncle So-and-So this year!' The thought of killing them is abhorrent. But most people along Tysfjord loudly support the approximately two hundred whalers killing minke whales. Their culture is in many ways a hunter and gatherer culture, and locals and experts agree that there are about 80,000 minke whales. The point I want to emphasize here is that in the next century the norm will be abandoned that we humans have the right to go on killing as long as the existence of a species of marine mammals is not threatened. Abundance, please! Marine mammals might be the first wide class of mammals the abundance of which will be well protected through international regimes.

Two kinds of conflicts emerge: when a concept of ecological sustainability was introduced, to what extent could it be expected that countries with grave unfulfilled material needs would follow a course of ecologically sustainable development? The answer by economic and political theorists has often been that fulfilling the country's grave needs must have priority: first development, then environmental regulations. Others answered that both must be seriously considered together. I agree with the latter. But evidently some of the 'expensive' rules which the rich countries can afford to introduce cannot be exported to the economically poor countries. Strong measures to moderate the increasing ecological unsustainability, yes, but together with serious attention to other problems. If a father can only rescue his children from starvation by killing the last tigers, it is, in my view, his duty to kill the tigers, provided he has the necessary (exceptional) power to do so. *But shame on the government of his country, and shame on us!*

The interference of the materially richest countries in the planetary ecosystem takes place in a per capita excessive and unfair way. As the governments of these countries do not have the intention to support (in ethically fully acceptable ways) a slow decrease of population, they must significantly reduce their standard of living. I do not narrow this down to 'material standard of living', because there is very significant waste, for instance when teaching students of chemistry. Institutional waste. How can people in the poor countries believe in the importance of ecologically sustainable development when they see how the rich live, and intend to continue to live?

Second, it is said that pollution knows no borders. But the rich countries may perhaps continue their unecological policies for several generations, and also profit in terms of long-distance trade from a destructive global free market, at the same time as ecological conditions in the poorest countries become desperate. Unfortunately, before the powerful rich countries are hit by tragedies far greater than that of Chernobyl, conditions are likely to worsen. When I am asked whether I am an optimist or a pessimist, I answer, 'optimist!', adding, 'on behalf of the twenty-second century'. If there is a large audience listening, I point to individuals, shouting, 'How far down we go in the next century depends on YOU, YOU, YOU!'

How can it depend upon each of us? I shall answer with one of my favourite slogans: 'The frontier is long!' By this four-word formulation I refer to the very great diversity of jobs available for those who wish to be 'activists'. There is a tendency among activists to say, 'What *you* are doing cannot have the highest priority (or does not hit the core of the problems), come over to where *I* work.' This is mostly counter-effective. We should help people to find what would be most interesting to them, pointing to very different, but important, issues. And we should also insist that they stand up and talk and write, and perhaps, or perhaps not, occasionally take part in direct actions. Personally I have only been arrested and carried away by the police twice, but I cherished my close contact with them. After all, green societies will also presumably have police.

A factor that is sometimes underrated is the tone of the communication between conflicting groups. The heat of the debate tends to reduce the value of the communication. In order to avoid this reduction it is important first of all not to distort the views of the opponent. We have to shorten and simplify the views of our opponents as well as our own. When the views of the opponent are rendered in a way that makes them less tenable, or less ethically justifiable, or when we attribute views to the opponent which the opponent does not have, or if consequences are invented which do not strictly follow, then no agreement can arise, and the atmosphere of the debate is poisoned. A solution for the conflict is impossible to reach.

Non-violence in debate today is more relevant than ever because of the ever-increasing role of communication between the increasing number of people on an increasing number of crises. This is why I have mentioned the role of non-violent communication in environmental ethics. But there is a second reason. We ought to talk much more to people who despise what we are doing, or to business people who find the usual pictures of green societies unbearably boring. Taskmindedness requires less contact with in-groups, more contact with out-groups, business people tell me.

Fair distribution of and fair profit from non-renewable resources

From the point of view of the rich industrial countries (sometimes abbreviated to the 'Rs') there are a number of areas of concern, all of which have philosophical relevance:

1   The Rs maintain a standard of living that is unsustainable on a global scale.
2   The Rs profit unduly from trade with the countries with low material standard (the Ls) and, especially through the behaviour of corporations, induce people to try to copy the unsustainable way of life in the rich countries.
3   It is an obligation of the Rs to cooperate with the minority in the Ls who are aware of the dangers of uncritical development to rectify their direction. The cooperation must be based on an attitude of self-criticism on the part of the Rs, and without the slightest tone of arrogance.
4   The huge old loans which the Rs furnished for development of the Ls served only in part the long-range independent development of the Ls. The interests on the loans which the Ls pay today is a burden and must be eliminated.
5   Until the Rs adopt a globally sustainable way of production and consumption per capita, it is the obligation of the Rs to leave the industrial utilization of the natural resources of the Ls for *their* sustainable development

mainly in their own hands. This implies a control of the way by which the Rs take over an unduly large fraction of the industrial processing.

6   The technical questions are very moderate compared with the political, both in the Rs and the Ls. The present trend is towardss *centralization* of trade; that is, inventing a global liberalization of trade, constructing an immensely strong global market. In some Rs such as the Scandinavian countries, the governments have in the last half of the twentieth century proclaimed their support for economic decentralization and therefore local markets. The price of milk in Norway, to mention a concrete example, is much higher than in the world market. This is done in order to protect what is left of the fairly sustainable agricultural culture, avoiding large-scale industrialization of the production of life necessities. Free markets combined with ecological protectionism!

There are in the Rs fortunately no strong opinions in favour of authoritarian regimes, or belief that such regimes could ensure responsible economic development. The main problems will have to be solved within the framework of democracies. The outlook today is grim because the public is interested in short-range problems, whereas environmental justice is a long-range issue. Injustice may not result in any violent crisis that seriously affects the Rs. The very few doomsday prophets have evidently been wrong in their predictions of great headline catastrophes.

Let us, as an example of problems of environmental justice, inspect the distribution of sources of oil and natural gas. By chance Britain and Norway are placed near enormous sources of oil and gas. In the early 1960s these countries had the necessary scientific and technical know-how to make large-scale use of these resources. The international rules at that time allowed them to have a complete monopoly. They could do what they wanted: to 'produce' millions of barrels of oil a day and a comparable amount of natural gas. Private and state corporations could make gigantic profits from selling the oil and gas, within their own market, but preferably for export.

In Norway, a minority in the 1960s found this arrangement unjust: unfair to the economically poor countries and unfair to future generations. We ought to hand over a part of it to the Third World, they said, and to limit the extractions of oil and natural gas so that future generations could have a fair amount at their disposal, or if necessary simply leave the gas and oil untouched. The industrial extraction causes great CO_2 pollution. The supporters of fair play have not changed their position.

Ecologically, the use of oil and gas for heating and transport should and probably will decrease, but as long as the price is very low, other means cannot compete. The pressure to use very large amounts is great within the present non-ecological system of economy. Norwegian production might within one generation (thirty years) be cut down by about 80 per cent, and some of the reduced volume should be available to those countries that need the resources today. If, as seems likely, the technology of renewable energy sources will be used on a dominant scale all over the planet within one or two generations, the resources of oil and gas can and should be capable of being used at a level which would make them available practically indefinitely.

To decide on policies, an international energy commission is required, and today it might be set up by the United Nations. The technology required is 'advanced' and costly, and it should not be necessary for users of oil and gas to build the great platforms and do the extraction themselves. The main thing is to view the resources of oil and gas, and primarily those in the oceans, as common resources for humanity, and to avoid obviously unecological ways of using them, and especially to avoid the use of oil and gas as a means of heating and transport. The present situation is ethically, ecologically and politically irresponsible. But a satisfactory solution to the problems involved cannot be expected to materialize within ten years, perhaps not even within a generation or two. But something could be done immediately to propose practical changes on a minor scale. It is not my job in this opening address to suggest steps to be taken, but the access to sources of energy seems to me today an issue requiring immediate action at the level of the United Nations.

Today a product we buy on the market has mostly a very complicated genesis. The necessary natural resources have been collected and then a long series of processes, let us say ten processes, commence, ending with a 'finished' product to be sold at a definite place at a definite time. But today the poor countries do not take part in, say, the eight last processes. The rich take over, and the corporations take over the 80 per cent *profit*. The more processes of the industrial kind, the more profit – roughly speaking, of course. How could this be changed? Some, but not enough, people try to find out how to proceed. I cannot do more than remind us of the existence of the problems which fairness implies.

From what I have said, it does not follow that there should be the same level of material production and consumption per capita in every country. But today the difference between rich and poor both within countries and between countries is clearly unacceptable and may cause future violent conflicts if the difference continues to increase.

Perhaps I have preached too much already to the converted. I shall, in my survey, mention an area which is emotionally rather touchy. It is generally conceded among fairly well-informed people that any population increase in the rich consumerist and wasteful countries gravely contributes to the ecological crisis. We in the rich countries should in the near future be able to tell the rest of the world that our governments and public institutions do nothing to maintain the population level. We should be able to tell of how much we try to inform our people of the grave responsibility we incur as the situation is today. With a half per cent fewer births a year – that is, 199 instead of 200 births in an area – a process would start which with time would eliminate serious population pressure on areas where people prefer to live (for instance, certain coastlines). As it is now, such areas tend to lose their special qualities because of over-crowding. But, of course, the main problem is perhaps not the unpleasant population pressures in the rich countries, but the unwarranted signals the rich countries send to the Third World: 'Try to reach our level of material standard of living and waste. We see nothing wrong with continuing to increase our population without changing our way of life.'

As long as some countries such as the United States do not work to stabilize or reduce the population, it is completely unacceptable to ask the poor countries to do so. But in any case, to talk about 'overpopulation' is unnecessary. People tend to interpret this as a violent threat; and they also tend to forget we are talking about a time span of perhaps hundreds of years. Our goal is to secure sustainable high quality of life for everybody, and with free choice to decide where to live. This implies reduction of population pressures.

Some people are active within environmentalism on the basis of, or are strongly motivated by, their life philosophy. I would, as a supporter of the deep ecology movement, say they act in part on the basis of their 'ecosophy'. Others call themselves supporters of the social ecology movement (Bookchin, 1980, 1982). The latter tend to think that domination over nature (as a goal) is caused by humans' domination over humans. Environmental justice may from this point of view be realized through giving up hierarchical thinking of every kind. In Uruguay, social ecologists have tried to empower and emancipate fishermen and make them preserve old, ecologically sustainable methods, rejecting the destructive new methods of the industrial societies. Domination implies injustice. Applied to humankind's domination of other beings, a generalized Rawlsian fiction is relevant: Forget completely who and what kind of being you are, and ask, 'what social and interspecies relation would you consider just?' (cf. 'veil of ignorance'; Rawls, 1971). If you thought of yourself as a rattlesnake, you would find it unjust of humans to kill you. You are dangerous only to people who walk carelessly and hastily within your territory. You would tell humans, 'As a rattlesnake, I am entitled to live where I was born. Humans can scarcely maintain the view that there is an overpopulation of rattlesnakes.'

Opinion surveys of how people feel about environmental justice and, more generally, about environmentalism suggest marked positive attitudes. But when they vote, people tend to give their votes to the politicians and political parties they criticize for unsatisfactory and weak environmental policies. The strong term 'schizophrenia' is often used in this connection. People feel an environmentally deep commitment, but also a deeply seated habit of choosing politicians who do not have any concrete proposals. Under these circumstances politicians claim that they cannot propose strong ecological measures without losing votes. They must get a minimum of feedback from the voters. In Norway, two attempts by established parties to put strong measures into their programmes clearly resulted in a decrease of votes. It is obvious that environmental justice proponents must hail any courageous initiative by any leaders of a political party – by fan mail and otherwise. I am speaking about the situation in democracies where the attitudes of individual voters are said to have some influence, however small. And surveys suggest that there is complete agreement that if democratic governments are unable to realize environmental justice, authoritarianism or dictatorships will not have a better chance. On the other hand, it is an open question whether or not undemocratic regimes will appear in the next century if the environmental situation gets much more threatening. It is unreasonable to believe that authoritarians would be highly motivated by considerations of justice as fairness.

The Europeans have had the supreme 'advantage' of having colonized the world. I place 'advantage' in inverted commas because, in the long run, this advantage may prove to have been a disadvantage not only for the rest of the vast population of the world, but for themselves. Now they are afraid of the increasing economic competition, and feel compelled to unify themselves within one immense market, the European Union (EU), with 'free flow' of international capital and a vast increase in long-distance trade. Ecological protectionism – that is, protection, for instance, against import of cheap products made cheap by disregard of pollution and energy waste – is prohibited. The increasing waste industry helps economic growth, and the EU is in part motivated by the belief in such growth. It is a bad sign that opinion in the United States about 'closer and closer' European union is very favourable. (All through the twentieth century Americans have tended to say to us Scandinavians, 'Your countries are too small, much too small! Integrate! It is very harmful not to have a common market. You will soon be unable to compete!' But competition is not everything, and 'big is beautiful' has only limited applicability.)

The EU represents a formidable step towards globalization of a definite kind of liberal economy. The Foundation for Deep Ecology in San Francisco arranged in New York the first international conference against this globalization. It is increasingly realized in the United States that the liberal economy favours the strongest – that is, the United States – but not the weakest.

The popularity of gradually decreasing environmental injustice is today only moderate in the Rs, and knowledge about what is going on is also moderate. But the activism of a small minority may ultimately have a sufficient impact to change policies. Slavery and certain other major evils have eventually been eliminated from large areas of the world through the persistent work of small minorities. The majority feel the rightness of the cause, but it takes time before they join.

Tens of thousands of young people in the Rs would gladly cooperate with the small, environmentally active minority in the Ls, but they must be assured of getting work in the Rs after having spent one or more years in the Third World. As it is now, they are threatened with unemployment, and they cannot take the risk of 'losing' time by doing even important and meaningful work abroad. Many of those who come back to their Rs have indispensable knowledge, including proficiency in one or more languages of the Third World. It would be a crime to let them go unemployed or let them try to get a job in advertising or some other flourishing industry. They are needed in the institutions in the First World dealing with relations with the Third World, and with the Fourth. (I am sorry about using these misleading numbers. But the First World is the first that must change its unsustainability!)

The ethically unacceptably high level of unemployment in the European Union might be lowered if Green economics and Green political theory were taken more seriously. What do politicians tend to answer when they are told this? They tend to say, at least in Scandinavia, that they largely agree with the Greens but that they are dependent on the voters. Again and again, it is clear that the public are not ready to accept the 'burdens' they think are implied by responsible environmental policies. The public resist any major political move.

We read in our newspapers, especially during the 1970s and 1980s, about environmentalists who *predicted* vast environmental catastrophes. They were called ecological 'doomsday prophets'. But who were they and who are they today? A great many professional ecologists have asserted that *if* such and such trends continue unhindered, the state of affairs will approach a major catastrophe. But they are evidently *not* doomsday prophets. The usefulness of their warnings is clear: something must be done, and the sooner the better. The latter words are important. Unsustainability does not increase linearly, but exponentially, perhaps, very roughly, by 3 per cent annually. If 3 per cent, then unsustainability doubles within 24 years. But quantification is rather unsuitable here.

Unhappily, some serious ecologists have greatly underrated in their publications the richness of the Earth's resources. The most powerful countries will be able to continue their lifestyle, unsustainable in the long term, for many years at a time when there are very serious crises elsewhere, for instance lack of clean water and very high frequency of environmentally induced illnesses. In certain areas at least 20 per cent of illnesses are today considered to be environmentally caused. But no prominent ecologists are 'doomsday prophets'. The term is now used mainly in the literature characteristic of the international 'ecological backlash': the worldwide attempt to discredit environmental movements.

Less talk should have been devoted to discussions about resources and more devoted to environmental justice as fairness. Until about twenty years ago, fair distribution and fair access to *appropriate* technology could have assured satisfaction of the vital needs of a rapidly increasing population. By the end of the twentieth century the general level of material aspiration has increased significantly, and consequently what was in 1970 considered adequate to satisfy vital needs is now considered unsatisfactory. Here we are touching on a very complex and delicate situation. Environmental justice cannot be defined in terms of a global approximation towards the production and consumption pattern of the rich industrialized countries. That pattern cannot be universalized. To simply continue that pattern in the rich countries is in itself unjust, unfair. It violates the high-ranking norm of universalizability, it accelerates the rate of decrease of life conditions of the planet, and it neglects future generations.

The impact of the maxim 'increase environmental justice!' depends upon the way the sufferers of unfair arrangements *define* the status of satisfactory environmental justice. It must be defined not in terms of material standard of living, but in terms of access to resources, access to a level of technology which makes industrial and other use of the resources practicable. All the time, a concept of vital *need* is relevant: the ocean of human desires seems never to diminish, whatever the level reached. But the mere existence of a group which suffers from lack of satisfaction of vital needs does not entitle this group to claim environmental unfairness. There must be other groups or countries which have been unfair in their relation to the sufferers.

Green economics and political theory

Since the 1960s there has been a mounting stream of publications conveniently called contributions to 'Green economics and Green political theory'. The capital G indicates that not only selected practical reforms are considered, but also significant, substantial theoretical changes involving changes of attitudes among people. It is a difference comparable to that between the 'shallow' and the 'deep' ecology movements. It is not my task to express any opinions within this extended field of economic and political discourses, but a few points may be tentatively formulated.

Today's strict market economy makes it extremely difficult to avoid substantial unemployment. Highly competent economists who are at the same time active within environmentalism are often called 'Green economists' again with a capital G. They differ from green economists – with an ordinary g – in much the same way that supporters of the deep ecology movement differ from supporters of the ecology movement who do not envisage substantial changes of a social and political kind. The fundamental point I am making may be formulated thus. Green economics is a labour-intensive economy. The kind of liberal economy we have in the 'overdeveloped' world today is capital intensive, and unemployment is difficult to keep down. It is a formidable case of injustice to deprive people of a job that enables them to support themselves. Green economists warn against just closing factories here and there because of clearly unacceptable levels of pollution.There must be a consistent Green economy in order significantly to reduce ethically unacceptable unemployment.

Corporations calculate what it costs for them to reduce CO_2 emissions by 20 or more percent. Governments must somehow arrange 'the rules of the game' in such a way that makes it profitable to change ways of production in a Green direction. Green economic principles require free markets, but also a kind of ecological 'protectionism' such as ecological costs being included in prices. People look for 'green products'. But deeper green – that is, Green – products seem to require a system of economics which relies on *incentives* provided by public institutions.

The so-called Green political theorists all work within a democratic framework. This is compatible with a critical attitude towards the operation of present democracies. How can it be avoided that there are pressure groups so powerful that it is in practice more or less impossible not to submit to their special interests? There are many other areas where critical discussion is lively, and hopefully creative.

These remarks on economics and politics have centred on conditions in the 'overdeveloped' countries. It is clear that the kind of Green, or even green, policies under discussion in the rich countries are in part out of reach for the economically poor. Must we conclude 'Development first, then environment!'? Is it a form of environmental injustice to advise economists and politicians in the Third World to ask for Greenness? On the contrary. In order to protect resources for future generations *and* develop profitable use of resources, a multitude of ecologically motivated policies *within* their economic reach must be realized. Cooperation with well-trained, well-educated Westerners is also a must within a number of fields of economic production. There is work to be done by young Westerners with knowledge of relevant local languages, who will stay where they can cooperate for at least a year, and preferably many more years. The Third World's relations with the rich nations suffer seriously from the tendency of people from the latter who stay only for a few months or a single year. Much more would be accomplished if people stayed for several years to establish trust and gain a deeper knowledge of the relevant social and cultural affairs. Young people with specialized competence must be guaranteed that they will not come home to face unemployment. The most satisfactory way this can be achieved is by using their competence within permanent institutions of international cooperation.

But all this requires political goodwill and informed opinions. We are back to the problem of how to increase people's awareness of the environmental justice challenges; how to activate the constructive imagination of young people, appealing to their taste for the big, global questions. We have to help more people who travel as tourists to combine their travel with a study of the relevant cultures, to report on how people in the countries they visit tend to experience themselves and the world. In short, we must concentrate on feelings and attitudes. More travellers could be induced to write about their experiences, to articulate reflections that can be useful for those who seriously wish to take an active part in the great movements of the next century: the movement for peace, the movement to eradicate unacceptable poverty completely, and the movement to establish ecological sustainability and justice.

In some rich countries, including Norway, the government seems to avoid taking environmental justice seriously by telling us that *first* the rich countries must help the rest of the world to get rid of poverty, *then* ecological unsustainability should get a high priority. 'There can be no green society of poor people.' Yes, we may answer, but to eradicate poverty will take a lot of time, and we must not neglect the yearly environmental degradation and the injustices. We shall have to spend time, work and money on both factors *now*.

Unfortunately, the reason for goverments to propose elimination of poverty *first* is politically a very strong one: to spend, say, half of one per cent of GDP on poverty is politically very much easier than to get voters to accept a substantial reduction of CO_2 emission. 'People feel the time is not ripe for such new policies.' They feel virtuous spending a little more on poverty, and it does not harm business.

The relationship between population increase and environmental concerns has to be discussed. There are serious population pressures of different kinds in specific areas. But many people tend to feel that the word 'overpopulation' suggests a threat to existing humans. 'You are the ones who are too many!' Fortunately there is no need to speak of 'overpopulation'.

The easy and safe access by small children to patches of free nature not dominated by humans is decreasing drastically in many urban areas. Such access has played an important role until recently. But pressures to 'develop' the patches – for instance, turning them into commercial areas – have been immensely strong, and such pressures grow with population growth. The astonishing argument that the 'value' of the patches is too great for them to be left to children has been used implicitly or even explicitly. This is evidently environmental injustice towards children, and destructive use of 'value' as a synonym for 'market value'.

'Population pressure' may for some people sound misanthropic ('do people press me?'), but even in Norway, with only 4.3 million people on 325,000 square kilometres – less than 10 people per square kilometre – there are population pressures. People relish living along the fjords, but there are 'too many' competing for the beautiful areas. Prices of real estate are sky-high and it has been necessary to introduce laws defending the right to walk along the shoreline. No private property all the way down to the edge of the water!

People who have a desire or need for 'elbow room' may try to escape to a roomier environment, even leaving a place they love if they feel this place has 'developed' into something resembling an anthill. On the other hand, a family of six living in one small room in a skyscraper in downtown Hong Kong may prefer to stay there even if they have the opportunity to exchange it for a two-bedroom apartment in a much less crowded area, but with not nearly as good access to their workplace. The term 'crowding' is often used, but some people are perfectly adapted to what others call 'insufferable crowding'. In short, by 'population pressure' should be meant a situation where pressures are *felt* because of the density of people, not as a term for mere situations of very high density. Fortunately, many people love high density.

When it is argued that a smaller population might in some areas make it less difficult to decrease environmental injustice, including that towards children, people tend to think in terms of significant population reduction within a hundred years or even within a smaller time limit. But it seems to be forgotten that even a half a percent annual reduction makes a significant difference in the very long term, say in five hundred years. To many people the very *word* 'reduction' feels threatening, but mainly because they think in terms of ethically unacceptable measures, like those in China.

I mention population mainly to remind us of the need to debate population calmly and not pay much attention to extreme theoretical views, or extreme proposals. Personally I am interested in the creative imagination fostered by children with easy and safe access to patches of free nature. They have less need of the gadgets made by adults. Every development restricting free movement and play by children is misused, making them more dependent on objects bought on the market.

A generally accepted norm is severely violated in the rich countries: 'Do not live in such a way that you cannot seriously wish that others also would live, should they wish to do so.' This is the ethical norm of universalizability. Because of the seriously aggravated environmental conditions if an additional 1 billion Chinese and 1 billion Indians turned to the American (or Norwegian) way of life, we cannot seriously wish them to do so. Of course, we grant their right to live as we do, but it will not be possible without very grave consequences.

The above discussion of population problems does not imply that I consider population problems to have a high priority, but they should not be neglected when speaking about environmental justice. But we also have to look at the influence of various economic systems and ask, for instance, whether a liberal, capitalist global economy is an asset or an obstacle, or neither. And how can indigenous cultures be economically protected?

All the areas we touch upon have in common the need for high-quality mediators to soften polarization of conflicting points of view. We are all here as fallible humans, with more or less strong convictions – but nevertheless fallible. Now I am full of anticipation. I shall close with the formulation of two maxims or slogans:

The first is: *the frontier is long*. What I mean to say is that the problems we face are so many, and the kinds of activity required are of so many kinds, that there is room for everybody, whatever their inclinations and interests. Everyone will find something very meaningful to do. But we should not pressure others to do the same thing as we do; we should not insist that a definite problem or kind of activity is *the* most essential.

The other maxim is: *concepts of justice are very similar worldwide, but opinions on matters of fact differ widely.* Optimism is to a large extent dependent on the similarity of concepts, because people can be more easily made to change their opinions about facts than their basic views about what constitutes justice. An extreme example: millions of young Austrians and Germans supported Hitler in the early 1930s because Hitler convinced them that treason by communists and Jews was the cause of the defeat in the First World War, that the terrible economic crisis in the 1920s, when it cost millions of marks to send a letter, was due to the perfidy of the Jews, that the biological make-up of the Aryans was superior to that of any other race, and that of the Jews was sinister. The young Nazis in the early 1930s did not have a remarkably different concept of justice from us, but they were convinced about the rightness of certain factual matters which turned out to be wrong. Few tried to convince them that they were wrong.

In short, there is an immense need patiently to disseminate information, to dwell repeatedly on the concrete cases of injustice and on the concrete cases of ecological unsustainability. Let us not hurry to complain about basically different concepts of justice!

This chapter is based on the Opening Address given at the conference 'Environmental Justice and Global Ethics for the 21st Century', the University of Melbourne, Australia, 1–3 October 1997. It has been somewhat modified in the light of contributions by other participants.

References

Bookchin, M. (1980) *Toward an Ecological Society*, Montreal: Black Rose Books.
—— (1982) *The Ecology of Freedom: The Emergence and Dissolution of Hierarchy*, Palo Alto: Cheshire Books.
Rawls, J. (1971) *A Theory of Justice*, Cambridge, MA: Harvard University Press.

Part I

Environmental justice challenges

3 Environmental justice challenges at home and abroad

Robert Bullard

Despite significant improvements in environmental protection over the past several decades, millions of Americans continue to live in unsafe and unhealthy physical environments. Many economically impoverished communities and their inhabitants are exposed to greater health hazards in their homes, their jobs and in their neighbourhoods when compared to their more affluent counterparts (Bullard, 1990; US Environmental Protection Agency, 1992; Bryant and Mohai, 1992). Much of the world does not get to share in the *benefits* of the United States' high standard of living. From energy consumption to the production and export of chemicals, pesticides and other toxic products (including tobacco), more and more of the world's peoples are sharing the health and environmental *burden* of the United States' wasteful consumer-driven throwaway society.

Hazardous wastes and 'dirty' industries have followed the path of least resistance. Transnational corporations and governments (including the military) have often exploited the economic vulnerability of poor communities, poor states, poor nations and poor regions for environmentally unsound, unhealthy and 'risky' and unsustainable operations. For years, economically and politically disenfranchised populations watched helplessly as their communities gradually became the dumping grounds for all types of locally unwanted land uses (LULUs – Bullard, 1990). From urban ghettos, *barrios*, Native American reservations to rural 'poverty pockets' in the United States, unequal protection is creating endangered communities, people and environments. In the southern United States, for example, 'Jim Crow' (i.e. apartheid American style) racial discrimination, institutionalized in housing, employment and education, buttressed this process.

A new form of activism emerged out of the struggles against disparate and unequal enforcement of environmental protection laws in black and white communities. This new environmental activism was an extension of the modern anti-racist civil rights movement. Environmental racism may be difficult to prove in a court of law. Nevertheless, it is as real as the racism found in housing, employment, education and voting (Bullard, 1993a). *Environmental racism refers to any policy, practice or directive that differentially affects or disadvantages individuals, groups or communities on the basis of race or colour, whether the differential effect is intended or unintended.* Environmental racism is just one form of environmental injustice and is reinforced by government, legal, economic, political and military institutions. Environmental racism combines with public policies and industry practices to provide benefits for whites while shifting costs to people of colour (Godsil, 1990; Colquette and Robertson, 1991; Collin, 1992; Chase, 1993; Bullard, 1993a; Coleman, 1993; Colopy, 1994; Westra and Wenz, 1995).

The environmental justice movement – as is true of most other social movements in the United States and elsewhere – emerged in response to practices, policies and conditions that were judged to be unjust, unfair and illegal. Some of these practices, policies and conditions include (a) unequal enforcement of environmental, civil rights, and public health laws; (b) differential exposure of some populations to harmful chemicals, pesticides and other toxins in the home, school, neighbourhood and workplace; (c) faulty assumptions in calculating, assessing and managing risks; (d) discriminatory zoning and land-use practices; (e) disparate siting of polluting facilities; and (f) exclusionary practices that limit some individuals and groups from participation in decision-making (Lee, 1992; Bullard, 1993b, 1994).

The environmental justice paradigm

The current environmental protection apparatus is broken and needs to be fixed. The current apparatus manages, regulates and distributes risks. The dominant environmental protection paradigm institutionalizes unequal enforcement, trades human health for profit, places the burden of proof on the 'victims' and not the polluting industry, legitimates human exposure to harmful chemicals, pesticides and hazardous substances, promotes 'risky' technologies, exploits the vulnerability of economically and politically disenfranchised communities, subsidizes ecological destruction, creates an industry around risk assessment and risk management, delays clean-up actions, and fails to develop pollution prevention as the overarching and dominant strategy (Austin and Schill, 1991; Bullard, 1993c).

On the other hand, the environmental justice paradigm embraces a holistic approach to formulating health policies and regulations, developing risk reduction strategies for multiple, cumulative and synergistic risks, ensuring public health, enhancing public participation in environmental decision-making, promoting community empowerment, building infrastructure for achieving environmental justice and sustainable communities, ensuring interagency co-operation and coordination, developing innovative public–private partnerships and collaboratives, enhancing community-based pollution prevention strategies, ensuring community-based sustainable economic development, and developing geographically oriented community-wide programming.

The question of environmental justice is not anchored in a debate about whether or not decision-makers should tinker with risk assessment and risk management. The environmental justice framework rests on an ethical analysis of strategies to eliminate unfair, unjust and inequitable conditions and decisions. The framework attempts to uncover the underlying assumptions that may contribute to and produce differential exposure and unequal protection. It also brings to the surface the *ethical* and *political* questions of 'who gets what, when, why, and how much'. Some general characteristics of this framework include:

- the principle of the 'right' of all individuals to be protected from environmental degradation;
- a public health model of prevention – i.e. elimination of the threat before harm occurs – as the preferred strategy;
- the burden of proof being shifted to polluters/dischargers who do harm to, discriminate against or do not give equal protection to people of colour, low-income persons, and other 'protected' classes;
- the use of disparate impact and statistical weight or an 'effect' test, as opposed to 'intent', to infer discrimination;
- redressing disproportionate impact through 'targeted' action and resources. In general, this strategy would target resources where environmental and health problems are greatest (as determined by some ranking scheme but not limited to risk assessment).

Endangered communities

Numerous studies reveal that low-income persons and people of colour have borne greater health and environmental risk burdens than society at large (Goldman, 1992; Goldman and Fitton, 1994). Elevated public health risks have been found in some populations even when social class is held constant. For example, race has been found to be independent of class in the distribution of air pollution, contaminated fish consumption, location of municipal landfill sites and incinerators, abandoned toxic waste dumps, clean-up of superfund sites, and lead poisoning in children (Commission for Racial Justice, 1987; Agency for Toxic Substances and Disease Registry, 1988; West *et al.*, 1992; Bryant and Mohai, 1992; Geddicks, 1993; Goldman and Fitton, 1994; Lavelle and Coyle, 1992; Pirkle *et al.*, 1994).

Asthma is a classic example of an environmental health problem that disproportionately impacts African American and Latino children and the poor (Schwartz *et al.*, 1990). From 1982 to 1991, the age-adjusted death rate for asthma for persons aged 5 to 34 was approximately five times higher among African Americans than whites. Between 4 and 5 million children under age 18 suffer from asthma, the most common chronic disease among children. It is the fourth leading cause of disability in children. Poor children are at special risk from air pollution (Thurston *et al.*, 1992). Asthma is 26 per cent higher among African American children than among white children.

Persons suffering from asthma are particularly sensitive to the effects of carbon monoxide, sulphur dioxides, particulate matter, ozone and nitrogen oxides. Hospitalization and mortality due to asthma exhibit wide racial differences. The US federal Centers for Disease Control found that African Americans are two to three times more likely than whites to be hospitalized for or die from asthma (Centers for Disease Control, 1992). In Atlanta, for example, ozone pollution appears to exacerbate childhood asthma problems. The average number of hospital visits for asthma or reactive airway disease was 37 per cent higher on the days after the high ozone pollution (White *et al.*, 1994).

Childhood lead poisoning is another preventable disease that dispro-portionately affects poor children and children of colour. Figures reported in the July 1994 *Journal of the American Medical Association* on the Third National Health and Nutrition Examination Survey (NHANES III) revealed that 1.7 million children (8.9 per cent of children aged 1 to 5) are lead poisoned, defined as having blood lead levels equal to or above 10 micrograms per decilitre. The NHANES III data found African American children to be lead poisoned at more than twice the rate of white children at every income level (Pirkle *et al.*, 1994). Over 28.4 per cent of all low-income African American children were lead poisoned compared to 9.8 per cent of low-income white children. Between 1976 and 1991, the decrease in blood lead levels for African American and Mexican American children lagged far behind that of white children.

In California, a coalition of environmental, social justice and civil libertarian groups joined forces to challenge the way the state carried out its lead screening of poor children. The Natural Resources Defense Council, the National Association for the Advancement of Colored People (NAACP) Legal Defense and Education Fund, the American Civil Liberties Union and the Legal Aid Society of Alameda County, California, won an out-of-court settlement worth $15 million–$20 million for a blood lead-testing programme. The lawsuit *Matthews v. Cove* involved the failure of the State of California to conduct federally mandated testing for lead of some 557,000 poor children who receive Medicaid (Lee, 1992). This historic agreement triggered similar lawsuits and actions in several other states that failed to live up to the mandates.

Impetus for policy shift

The impetus behind the environmental justice movement did not come from within government, from academia or from within largely white, middle-class, nationally based environmental and conservation groups. The impetus for change came from people of colour, grassroots activists and their 'bottom-up' leadership approach. Grassroots groups organized themselves, educated them-selves and empowered themselves to make fundamental change in the way environmental protection is performed in their communities.

The environmental justice movement has come a long way since its humble beginning in rural, predominantly African American Warren County, North Carolina, where a PCB landfill site ignited protests and over five hundred arrests. The Warren County protests provided the impetus for a US General Accounting Office (1983) study, *Siting of Hazardous Waste Landfills and Their Correlation with Racial and Economic Status of Surrounding Communities*. That study revealed that three out of four of the off-site, commercial hazardous waste landfill sites in Region 4 (which comprises eight states in the South) happened to be located in predominantly African American communities, although African Americans made up only 20 per cent of the region's population. In 1997, both of the operating commercial offsite hazardous waste landfill sites in the region were located in mostly African American communities.

The protests also led the Commission for Racial Justice (1987) to produce *Toxic Waste and Race*, the first national study to correlate waste facility sites and demographic characteristics. Race was found to be the most potent variable in predicting where these facilities were located – more powerful than poverty, land values and home ownership. In 1990, *Dumping in Dixie: Race, Class and Environmental Quality* chronicled the convergence of two social movements – social justice and environmental movements – into the environmental justice movement (Bullard, 1990). This book highlighted African Americans' environmental activism in the South, the same region that gave birth to the modern civil rights movement. What started out as local and often isolated community-based struggles against toxic waste and landfill facility siting blossomed into a multi-issue, multiethnic and multiregional movement.

The 1991 First National People of Color Environmental Leadership Summit was probably the most important single event in the movement's history. The summit broadened the environmental justice movement beyond its anti-toxic waste focus to include issues of public health, worker safety, land use, transportation, housing, resource allocation and community empowerment (Lee, 1992). The meeting, organized by and for people of colour, demonstrated that it is possible to build a multiracial grassroots movement around environmental and economic justice (Alston, 1992).

Held in Washington, DC, the four-day summit was attended by over 650 grassroots and national leaders from around the world. Delegates came from all fifty states including Alaska and Hawaii, from Latin America including Puerto Rico, Chile, Mexico, and from as far away as the Marshall Islands. People attended the summit to share their action strategies, redefine the environmental movement, and develop common plans for addressing environmental problems affecting people of colour in the United States and around the world (Alston and Brown, 1993).

On 27 October 1991, summit delegates adopted seventeen 'Principles of Environmental Justice'. These principles were developed as a guide for organizing, networking, and relating to government and non-governmental organizations (NGOs). By June 1992, Spanish and Portuguese translations of the principles were being used and circulated by NGOs and community groups at the Earth Summit in Rio de Janeiro.

Federal, state and local policies and practices have contributed to residential segmentation and unhealthy living conditions in poor, working-class, and people-of-colour communities (Bullard and Johnson, 1997). Several recent California cases bring this point to life (Lee, 1995). Disparate highway siting and mitigation plans were challenged by community residents, churches and the NAACP Legal Defense and Education Fund. A case was brought against the US Department of Transportation (*Clean Air Alternative Coalition v. United States Department of Transportation*, N.D. Cal. C-93-0721-VRW), involving the reconstruction of the earthquake-damaged Cypress Freeway in West Oakland. The plaintiffs wanted the downed Cypress Freeway (which split their community in half) rebuilt further away. Although the plaintiffs were not able to get their plan implemented, they did change the course of the freeway in their out-of-court settlement.

The NAACP LDF filed an administrative complaint, challenging the construction of the 4.5-mile extension of the Long Beach Freeway in East Los Angeles through El Sereno, Pasadena and South Pasadena (*Mothers of East Los Angeles, El Sereno Neighborhood Action Committee, El Sereno Organizing Committee, et al. v. California Transportation Commission, et al.*, before the US Department of Transportation and US Housing and Urban Development Department). The plaintiffs argued that mitigation measures to address noise, air and visual pollution proposed by the state agencies discriminated against the mostly Latino El Sereno community. For example, all of the freeway in Pasadena and 80 per cent of it in South Pasadena will be below ground level. On the other hand, most of the freeway in El Sereno will be above ground. White areas were favoured over the mostly Latino El Sereno in allocation of covered freeway, historic preservation measures and accommodation to local schools (Lee, 1995; Bullard and Johnson, 1997).

Los Angeles residents and the NAACP LDF have also challenged the inequitable funding and operation of bus transportation, used primarily by low-income residents and people of colour. A class-action lawsuit was filed on behalf of 350,000 bus riders represented by the Labor/Community Strategy Center, the Bus Riders Union, the Southern Christian Leadership Conference, Korean Immigrant Workers Advocates and individual bus riders. The plaintiffs argued that the Los Angeles Metropolitan Transportation Authority had used federal funds to pursue a policy of raising costs of bus riders and reducing quality of service in order to fund rail and other projects in predominately white, suburban areas (*Labor/Community Strategy Center v. Los Angeles Metropolitan Transportation Authority* (Cal. CV 94-5936 TJH Mcx). In September, 1996, the Labor/Community Strategy Center and their lawyers won a historic out-of-court settlement against the MTA (Bullard and Johnson, 1997).

Making government more responsive

Many of the United States' environmental policies distribute costs in a regressive pattern while providing disproportionate benefits for whites and individuals who fall at the upper end of the education and income scale. A 1992 study reported in the *National Law Journal* uncovered glaring inequities in the way the federal Environmental Protection Agency (EPA) enforces its laws:

> There is a racial divide in the way the U.S. government cleans up toxic waste sites and punishes polluters. White communities see faster action, better results and stiffer penalties than communities where blacks, Hispanics and other minorities live. This unequal protection often occurs whether the community is wealthy or poor.
>
> (Lavelle and Coyle, 1992)

The *National Law Journal* study reinforced what many grassroots activists have known for decades: all communities are not treated the same. Communities that are located on the 'wrong side of the tracks' are at greater risk from exposure to lead, pesticides (in the home and workplace), air pollution, toxic releases, water pollution, solid and hazardous waste, raw sewage and pollution from industries (Goldman, 1992).

Government has been slow to ask questions concerning who gets help and who does not, who can afford help and who cannot, why some contaminated communities are studied while others get left off the research agenda, why industry poisons some communities and not others, why some contaminated communities get cleaned up while others do not, why some populations are protected and others are not protected, why unjust, unfair and illegal policies and practices are allowed to go unpunished. In 1990 the Agency for Toxic Substances and Disease Registry (ATSDR) held a historic conference in Atlanta. The National Minority Health Conference focused on contamination in people-of-colour communities (Johnson *et al.*, 1992).

In 1992, after meeting with community leaders, academicians and civil rights leaders, the EPA (under the leadership of William Reilly) admitted there was a problem, and established the Office of Environmental Equity. The name was changed to the Office of Environmental Justice under the Clinton administration. In the same year, the EPA produced one of the first comprehensive documents to examine the whole question of risk and environmental hazard (US Environmental Protection Agency, 1992). The report and its Office of Environmental Equity were initiated only after prodding from people-of-colour environmental justice leaders, activists, and a few academicians.

The EPA also established a twenty-five-member National Environmental Justice Advisory Council (NEJAC) under the Federal Advisory Committee Act (FACA). NEJAC divided its environmental justice work into six subcommittees: Health and Research, Waste and Facility Siting, Enforcement, Public Participation and Accountability, Native American and Indigenous Issues, and International Issues. NEJAC is made up of stakeholders representing grassroots community groups, environmental groups, NGOs, state, local and tribal governments, academia and industry.

In February 1994 seven federal agencies, including the Agency for Toxic Substances and Disease Registry (ATSDR), the National Institute for

Environmental Health Sciences (NIEHS), the EPA, the National Institute of Occupational Safety and Health (NIOSH), the National Institutes of Health (NIH), the Department of Energy (DOE) and the Centers for Disease Control (CDC) sponsored a national health symposium, 'Health and Research Needs to Ensure Environmental Justice'. The conference planning committee was unique in that it included grassroots organization leaders, affected community residents, and federal agency representatives. The goal of this symposium was to bring diverse stakeholders and those most affected by environmental problems to the decision-making table (National Institute for Environmental Health Sciences, 1995). Some of the recommendations from the symposium included:

- conduct meaningful health research in support of people of colour and low-income communities;
- promote disease prevention and pollution prevention strategies;
- promote interagency coordination to ensure environmental justice;
- provide effective outreach, education and communications programmes; and
- design legislative and legal remedies.

In response to growing public concern, President Clinton on 11 February 1994 (i.e. the second day of the health symposium) issued Executive Order 12898, 'Federal Actions to Address Environmental Justice in Minority Populations and Low-Income Populations'. This Order is *not* a new law but an attempt to address environmental injustice within existing federal laws and regulations. Executive Order 12898 reinforces the thirty-year-old Civil Rights Act of 1964, Title VI, which prohibits discriminatory practices in programmes receiving federal funds. The Order also focuses the spotlight back on the National Environmental Policy Act (NEPA), a twenty-five-year-old law that set policy goals for the protection, maintenance and enhancement of the environment. NEPA's goal is to ensure for all Americans a safe, healthful, productive, and aesthetically and culturally pleasing environment. NEPA requires federal agencies to prepare a detailed statement on the environmental effects of proposed federal actions that significantly affect the quality of human health.

The Executive Order calls for improved methodologies for assessing and mitigating impacts, health effects from multiple and cumulative exposure, collection of data on low-income and minority populations who may be disproportionately at risk, and impacts on subsistence fishers and wildlife consumers. It also encourages participation of the affected populations in the various phases of assessing impacts – including scoping, data gathering, alternatives, analysis, mitigation and monitoring.

The Executive Order thus focuses on 'subsistence' fishers and wildlife consumers. Not everybody buys their fish at the supermarket. There are many people who are subsistence fishers, who fish for protein, who basically subsidize their budgets, and their diets, by fishing from rivers, streams and lakes that happen to be polluted. These subpopulations may be underprotected if standard assumptions are made using the dominant risk paradigm. The Executive Order is one attempt to fix this problem.

Winning in the courts: the case of Citizens Against Nuclear Trash v. Louisiana Energy Services (*CANT v. LES*)

The first lawsuit to challenge environmental racism on civil rights grounds in the courts was *Bean v. Southwestern Waste Management*, filed in 1979 by Linda McKeever Bullard on behalf of the Northeast Community Action Group (NECAG), a Houston-based neighbourhood organization (Bullard, 1990). Dozens of lawsuits have been filed, argued and lost since *Bean*. In the real world, all people and communities are *not* created equal. Some populations and interests are 'more equal than others'. Unequal interests and power arrangements have allowed treatment of the poisons of the rich to be offered as short-term remedies for the poverty of the poor (Bryant and Mohai, 1992). Institutional racism creates 'invisible' communities – whose people are not seen and whose voices are muted. This is not fair or just, *but it is often judged to be legal*. Few low-income communities have the financial resources to hire lawyers and experts. Generally, resource-rich organizations win over poor, unorganized communities. However, there are exceptions.

It has taken nearly two decades since *Bean* for a court to rule in favour of a community's charge of environmental discrimination. Since 1991 the Nuclear Regulatory Commission had under review a proposal from Louisiana Energy Services (LES) to build the nation's first privately owned uranium enrichment plant. A national search was undertaken by LES to find the 'best' site for a plant that would produce 17 per cent of the nation's enriched uranium. LES supposedly used the Kepner–Tregoe (1981) method in designing its site selection process. The Kepner–Tregoe (1981) decision analysis method is a widely used means for comparing alternatives on the basis of multiple criteria using a 10-point weighted scoring system in which criteria are divided into those that *must* be met ('musts') and those that are *desirable* ('wants'), with the 'wants' weighted according to relative importance.

The southern United States, Louisiana and Claiborne Parish ended up being the dubious 'winners' of the site selection process. The Claiborne Enrichment Center was slated to be built in the midst of two African American communities: Forest Grove and Center Springs. Many of the residents of these two rural communities hunt, fish and grow gardens to subsidize their household budgets. A sizeable share of households use wells for drinking water. There are no industries, stores, shops, schools or hospitals in these two rural, racially segregated communities.

Forest Grove and Center Springs residents and citizens from Homer disagreed with the site selection process and outcome. They organized themselves into a group called Citizens Against Nuclear Trash, or CANT. CANT and the community residents saw the LES proposal as unjust. Local residents doubted the promise of jobs. They distrusted assurances that the 100 tons of radioactive debris that would be stored at the site would be safe (Bullard, 1995). They charged LES and the NRC staff with practising environmental racism.

In 1991, CANT hired the Sierra Club Legal Defense Fund and sued LES. The *CANT v. LES* lawsuit dragged on for more than seven years. CANT also hired an environmental sociologist to provide expert testimony on the conclusions of the environmental impact statement (EIS) and environmental report (ER) prepared for the proposed uranium enrichment plant and their consideration of environmental justice implications (Bullard, 1995).

On 1 May 1997 a three-judge panel of the NRC Atomic Safety and Licensing Board finally issued an initial decision on the case. The judges denied the permit and concluded that 'racial bias played a role in the selection process' (US Nuclear Regulatory Commission, 1997). The precedent-setting federal court ruling was handed down some two years after President Clinton signed Executive Order 12898. The judges in a thirty-eight-page written decision also chastised the NRC staff for not addressing the provision called for under Executive Order 12898 (US Nuclear Regulatory Commission, 1997).

A clear racial pattern emerged during the search and multistage screening process (Bullard, 1995). For example, African Americans comprise about 13 per cent of the US population, 20 per cent of the Southern states' population, 31 per cent of Louisiana's population, 35 per cent of Louisiana's northern parishes, and 46 per cent of Claiborne Parish. This progressive trend, involving the narrowing of the site selection process to areas of increasingly high poverty and African American representation, is also evident from an evaluation of the actual sites that were considered in the 'Intermediate' and 'Fine' Screening stages of the site selection process.

The aggregate average percentage of black population for a one-mile radius around all of the seventy-eight sites examined (in sixteen parishes) is 28.4 per cent. When LES completed its initial site cuts, and reduced the list to thirty-seven sites within nine parishes, the aggregate percentage of black population rose to 36.8 per cent. When LES then further limited its focus to six sites in Claiborne Parish, the aggregate average percentage black population rose again, to 64.8 per cent. The final site selected, the 'LeSage' site, has a 97.1 per cent black population within a one-mile radius (Table 3.1). The plant was

Table 3.1 Population by race living within a one-mile radius of Louisiana Energy Services candidate sites for a uranium enrichment plant

Candidate sites	Total population	Black population	Percentage
Initial 78 sites	18,722	5,321	28.4
Intermediate 37 sites	8,380	3,082	36.8
Fine-screening, 6 sites	1,160	752	64.8
Final selection, 1 site	138	134	97.1

Source: US Bureau of the Census, PL94-171

proposed to be located on Parish Road 39 – a road that connects two African American communities, namely Forest Grove and Center Springs. The proposed site was just a quarter of a mile from Center Springs (founded in 1910) and one and a quarter miles from Forest Grove (founded in the 1860s just after the abolition of slavery). In Claiborne Parish per capita income is only $5,800 per year, just 45 per cent of the national average; over 58 per cent of the African American population is below the poverty line.

The two rural communities of Forest Grove and Center Springs were rendered 'invisible' since they were not even mentioned or shown on any maps in the Nuclear Regulatory Commission Draft Environmental Impact Statement (US Nuclear Regulatory Commission, 1993). Their population was severely undercounted. But more important, risk calculations and assumptions for the permit were based on the populated area of Homer – a community that is five miles from the proposed site. Only after intense public criticism did the NRC staff attempt to address environmental justice and disproportionate impact implications as required under NEPA and called for under the Environmental Justice Executive Order 12898. For example, NEPA required that the government consider the environmental impacts and weigh the costs and benefits of the proposed action. These include health and environmental effects, the risk of accidental but foreseeable adverse health and environmental effects, and socio-economic impacts.

The NRC devoted less than a page to addressing environmental justice concerns of the proposed uranium enrichment plant in its Final Environmental Impact Statement (US Nuclear Regulatory Commission, 1994). Overall, the Final Environmental Impact Statement (EIS) and Environmental Report (ER) are inadequate in the following respects: (a) they inaccurately assessed the costs and benefits of the proposed plant; (b) they failed to consider the inequitable distribution of costs and benefits of the proposed plant to white and African American populations; and (c) they failed to consider the fact that the siting of the plant in a community of colour follows a national pattern in which institutionally biased decision-making leads to the siting of hazardous facilities in communities of colour, and results in the inequitable distribution of costs and benefits to those communities (Bullard, 1995).

The distributive costs not analysed in relationship to Forest Grove and Center Springs included the disproportionate burden of health and safety, the impact on property values, risk of fire and accidents, noise, traffic, radioactive dust in the air and water, and dislocation resulting from closure of a road that connects the two communities. The EIS and ER failed to live up to the mandates and requirement under NEPA. Overall, the CANT victory points to the utility of *combining environmental and civil rights laws* and the President's Executive Order on Environmental Justice to prevent future injustices. The NRC did the right and just thing in denying a permit for the LES uranium enrichment plant.

Conclusion

The environmental protection apparatus in the United States is *broken* and needs to be *fixed*. Governments must live up to their mandate of protecting all peoples and the environment. The call for environmental and economic justice does not stop at the United States borders but extends to communities and nations that are threatened by hazardous wastes, toxic products, environmentally unsound technology and non-sustainable development models (see this volume, Chapter 4 by Shiva). All communities are not created equal. Some communities and their inhabitants are more vulnerable than others. Many environmental inequities result from land use, housing and development practices that disproportionately and adversely affect the poor and people of colour. Racism renders some people and communities 'invisible' and vulnerable to environmental exploitation.

The environmental justice movement has set out clear goals of eliminating unequal enforcement of environmental rights, civil rights, and laws on housing, transportation, facility siting and public health. In the United States, community residents are fighting to end their exposure to harmful chemicals, pesticides and other toxins in their homes, schools, neighbourhoods and workplaces. They are challenging the so-called 'science' and the faulty assumptions in selecting sites for polluting facilities, and for calculating, assessing and managing risks. In many cases the only science involved in the siting of locally unwanted land uses or LULUs is political science. Environmental justice advocates are demanding an end to discriminatory zoning and land-use practices, and exclusionary policies that limit the participation of poor people and people of colour in decision-making.

Finally, the solution to environmental injustice lies in the realm of equal protection of all individuals, groups and communities. The key is prevention. Many of these problems could be eliminated if existing environmental, health, housing and civil rights laws were vigorously enforced in a non-discriminatory way. No community, state or nation, rich or poor, urban or suburban, black or white, should be allowed to become a 'sacrifice zone' or dumping ground.

References

Agency for Toxic Substances and Disease Registry (1988) *The Nature and Extent of Lead Poisoning in Children in the United States: A Report to Congress*, Atlanta: US Department of Health and Human Services.

Alston, D. (1992) 'Transforming a Movement: People of Color Unite at Summit against Environmental Racism', *Sojourner* 21: 30–1.

Alston, D. and Brown, N. (1993) 'Global Threats to People of Color', in R. D. Bullard (ed.) *Confronting Environmental Racism: Voices from the Grassroots*, Boston: South End Press, pp. 179–94.

Austin, R. and Schill, M. (1991) 'Black, Brown, Poor, and Poisoned: Minority Grassroots Environmentalism and the Quest for Eco-Justice', *Kansas Journal of Law and Public Policy* 1(1): 69–82.

Bryant, B. and Mohai, P. (1992) *Race and the Incidence of Environmental Hazards*, Boulder, CO: Westview Press.

Bullard, R. D. (1990) *Dumping in Dixie: Race, Class and Environmental Quality*. Boulder, CO: Westview Press.

—— (ed.) (1993a) *Confronting Environmental Racism: Voices from the Grassroots*, Boston: South End Press.

—— (1993b) 'Race and Environmental Justice in the United States', *Yale Journal of International Law* 18: 319–35.

—— (1993c) 'Environmental Racism and Land Use', *Land Use Forum: A Journal of Law, Policy and Practice* 2: 6–11.

—— (1994) *Unequal Protection: Environmental Justice and Communities of Color*, San Francisco: Sierra Club Books.

—— (1995) 'Pre-filed Written Testimony at the *CANT vs. LES* Hearing', Shreveport, LA.

Bullard, R. D. and Johnson, G. S. (1997) *Just Transportation: Dismantling Race and Class Barriers*, Gabriola Island, BC: New Society Publishers.

Centers for Disease Control (1992) 'Asthma: United States, 1980–1990', *MMWR* 39: 733-735.

Chase, A. (1993) 'Assessing and Addressing Problems Posed by Environmental Racism', *Rutgers University Law Review* 45: 385–89.

Coleman, L. A. (1993) 'It's the Thought That Counts: The Intent Requirement in Environmental Racism Claims', *St Mary's Law Journal* 25: 447–92.

Collin, R. W. (1992) 'Environmental Equity: A Law and Planning Approach to Environmental Racism', *Virginia Environmental Law Journal* 11: 495–546.

Colopy, J. H. (1994) 'The Road Less Traveled: Pursuing Environmental Justice through Title VI of the Civil Rights Act of 1964', *Stanford Environmental Law Journal* 13: 126–89.

Colquette, K. C. and Robertson, E. A. H. (1991) 'Environmental Racism: The Causes, Consequences, and Commendations', *Tulane Environmental Law Journal* 5(1): 153–207.

Commission for Racial Justice (1987) *Toxic Waste and Race in the United States*, New York: United Church of Christ.

Geddicks, A. (1993) *The New Resource Wars: Native and Environmental Struggles against Multinational Corporations*, Boston: South End Press.

Godsil, R. D. (1990) 'Remedying Environmental Racism', *Michigan Law Review* 90/394: 394–427.

Goldman, B. (1992) *The Truth about Where You Live: An Atlas for Action on Toxins and Mortality*, New York: Random House.

Goldman, B. and Fitton, L. J. (1994). *Toxic Wastes and Race Revisited*, Washington, DC: Center for Policy Alternatives, National Association for the Advancement of Colored People, United Church of Christ.

Johnson, B. L., Williams, R. C. and Harris, C. M. (1992) *Proceedings of the 1990 National Minority Health Conference: Focus on Environmental Contamination*, Princeton, NJ: Scientific Publishing Co.

Kepner, C. H. and Tregoe, B. B. (1981) *The New Rational Manager*, Princeton, NJ: Princeton Research Press.

Lavelle, M. and Coyle, M. (1992) 'Unequal Protection', *National Law Journal*, 21 September: 1–2.

Lee, B. L. (1992) 'Environmental Litigation on Behalf of Poor, Minority Children: *Matthews v. Coye*: A Case Study', Paper presented at the Annual Meeting of the American Association for the Advancement of Science, Chicago, 9 February.

—— (1995) 'Civil Rights Remedies for Environmental Injustice', Paper presented at the Transportation and Environmental Justice: Building Model Partnerships Conference, Atlanta, GA, 11 May.

Lee, C. (1992) *Proceedings: The First National People of Color Environmental Leadership Summit*, New York: United Church of Christ Commission for Racial Justice.

National Institute for Environmental Health Sciences (1995) *Proceedings of the Health and Research Needs to Ensure Environmental Justice Symposium*, Research Triangle Park, NC: NIEHS.

Pirkle, J. L., Brody, D. J., Gunter, E. W., Kramer R. A., Paschal, D. C., Glegal, K. M. and Matte, T. D. (1994) 'The Decline in Blood Lead Levels in the United States: The National Health and Nutrition Examination Survey – HANES III', *Journal of the American Medical Association* 272: 284–91.

Schwartz, J., Gold, D., Dockey, D. W., Weiss, S. T. and Speizer, F. E. (1990) 'Predictors of Asthma and Persistent Wheeze in a National Sample of Children in the United States', *American Review of Respiratory Disease* 142: 555–62.

Thurston, G. D., Ito, K., Kinney, P. L. and Lippmann, M. (1992) 'A Multi-year Study of Air Pollution and Respiratory Hospital Admission in Three New York State Metropolitan Areas: Results for the 1988 and 1989 Summers', *Journal of Analytical Environmental Epidemiology* 2: 429–50.

US Environmental Protection Agency (1992) *Environmental Equity: Reducing Risk for All Communities*, Washington, DC: US EPA.

US General Accounting Office (1993) *Siting of Hazardous Waste Landfills and Their Correlation with Racial and Economic Status of Surrounding Communities*, Washington, DC: Government Printing Office.

US Nuclear Regulatory Commission (1993) *Draft Environmental Impact Statement for the Construction and Operation of Claiborne Enrichment Center, Homer, Louisiana*, Washington, DC: NRC.

—— (1994) *Final Environmental Impact Statement for the Construction and Operation of Claiborne Enrichment Center, Homer, Louisiana*, Washington, DC: NRC.

—— (1997) 'Final Initial Decision – Louisiana Energy Services', US Nuclear Regulatory Commission, Atomic Safety and Licensing Board, Docket No. 70-3070-ML (1 May).

West, P., Fly, J. M., Larkin, F. and Marans, P. (1990) 'Minority Anglers and Toxic Fish Consumption: Evidence of the State-Wide Survey of Michigan', in B. Bryant and P. Mohai (eds) *Race and the Incidence of Environmental Hazards*, Boulder, CO: Westview Press, pp.100–13.

Westra, L. and Wenz, P. S. (eds) (1995) *Faces of Environmental Racism: Confronting Issues of Global Justice*, Lanham, MD: Rowman & Littlefield.

White, M. C., Etzel, R. A., Wilcox, W. D., and Lloyd, C. (1994) 'Exacerbations of childhood asthma and ozone pollution in Atlanta', *Environmental Research* 65: 56–68.

4 Ecological balance in an era of globalization

Vandana Shiva

Introduction: globalization as a political phenomenon

In 1992, the Earth Summit in Rio marked the maturing of ecological awareness on a global scale. The world was poised to make a shift to ecological sustainability. However, the Rio process and the sustainability agenda were subverted by the free trade agenda. In 1993 the Uruguay Round of the General Agreement on Tariffs and Trade (GATT) was completed; in 1995 the World Trade Organization (WTO) was established and world affairs started to be dictated by trade and commerce. The normative political commitment to sustainability and justice was replaced by the rule of trade and the elevation of exploitation, greed and profit maximization as the organizing principles of the market, the state and society. Instead of the state regulating the market for the good of society, global economic powers and commercial forces are now regulating the state and society for the benefit of corporations. Instead of commerce being accountable to state and society, economic globalization is making citizens and their governments accountable to corporations and global economic bodies.

Economic globalization is not merely an economic phenomenon related to the reduction of tariff barriers and the removal of 'protectionist' policies. It is a normative process which replaces all value by commercial value. Free trade is in reality the rule of commerce. GATT and the WTO basically undo the Rio agenda. Five years after Rio we do not have Rio plus five but Rio minus five. The search for ecological balance in an era of globalization requires on the one hand an assessment of the social and ecological impact of globalization. On the other hand it requires the imagination and realization of an alternative order which puts ecological balance and social and economic justice rather than trade at the centre of economic policy.

Globalization is not a natural, evolutionary or inevitable phenomenon, as is often argued. It is a political process which has been forced on the weak by the powerful. Globalization is not the cross-cultural interaction of diverse societies. It is the imposition of a particular culture on all others. Nor is globalization the search for ecological balance on a planetary scale. It is the predation of one class, one race and often one gender of a single species on all others. The 'global' in the dominant discourse is the political space in which the dominant *local* seeks global control, and tries to free itself from natural limits arising from the imperatives of ecological sustainability and social justice. The global in this sense does not represent the universal human interest; it represents a particular *local* and *parochial* interest and culture, which has been globalized through its reach and control, its irresponsibility and lack of reciprocity.

Three waves of globalization

Globalization has come in three waves. The first wave was the colonization of the Americas, Africa, Asia and Australia by European powers over a period of five hundred years. The second wave was the imposition of the West's idea of 'development' on non-Western cultures in the post-colonial era of the past five decades. The third wave of globalization was unleashed approximately five years ago as the era of 'free trade', which for some commentators implies an end to history but for us in the Third World is a repeat of history through recolonization. Each wave of globalization is cumulative in its impact, even while it creates a discontinuity in the dominant metaphors and actors. Each wave of globalization has served Western interests, and each wave has created deeper colonization of other cultures and of the planet's life.

Each time a global order has tried to wipe out diversity and impose homogeneity, disorder and disintegration have been induced, not removed. Globalization introduces violence when it replaces the self-organizing capacity and order of diverse social and natural systems with an externally controlled global order, which is unstable and is maintained only through coercion and force. Globalization and homogenization are now being carried out not by nation states, but by the global powers that control global markets. 'Free trade' is the ruling metaphor for globalization in our times. But far from protecting the freedom of citizens and countries, free trade negotiations and treaties have become the primary sites of the use of coercion and force. The Cold War era has ended and the era of trade wars has begun. Among examples of violence in the free trade era is the US Trade Act, especially the Super and Special 301 clauses that allow the United States to take unilateral action against any country that does not open up its market to US corporations. Super 301 is used to force freedom for investment. Special 301 is used to force freedom for monopoly control of markets through the protection of intellectual property rights. Free trade is in fact an asymmetric arrangement that combines liberalization with protection for Northern interests.

Third World countries resisted the expansion of GATT into new areas like services, investments and intellectual property rights. By merely affixing the term 'trade-related' to issues that have been decided domestically, GATT, through its successor organization the WTO, will not merely regulate international trade, but will in essence determine domestic policy. Brute force continued to be used against the Third World even in multilateral negotiations of the Uruguay Round of GATT. In a speech on 14 January 1994, Fernando Jaramillo, chairman of the Group of 77 and Colombia's permanent representative in the

UN, said, 'The Uruguay Round is proof again that the developing world continues to be sidelined and rejected when it comes to defining areas of vital importance to their survival' (Raghavan, 1995).

Because of the process through which they have been arrived at, free trade treaties like GATT are not expressions of freedom for all. They have been forced on citizens and on weaker trading partners, namely the countries of the Third World. Multilateral treaties like GATT are not really multilateral. Nothing makes this clearer than delivery in 1991 of a 'take it or leave it' draft prepared by the GATT Secretary-General, Arthur Dunkel – which in India has acquired the not so pleasant acronym of DDT (Dunkel Draft Text). An even more blatant expression of both undemocratic decision-making and non-transparency was the process in the final stages of GATT negotiations in which two men, Micky Kantor, the US Trade Representative, and Leon Brittan, the European Union negotiator, sat behind closed doors and then presented the world with a 'free trade' treaty which was supposed to have been negotiated multilaterally. This is far from multilateralism and even further from global democracy.

Despite insisting that the negotiations were global, the countries of the North refused in the end to accept any discussions, even bilateral ones, with the countries of the Third World, illustrating once again that globalization is the imposition of the will of the economic powers of the North on the rest on the world. A new authoritarian structure is emerging, as Ambassador Jaramillo of Colombia observed in his speech in Geneva:

> The Bretton Woods Institutions continue to be made the centre of gravity for the principal economic decisions that affect the developing countries. We have all been witnesses to the conditionalities of the World Bank and the IMF. We all know the nature of the decision-making system in such institutions; their undemocratic character, their lack of transparency, their dogmatic principles, their lack of pluralism in the debate of ideas, and their impotence to influence the policies of the industrialised countries. This also seems to be applicable to the new World Trade Organization. The terms of its creation suggest that this [institution] will be dominated by the industrialised countries and that its fate will be to align itself with the World Bank and IMF. We could announce in advance the birth of a New Institutional Trinity which would have as its specific function to control and dominate the economic relations that commit the developing world.
>
> (Raghavan, 1995)

Free trade has come to mean in reality the vastly expanded freedom of transnational corporations to trade and invest, while national governments have significantly reduced powers to restrict their operations. Multinational corporations, which were the real power behind the Uruguay Round, have gained new rights and have given up old obligations to protect workers' rights and environmental rights.

The community, the state and the corporation

Globalization has distorted the relationship between the community, the state and the economy, or, to use Marc Nerfin's more colourful categories, the relationship between the citizen, the prince and the merchant. Globalization privileges the economy and its key actor, the corporation, inasmuch as the state and the community are increasingly becoming mere instruments of global capital. The appeal of globalization is usually based on the idea that it implies less red tape, less centralization and less bureaucratic control. It is celebrated because it implies the erosion of the power of the state, and a reduction in the role of bureaucratic institutions in the lives of people. Whereas over the previous fifty years the state had increasingly taken over the functions of the community and the self-organizing capacity of citizens, now, through globalization, corporations are taking over the functions of both the state and citizens. Food provisioning, health care, education and social security are all being commodified and transformed into corporate monopolies under the rubric of 'competitiveness' and 'efficiency'. People's rights and the public domain are being eroded by exporting the economic label of 'protectionism' to cover all domains – ethical, social and political. The protection of the environment and the protection of people's security are treated as non-tariff trade barriers which need to be dismantled.

While the state is being required to step back from the regulation of trade and commerce, it is also increasingly being called in to regulate citizens and remove communities which are an 'obstruction' to free trade. Thus the state is becoming *leaner* in dealing with big business and global industry, and *meaner* in dealing with people. In the South and in the North, the principle of 'eminent domain' is still applied to state take-over of people's land and resources, which are then handed over to global corporations. For example, in India under the new infrastructure policies, foreign companies can have 100 per cent equity participation, but the government will acquire the land, displace the people and deal with 'law and order' problems created by the displacements. In the United States, federal, state and local governments are appropriating citizens' homes and farms to hand over to large corporations. In Hurst, Texas, a suburb of Fort Worth, the government appropriated the land of more than a hundred homeowners to hand it over to its biggest taxpayer, the North East Mall. Four thousand two hundred residences were destroyed in Detroit so that General Motors could build a new plant (see *US News World Report*, 15 September 1997). Quite clearly, it is the prosperity of the powerful corporations that is being protected by the state in every part of the world under the new free trade regimes, while the property of the ordinary citizen has no protection.

Another area where the role of the state is also increasing is in 'intellectual property rights' (IPRs) – the privatization of knowledge (see Fourmile, this volume, Chapter 13). As larger and larger domains of knowledge are being converted into 'intellectual property' through patents (from microbes to mice, from seeds to human cell lines), the state is being increasingly called upon to police citizens to prevent them from engaging in everyday activities such as saving seeds and exchanging knowledge. Through intellectual property legislation our most human acts have been criminalized in relationship to ourselves, to each other and to other species.

The inverted role of the state with respect to citizens is exemplified in the concrete cases of John Moore, Peter Toborsky, Josef Albrecht and Dennis and Becky Winterboer. John Moore, a cancer patient, went to court when he discovered that his own doctor, Dr Golde, had patented his cell line as the Mo-cell line. The Supreme Court of California ruled that Moore could not have autonomy and integrity with respect to his own body since this would interfere with trade and scientific progress (*Bija Newsletter*, 1995). Peter Taborsky worked as a laboratory assistant in the University of Florida on a project funded by the Progressive Technologies Corporation. Outside his scheduled work hours Peter did research of his own, for which he obtained a patent. He was accused of 'theft' by the corporation and was arrested. Peter's arrest dramatizes the problems of IPRs linked to the private funding of public institutions. Most laboratories and research facilities have been built by public funds. When a corporation finances a project, and the research product becomes its intellectual property, it is forgotten that the facilities that make knowledge production possible were built up as a public resource. Later, when someone uses that public resource to generate new ideas, it is treated as theft, as in the case of Peter Taborsky (Shiva, 1996).

The case of farmer Josef Albrecht in Germany and potato seed farmers in Scotland are examples of how Seed Acts prevent farmers from engaging in their own seed production. Albrecht is an organic farmer in the village of Oberding in Bavaria. Not satisfied with commercially available seed, he developed his own ecological varieties of wheat. Ten other organic farmers from neighbouring villages took his wheat seeds. Farmer Albrecht was fined by the government of Upper Bavaria because he traded in uncertified seed. He has challenged the penalty and the Seed Act because he feels restricted by this law in freely exercising his occupation as an organic farmer. During the Leipzig conference on Plant Genetic Resources, Josef Albrecht initiated in Leipzig (in the same church from which the democracy movement against the erstwhile communist state of East Germany was organized) a non-cooperation movement against seed legislation that denies farmers the right freely to breed and exchange their seeds (*Bija Newsletter*, 1996).

Globalization certainly means 'less government' regulating business and commerce. But *less* government for commerce and corporations can go hand in hand with *more* government in the lives of people. As globalization allows increasing transfer of wealth and resources from the public domain, under the control either of communities or of the state, the result is increasing *poverty*, increasing *unemployment*, increasing *insecurity* (Martin and Schumann, 1996; Athanasiou, 1998). Discontent and dissent necessarily increase, leading to law and order problems. In such a situation even a minimalist state, restricted only to policing, will become enormously large and all-pervasive, devouring much of the wealth of society and intruding into every aspect of citizens' lives.

Most of the ideological projection of globalization has focused on the new relationship of the prince and the merchant, the state and the corporation, the government and the market. The state has been stepping back more and more from the regulation of commerce and capital. However, the shift from the rule of the nation state to the rule of the corporations does not imply more power to the people. If anything, it implies less power in the hands of people because corporations, especially transnational corporations, are more powerful than governments, they are more totalitarian than governments and they are less accountable than governments to democratic control.

The erosion of the power of the nation state from outside and from above leads to a concentration of power in the hands of corporations. It does not devolve power to the people. It does not move power downwards into the hands of communities. In fact, it takes power away from the local level and transforms institutions of the state from being protectors of the health and rights of people to protectors of the property and profits of corporations. This creates an inverted state, a state more committed to the protection of foreign investment and less to the protection of the citizens of the country. An inverted state that no longer protects the environment or jobs also becomes an unaccountable state – a state beyond the reach of citizens, a state captive to global commercial interests.

Globalization therefore runs into conflict with the democratic space of citizens to determine and influence the conditions for their health and well-being, and their right to elect governments which protect the public interest. In country after country, governments are elected on the basis of manifestos that expose the social and ecological destruction inherent to processes of globalization. In every country, however, governments of every shade in the political spectrum enforce the free trade agenda more forcefully than their predecessors and act against the citizens who have put them in power.

In India, the Bharatiya Janata Party won the Maharashtra elections by criticizing the Congress Party for giving clearance to the Enron power project. It came to power on this anti-Enron platform. But the first thing it did was to clear the power project, even though it involved human rights violations against the party's own supporters at the grassroots (Amnesty International, 1997). And it is not just in the Third World that democracy is being eroded by globalization. In Denmark people voted against joining the European Union but their vote was brushed aside and a second referendum was organized. In Austria 1.4 million people voted against the import of genetically engineered food, but the European Union ruled against the democratic decision. Again in the European Union, the parliament rejected a law that allowed patenting of life, but the bureaucracy of the European Commission bulldozed the patent directive through in spite of the vote (Raghavan, 1997). Globalization and democracy are therefore mutually exclusive, both in the deeper sense of democracy as people's control over their lives and in the shallower sense of people's control over their elected representatives.

The expansion of corporate control is often made to appear as the expansion of the democratic space for citizens on the basis of 'consumer choice'. However, such choice is based on ever-narrowing alternatives. Choice within a narrow, predetermined set of options of corporate rule is not freedom because it involves the surrender of the right to determine the context of living and the values that govern society. The apparent widening of individual consumer choice for the elite in matters of automobiles and junk foods is based on the shrinking of the rights of communities to control their local natural resources, the shrinking of work opportunities for large numbers of people and the shrinking of social and political choice through a democratic public process. Globalization is creating more freedom for corporations, but this is not translating into more freedom for citizens.

Globalization as environmental apartheid

Apartheid literally means separate development. However, in practice apartheid is more appropriately described as a regime of exclusion. It is based on legislation that protects a privileged minority and excludes the majority. It is characterized by the appropriation of the resources and wealth of society by a small minority based on privileges of race or class. The majority is then pushed into a marginalized existence, without access to the resources necessary for well-being and survival. Erstwhile South Africa is the most dramatic example of a society based on racial apartheid. Globalization has in a deep sense been a globalization of apartheid (Kohler, 1982; Makhijani, 1992; Alexander, 1996). Global apartheid is particularly glaring in the context of the environment.

Globalization is restructuring the control over resources in such a way that the natural resources of the poor are systematically taken over by the rich, and the pollution of the rich is systematically dumped on the poor. In the pre-Rio period, it was the North that contributed most to the destruction of the environment. For example, 90 per cent of historic CO_2 emissions have come from the industrialized countries, and the developed countries generate 90 per cent of the hazardous wastes produced around the world every year. Global free trade has patterned this environmental destruction asymetrically (see Khor, 1993; Shiva, 1993). The economy is controlled by Northern corporations that are increasingly exploiting Third World resources for their global activities. It is the South which is disproportionately bearing the environmental burden of the globalized economy. Globalization is thus leading to an environmental apartheid.

The current environmental and social crisis demands that the world economy adjust to ecological limits and the needs of human survival. Instead, global institutions like the World Bank, the International Monetary Fund (IMF) and the WTO are forcing the costs of adjustment on to nature and women and the Third World. Across the Third World, structural adjustment and trade liberalization measures are becoming the most serious threat to survival of the people. Whereas the past five decades have been characterized by the 'globalization' of maldevelopment and the spread of the non-sustainable Western industrial paradigm in the name of development, the recent trends are towards an environmental apartheid in which, through global policy set by the holy trinity, the Western transnational corporations supported by the governments of the economically powerful countries attempt to maintain the North's economic power and the wasteful lifestyles of the rich by exporting environmental costs to the Third World. Resource- and pollution-intensive industries are being relocated in the South through the economics of free trade. Even the United Nations now acknowledges what has been obvious for years: 'the world's dominant consumers are overwhelmingly concentrated among the well-off – but the environmental damage from the world's consumption falls most severely on the poor' (United Nations Development Programme, 1998: 3).

As is now well known, Lawrence Summers, Secretary to the US Treasury and formerly the World Bank's chief economist, suggested that it makes economic sense to shift polluting industries to Third World countries. He was merely saying forthrightly what the economic logic of globalization demands. He justified his utilitarian logic on three grounds:

1   Since wages are low in the Third World, economic costs of pollution arising from increased illness and death are least in the poorest countries. According to Summers, 'relocation of pollutants to the lowest wage country is impeccable and we should face up to that'.
2   Since in large parts of the Third World pollution is still low, it makes economic sense to introduce pollution: 'I've always thought', he says, 'that countries in Africa are vastly underpolluted; their air quality is probably vastly inefficiently low compared to Los Angeles or Mexico City'.
3   Since the poor are poor, they cannot possibly worry about environmental problems: 'the concern over an agent that causes a one in a million change in the odds of prostate cancer is obviously going to be much higher in a country where people survive to get prostate cancer than in a country where under-five mortality is 200 per thousand'.

Thus Summers has recommended the relocation of hazardous and polluting industry to the Third World because in narrow economic terms, life is cheaper in the poorer countries. Utilitarian economic logic values life differentially in the rich North and the poor South – a conclusion which directly contradicts the principle of human rights, but is highly supportive of apartheid. Against this we must insist that all life is precious. It is equally precious to the rich and the poor, the white and the black, to men and women.

In this context, recent attempts of the North to link trade conditionalities with the environment in platforms like WTO need to be viewed as an attempt to build on environmental and economic apartheid. No Western country has stopped the export of its hazardous wastes and polluting industry to the South. Thus, the issue of export of domestically prohibited goods was never fully developed in GATT. The destruction of ecosystems and livelihoods as a result of trade liberalization is a major environmental and social subsidy to global trade and commerce and those who control it. The mantra of globalization is 'international competitiveness'. In the context of the environment this translates into the largest corporations competing for the natural resources that poor people in the Third World need for their survival. This competition is highly unequal not only because the corporations are powerful and the poor are not, but also because the rules of free trade allow corporations to use the machinery of the nation state to appropriate resources from the people, and prevent people from asserting and exercising their rights.

It is often argued that globalization will create more trade, that trade will create growth and that growth will remove poverty. What is overlooked in this myth is that globalization and liberalized trade and investment create growth by *destruction* of the environment and local, sustainable livelihoods. This is now well understood through the application of the second law of thermodynamics. Order – that is, the benign order of the environments of the rich – depends upon increasing disorder elsewhere. The market exhibits this tendency to entropy, as Wackernagel and Rees point out: 'beyond a certain point, the continuous growth of the economy . . . can be purchased only at the expense of increasing disorder (entropy) in the ecosphere' (Wackernagel and Rees, 1996: 43; and see Altvater, 1993, and this volume, Chapter 16). Globalization therefore creates poverty instead of removing it. The new globalization policies have accelerated and expanded environmental destruction and displaced millions of people from their homes and their sustenance base.

Importing pollution and toxics

Globalization implies that toxic waste, like anything else in the context of free trade, is treated as a commodity, which shifts around the world to those countries in which the economic costs of internalizing, processing and disposing of it are the lowest, and where the economic and political clout to resist is the lowest. Thus, under trade liberalization India has become a dumping ground for industrial toxic wastes from the North as well as a target for deceptive marketing, double standards and dangerous exports of rejected goods – or, as Beck (1995: 134) has put it, 'Supranational groups of regions and countries swallow poisons and waste on others' behalf' (see also Smith and Blowers, 1992; Edwards, 1995a, b; Low and Gleeson, 1998: 121–30).

Under the new economic trade recipes of the WTO, transnational companies are seeking freedom to relocate toxic and polluting industry to India. Drawn by the country's cheap raw materials, minimal labour costs, pliable bureaucracy, lax environmental standards and, very importantly, low processing and disposal costs for waste, transnational corporations find India an attractive option. Free trade translates into the right to free and unrestrained export of hazardous wastes, products and industry to countries where life is considered cheaper because people are poorer.

The Thapur–Du Pont case

Since 1985 Du Pont has been attempting to relocate a hazardous Nylon 6.6 manufacturing plant from the United States to the picturesque state of Goa. It was to be established at a cost of six billion rupees (Rs) to produce 18,500 tonnes of Nylon 6.6 per annum. However, the corporation was forced to give up its plans and move to Tamil Nadu owing to sustained agitation by the villagers of the area where the factory was to be located.

The house committee of the Goa Assembly set up to look at the environmental implications of the Nylon 6.6 plant stated that it would be an ill-advised move to allow large chemical industries to discharge even their 'treated' effluents in Goa's eco-rich and virgin rivers. The committee stated:

> We have to safeguard our rivers against any conceivable environmental and pollution threat. While deciding on the establishment of large chemical industries in a small, relatively densely populated and socio-economically rich state [such] as ours, we should consider not only the statistical probability of a possible industrial accident but also the disastrous danger to humans and ecology that may result from such an accident. It is also imperative to realise that in a tiny state like Goa, any large capital intensive industry is bound to consume and utilise a very significant percentage of available natural resources and infrastructural facilities, but in turn, contribute negligibly to the local economy.

The Federal Ministry of Environment also did not give clearance to Du Pont. However, in mid-December 1991 Gaza Feketekutty, the US Trade Representative, made a special trip to India to put pressure on the government to give clearance to Du Pont. Key officials in the US government gave hints that Du Pont was being treated as a test case for the new trade liberalization policies. On 1 January 1992 the Cabinet cleared the project. Then, under the guise of public purpose, the central government acquired grazing land for the Du Pont plant and authorized building work to commence without local Panchayat (community council) consultation. This was possible under the Industrial Development Act, which empowers governments to grant special favours to industry in the interests of development. However, people did not accept the Du Pont plant. Protests were initiated and an Anti-Enron Committee was formed. People were arrested repeatedly during protests. One youth, Nilesh Naik, was shot at point-blank range and many others were injured in clashes with the police. Even the Chief Minister was forced to admit, 'If the people do not want the project, how can we force it?'

The Panchayati Raj Act only became effective in Goa subsequent to this land acquisition. At its monthly meeting the Panchayat resolved unanimously that local public opinion should be sought prior to granting Du Pont permission to continue with its development. In brazen defiance of the Panchayat's decision, Du Pont's construction programme continued unabated while the Panchayat was still considering the pending application. Anticipating that the Director of Panchayats, a government employee, would not uphold the local Panchayat's decision, villagers clashed violently with police and Du Pont representatives (Alvares, 1995). However, the decision of the Panchayat, which subsequently and unanimously resolved that the planning application should be quashed, was honoured by the High Court. Even before the High Court decision was made, it was obvious that the site acquired by the government for the factory had returned to the possession of the villagers because of the anti-Du Pont movement. Village animals including goats and cows, prevented for several months from entering the area, were now seen once again freely browsing all over the plot.

This event represents an important precedent for the reversal of the logic of globalization and the establishment of local democracy. The rejection of planning permission by a Panchayat, following intense local opposition, even after the approval of central and state governments and in spite of the global power of Du Pont and the global pressure it had mobilized, is evidence that even in an age of globalization people's power is stronger than the power of multinational corporations.

Northern dumping in the South

The United States generates more than 275 million tonnes of toxic waste every year; it is the leading waste-exporting country in the world. The United States is one of the 161 countries which has signed the Basel Convention, but it has not ratified it (along with 58 other countries). Parties to the convention, such as India, are not allowed to trade in hazardous wastes with countries which are not parties to the convention, therefore the United States is blatantly violating international law in sending shipments of its waste to India.

The new industrial policy in India, announced in 1991, has had the effect of making the country more vulnerable to international market forces. The policy makes minimal the requirements of industrial licensing and regulation (though licensing is still required for hazardous chemicals). The new policy permits foreign equity participation up to 51 per cent in select areas, including chemical production. The implication of this is that important decision-making power on matters such as the health of workers, the community and the environment is given away to outsiders. What we are seeing here is an opening up of the market with a corresponding relinquishing of government control, but at the same time an increasing neglect of the environment by way of the dismantling and fragmenting of protective environmental legislation and regulatory mechanisms.

In the first half of 1996 approximately 1,500 tonnes of lead wastes was imported into India. Greenpeace findings state that the amount of toxic lead waste imported from industrialized countries into the country has doubled since the previous year. Imports from the United States, Australia, South Korea, Germany, the Netherlands, France, Japan and the United Kingdom account for about 67 per cent of the total exports of lead wastes to India. A UNEP report states that the OECD accounted for 98 per cent of the 400 million metric tonnes of toxic waste generated worldwide.

Toxic waste such as cyanide, mercury and arsenic is being shipped as 'recyclable waste' – a deliberate attempt to mislead, and disguise the true nature of the wastes. In reality there is no such use for or demand to recover these toxic chemicals; they are pure waste. The imported waste often ends up in backyard smelting units, not the commercial sector as stated by the government. Many of the importing units do not possess the technology or expertise to process the chemicals they are importing, therefore inadvertently cause more harm to the environment and their communities because of their ignorance concerning the chemicals they are dealing with. Eight thousand five hundred such units operate in Maharashtra alone.

Developed countries are offering lucrative prices (in Indian terms) to Indian 'recycling' companies to take their material for 'processing'. In reality India is being used as a dumping ground by the Northern industrialized countries, because the cost of treating and disposing of waste in a sustainable manner in the North has become very high. The cost of burying one tonne of hazardous waste in the United States rose from $15 in 1980 to $350 in 1992. In Germany it is US$2500 cheaper to ship a tonne of waste to a developing country than to dispose of it in Europe. Countries like Germany find it cheaper to export their waste to a landfill site than to recycle it themselves. Costs have become so high because of the stringent laws banning the dumping, burning or burying of waste in the country of origin. Because India does not charge the waste trafficker the cost of land filling, the profits made in the waste trade have made the industry even more attractive. Dumping in the developing world therefore becomes justified on grounds of economic efficiency.

In 1996 the Research Foundation for Science, Technology and Ecology (RFSTE) filed a Public Interest Litigation (PIL) seeking a ban on the importing of all hazardous and toxic wastes into the country. On 6 May 1997 the Supreme Court of India imposed such a blanket ban. A court statement established that 2,000 tonnes of toxic wastes was being generated every day in the country without adequate safe disposal sites. The court also directed the state governments to show cause why immediate orders should not be passed for the closure of more than 2,000 unauthorized waste-handling units identified by the central government in various parts of the country. The Supreme Court directed that no import should be permitted of any hazardous waste which is already banned under the Basel International Convention, or is to be banned in future with effect from the date specified (*Times of India*, 1997). This ban applies to state governments as well as the central government, preventing these governments from authorizing the importation of toxic wastes.

Today toxic waste dumping has become a national issue. Several NGOs are working specifically on the banning of toxic waste import and dumping and related issues. Srishti, Greenpeace, Toxics Link Exchange, the Public Interest Research Group, WWF-India and the RFSTE are Delhi-based movements which concern themselves with hazardous wastes and toxics issues, and in particular are opposing the importing of toxic wastes. Furthermore, some of us are involved in creating awareness within India about the actions of trans-national and local industries, which often openly defy existing environmental laws regarding importation, treatment, handling and disposal of hazardous wastes.

Exporting India's biological wealth

Cattle: the case of Al-Kabeer

The focus on meat production and meat exports in India's New Livestock Policy in fact concerns the export of the country's ecological capital in the form of animal wealth. It has led to a mushrooming of slaughterhouses. The biggest export-oriented slaughterhouse is the Al-Kabeer Slaughter House in Andhra Pradesh.

People's movements towards conserving livestock diversity and protecting the basis of sustainable agriculture are slowly emerging as the impact of slaughterhouses and meat exports becomes apparent. The courts are playing a significant role in this awareness. Following the severe decline in cattle and buffalo numbers in the vicinity of the Al-Kabeer Slaughter House over the past two years, a long-awaited judgment was received from the Supreme Court on 12 March 1997. The interim judgment called for Al-Kabeer Exports Ltd to reduce its operation to 50 per cent of its installed capacity by 1 April 1997.

Cattle dung is a major source of fuel for cooking for many Indian house-holds, as well as an important organic manure. An examination of the dung economy reveals the unsustainable nature of Indian reliance on imported fossil fuels. A buffalo produces around 12 kg of wet dung every day. This converts to 6 kg of dry dung. An average Indian family of five members needs 12 kg of dung cakes every day as cooking fuel, which translates into a pair of buffaloes. The 182,400 buffaloes that Al-Kabeer kills every year could satisfy the fuel needs of 91,200 families in India.

The depletion of cattle and buffaloes leads to a decline in the availability of dung. The government therefore has to supply kerosene or liquefied petroleum gas (LPG). The transport cost of this runs into tens of millions of rupees, which means that poor people pay vastly higher fuel costs which they cannot afford. Imports of LPG and kerosene increase every year. Kerosene costing Rs 5.475 billion was imported in 1987/88. By 1992/93 this increased to Rs 20.09 billion – an almost fourfold increase in five years. Thus the 91,200 families whose fuel requirements have been forcibly altered by the killing of

182,400 buffaloes a year by Al-Kabeer will now spend Rs 1,440 × 91,200. This amounts to Rs 131.3 million spent on buying fuel. This fuel has to be imported by the government paying foreign exchange.

If the animals were allowed to live rather than being slaughtered, the state of Andhra Pradesh would get 1,918,562 tonnes of farmyard manure with the help of their dung and urine every year. This farmyard manure would help to cultivate 383,712 hectares. In 1991, the average food grain produced per hectare was 1.382 tonnes. Therefore, the food grain produced would be 530,000 tonnes. If the animals were allowed to live out their natural lives instead of being slaughtered by Al-Kabeer Slaughter House, they would save foreign exchange worth Rs 314 billion for the state of Andhra Pradesh. The calculation based on data from Andhra Pradesh goes as follows. The annual availability of major nutrients in farmyard manure from the dung and urine of 1,924,000 buffaloes and 570,000 sheep per year works out to:

- 1,117,179 tonnes of nitrogen fertilizer.
  At world market prices for nitrogen (urea) of Rs 6453 per tonne = Rs 7.209 billion
- 216,415 tonnes of phosphorus.
  At world market prices for phosphate of Rs 6642 per tonne = Rs 1.437 billion
- 1,006,929 tonnes of potash.
  At world market prices for potash of Rs 3888 per tonne = Rs 3.914 billion (prices are based on 'World fertilizer indicator prices for 1996/97' ABARE, 1998: Table 90).

The value of nitrogen + phosphorus + potash = Rs 12.56 billion. All these items are now imported. Thus Andhra Pradesh would save foreign exchange worth Rs 12.56 billion per year from the animals which are going to be killed. Taking into account their average remaining life span of 5 years, they will save foreign exchange worth Rs 12.56 × 5 = Rs 62.8 billion. Following the same argument, if all the animals which are going to be killed during (say) 5 years of Al-Kabeer's operation live out their natural life span, then they will be able to save foreign exchange worth Rs 62.8 × 5 = Rs 314 billion. This means that against a projected export earning of Rs 200 million by Al-Kabeer through slaughter of the animals, the state can actually save Rs 314 billion in foreign exchange by not killing them. That calculation takes no account of the high shipping, port, and transportation costs of importing fertilizer.

Cattle have been sacred in India. Today, under the impact of globalization this conservation culture, which is necessary for the survival of the people, is being perceived as an obstruction to the production and trade of meat. Livestock policies are being rewritten with the objective of exporting cattle for meat. As the new livestock policy states, 'Religious sentiments (particularly in the Northern and Western parts of India) against cattle slaughter prevent the utilisation of a large number of surplus male calves.' In industrial livestock systems cattle are bred either for milk or for meat. The male calves of milk breeds are thus treated as 'surplus'. However, in India the diverse cattle breeds such as Ongole, Hallihar, Haryana, Tharparhar and Sindhi are dual-purpose breeds, bred for both milk and draught power. Both male and female offspring are therefore useful and indeed essential for sustainable agriculture. In the past decade there has been a significant decline of livestock in India, particularly the indigenous breeds known for their hardiness, milk production and draught power. The decline in livestock is primarily due to illegal slaughtering of cattle and buffalo for meat export.

The UN Food and Agriculture Organization (FAO) confirmed in 1996 that 'the diversity of domestic animal breeds is dwindling rapidly. Each variety that is lost takes with it irreplaceable genetic traits – traits that may hold the key to resisting disease or to productivity and survival under adverse conditions.' For example, some of the declining indigenous breeds today are Pangunur, Red Kandhari, Vechur, Bhangnari, Dhenani, Lohani, Rojhan, Bengal, Chittagong Red, Napalees Hill, Kachah, Siri, Tarai, Lulu and Sinhala. If measures to arrest the decline are not taken now, most of us will witness the extinction of live-stock within our own lifetime, and with it the foundation of sustainable agriculture will disappear.

In a judgment given in a Delhi court, the protection of cattle and the prevention of slaughter were upheld as a duty of citizens and the state. The judgment made the following statements:

> This fundamental Duty in the Constitution to have compassion for all living creatures, thus determines the legal relation between Indian Citizens and animals on Indian soil, whether small ones or large ones. This gives legal status to the view of ancient sages down the generations to cultivate a way of life to live in harmony with nature. Since animals are dumb and helpless and unable to exercise their rights, their rights have been expressed in terms of duties of citizens towards them.
>
> (S. 93)

> Their place in the Constitutional Law of the land, is thus a fountainhead of total rule of law for the protection of animals and provides not only against their ill treatment, but from it also springs a right to life in harmony with human beings.

If this enforceable obligation of the State is understood, certain results will follow. First, the Indian State cannot export live animals for killing: and second, cannot become a party to the killing of animals by sanctioning exports in the casings and cans stuffed with dead animals after slaughter. Avoidance of this is preserving the Indian Cultural Heritage, of which we claim proud by claiming India as land of Gandhi, Buddha and Mahavir. India can only export a message of compassion towards all living creatures of the world, as a beacon to preserve ecology, which is the true and common Dharma for all civilisations. This is in keeping with the culture of living in harmony with nature by showing respect to all life; and that is Vasudhaiv Kutumbakam referred to by our Minister of Environment, Mr. Kamal Nath, at the Environment Conference at Rio, June 1992.

(S.98)

Plants and biopiracy

India is a major region of biodiversity. India is recognized as a country which is uniquely rich in all aspects of biological diversity – from the ecosystem level to the species and genetic levels. It is estimated that over 75,000 species of fauna and 45,000 species of flora are found in India. Of the estimated 45,000 plant species, about 15,000 species of flowering plants are endemic to the country. Estimates of other plant taxa include 5,000 species of algae, 1,600 lichens, 20,000 fungi, 2,700 bryophytes and 600 pteridophytes. The 75,000 species of animals include 50,000 insects, 4,000 molluscs, 200 fish, 140 amphibians, 420 reptiles, 1,200 birds, 340 mammals, and other invertebrates. Thus, India is home to about 120,000 species of living organisms.

Most of the country's people derive their livelihood and meet their survival needs from the diversity of living resources – as forest dwellers, farmers, fisher-folk, healers or livestock owners. Indigenous knowledge systems in medicine, agriculture and fisheries are the primary base for meeting the food and health needs of the majority of the people (see Chapter 13 by Fourmile in this volume). This immense resource has been protected, preserved and conserved by India's indigenous people over the years, who have had a reverence for their natural heritage.

This indigenous knowledge also reflects the continuous, cumulative, collective innovation of the people of India in all their diversity. Conservation and utilization have been delicately, sensitively and equitably combined in the country's indigenous knowledge systems and cultures. The sharing and exchange of biodiversity and the knowledge of its properties and use has been the norm in all indigenous societies, and it continues to be the norm in most communities, including the modern scientific community.

Sharing and exchange, however, are converted to 'piracy' when individuals, organizations or corporations who freely receive biodiversity from indigenous communities, and knowledge convert these gifts into private property through intellectual property rights (IPR) claims. This blocks the continuity of free exchange, thus leading to an 'enclosure of the intellectual commons'. Biopiracy refers to the process through which the rights of indigenous cultures to these resources and knowledge are erased and replaced by monopoly rights for those who have exploited indigenous knowledge and biodiversity.

Biodiversity-based traditional knowledge systems of the forest dwellers, farmers and healers are fast becoming the private property of the transnational corporations (TNCs). The TNCs are usurping these systems from the domain of common knowledge through IPRs, which in essence promote resource piracy and intellectual piracy. In the present world market economy, where knowledge represents money, capital represents power, and profit is the sole aim, those who own capital seek IPRs to protect their 'discoveries', which are often based on the cumulative and collective innovation of traditional societies.

The IPR regime has expanded the domain of intellectual property to include biodiversity. However, these IPR regimes, as provided under TRIP (Agreement on Trade-Related Aspects of Intellectual Property Rights), recognize and provide protection only to the formal innovators, not to the informal indigenous innovators. The traditional knowledge of informal innovators (farmers, indigenous medical practitioners, forest dwellers) is being pirated by the formal innovators (scientists, plant breeders and technologists), who make minor modifications or advances and then seek patents, thereby claiming the knowledge as their private property. Biopiracy leaves the donors poorer, both materially and intellectually, as they are excluded from sharing in the benefits of their own resources and knowledge.

As we enter the third millennium we need to find ways to protect biological diversity and the intellectual heritage of India for the future of India's people. Emergent ecological concern for the conservation of biological diversity provides a new opportunity to value our biological and intellectual wealth, and to use and conserve it for the needs of our people.

Fish

The new economic policy states that 'India's fishery resources are being grossly underutilized; significant opportunities exist for corporate units in the areas of preservation and exports'. However, the National Federation of Fishworkers has successfully resisted foreign licences because India's fisheries are already overexploited; the small-scale traditional fisherfolk are getting declining catches and their traditional livelihoods are threatened.

The fish farms and shrimp industry promoted by the new policy and financed by the World Bank have already created ecological and social havoc along India's 7,500-km coastline. These impacts are dealt with in detail in an RFSTE report entitled *Towards Sustainable Aquaculture* (Research Foundation for Science, Technology and Ecology, 1997b). Movements have emerged in every state to resist the aquaculture industry. Industrial fish farming has destroyed mangroves and fertile agricultural land, created salinization of groundwater and thus a scarcity of drinking water, and polluted the sea. The Supreme Court even passed an order on the basis of a PIL filed by the movements to destroy farms within the Coastal Regulation Zone (CRZ) and regular farms outside the CRZ.

People's movements for the protection of biodiversity and collective rights

New social and environmental movements are emerging everywhere as a response to the widespread destruction of the environment and of livelihoods dependent upon biodiversity, and in response to piracy of indigenous resources and indigenous innovation. In India, the intricate link between people's livelihoods and biodiversity has evolved over centuries. Economic liberalization is threatening to sever this link by treating biodiversity as raw material for exploitation of life forms as property and people's livelihoods as an inevitable sacrifice to national economic growth and development.

Tribals' right to self-rule

In February 1995 the tribals from different parts of India came to Delhi on an indefinite fast to force the government to recognize their declaration of 'self-rule'. The National Front for Tribal Self-Rule, a national organization of tribal peoples, started a civil disobedience movement on 2 October 1995 for the establishment of self-rule. As they stated:

> We have carried the cross of virtual slavery for much too long in spite of independence. Other rural folks are also in a similar state. Yet, now that everything is clear and there is unanimity in the establishment as also amongst MPs and experts, the change must not be delayed. We will not tolerate this.

They claimed that self-governance is a natural right. In the hierarchy of Indian democratic institutions *gram-sabha* (village assembly) is above even Parliament. This is what Gandhi preached. They said, 'We will not obey any law which compromises the position of *gram-sabha*. In any case we resolve to establish self-rule with effect from 2 October 1995. We will have command over our resources and will manage our affairs thereafter.'

The struggle of the tribals was successful. The passing of the Provisions of the Panchayats (Extension to the Scheduled Areas) Act, which came into effect in December 1996, represents a landmark piece of legislation acknowledging the legal rights to self-rule of the tribals. Section 4(b) and (d) of the Act state that:

- A village shall ordinarily consist of a habitation or a group of habitations, or a hamlet or a group of hamlets comprising a community and managing its affairs in accordance with traditions and customs.
- Each *gram sabha* shall be competent to safeguard and preserve the traditions and customs of the people, their cultural identity, community resources and the customary mode of dispute resolution.

The implementation of the Panchayati Raj Act in Scheduled Areas had already set the precedent, as the building block of a decentralized democracy, for the recognition of communities as competent authorities for decision-making concerning resource use, cultural values and traditions, and community rights to common resources (Research Foundation for Science, Technology and Ecology, 1997a).

Over a hundred villages in and around the thick forests of Nagarhole in Karnataka have established self-rule to safeguard their livelihood, under the Provisions of the Panchayats (Extension to the Scheduled Areas) Act. However, this law has yet to be passed by the Karnataka Assembly to implement it in that state. The people have formed *gram-sabhas* and have established task forces to implement the self-rule programme. In some of the villages they have erected gates at the entrance and only the chief of the tribal community or village has been entrusted with the power to give permission to any outsider to enter the village. The villagers freely collect the minor forest produce and they even adjudicate legal disputes themselves rather than going to the police or courts.

The movement for declaration of community rights to biodiversity

People's movements against erosion, exploitation and usurpation of biodiversity are numerous and widespread throughout India. One example is a small community in the south of Kerala, which has taken a bold step to protect the area's biodiversity.

On 9 April 1997 in a remote part of Kerala, hundreds of local people gathered to declare their local biodiversity as a community-owned resource, which they will collectively protect, and which they will not allow to be privatized through patents on derived products, or varieties. The community is known as the Pattuvam Panchayat (the Panchayat is the locally elected body for village governance accountable to the village community who have elected it). The Panchayat has set up its own biodiversity register to record all species in the region. It has stated that no individual, TNC, state or central government can use the region's biodiversity without the permission of the Pattuvam Panchayat. The people of Pattuvam have thus taken a path-breaking step by declaring their biodiversity a community resource over which the community as a whole has rights. This step demonstrates a commitment to rejuvenating and protecting the area's biodiversity and knowledge systems from the exploitative forces of economic liberalization.

Similar movements are occurring in other parts of India whereby communities are declaring the biodiversity and knowledge as the common heritage of local communities. For example, in Dharward in Karnataka and in Chattisgarh, Madhya Pradesh, declaration ceremonies have been held announcing that biodiversity is a community resource and privatization of biodiversity and indigenous knowledge through patents is theft.

A People's Commission on Biodiversity and Indigenous Knowledge and People's Rights was launched on 17 March 1997, on the inititative of the

RFSTE under the chairmanship of Justice V. R. Krishna Iyer. The commission is to consist of Justice R. S. Sarkaria and Justice Kuldeep Singh, retired judges of the Supreme Court of India, as members, and Afsar H. Jafri (Research Officer at the RFSTE) as Secretary. The Commission's basic task is to inquire into the piracy of India's biodiversity so as to assist the country and the government to take appropriate measures to preserve India's rich biological and genetic resources.

The commission is carrying out wide-ranging consultations and holding public hearings in order to provide legal guidelines to ensure that the country's national laws protect the national and the public interests. It will submit its findings and recommendations to the government in an effort to frame new laws on patents and varieties protection. The commission held the first phase of the public hearings on 5, 6 and 7 August 1997 at India International Centre, New Delhi. The commission also requested the Indian government to wait for this democratic process to be completed before introducing any changes in the existing laws relating to IPRs. At present the commission has been engaged in the collection of relevant data and materials on the nation's biological wealth. The chairman also requested people to help and cooperate with the commission by providing relevant copies of books, articles, papers and data on India's natural and biological wealth and its loss.

Nationwide, people's movements have succeeded to date in stalling any legislation passing through Parliament that promotes intellectual property rights over biodiversity. Such opposition signifies the degree of democratic dissent being generated at the grassroots level to laws affecting people's livelihoods and rights over their resources. During the fiftieth anniversary of India's independence, Bija Yatras (Journey of the seed) are being organized throughout the country by Navdanya (see below) to create awareness of the nation's rich biodiversity and the rights of farmers over that biodiversity, and ways to rejuvenate and protect it.

Navdanya: seeds of freedom

I have started a national movement for the recovery of the biological and intellectual commons by saving native seeds from extinction. Seed is the first link in the food chain. It is also the first step towardss freedom in food (Shiva *et al.*, 1995). Globalization is leading to others' gaining total control over what we eat and what we grow. The tiny seed is becoming an instrument of freedom in this emerging era of total control. Our slogan is: native seed – indigenous agriculture – local markets. Through saving the native seed we are becoming free of chemicals. By practising a 'free' agriculture, we are saying no to patents on life and to biopiracy. Gandhi called such resistance 'Satyagraha' – the struggle for truth. Navdanya is a 'Seed Satyagraha' in which the most marginal and poor peasants are finding new hope.

A central part of the 'Seed Satyagraha' is to declare the 'common intellectual rights' of Third World communities, who have *gifted* the world the knowledge of the rich bounties of nature's diversity. The innovations of Third World communities might differ in process and objectives from the innovations in the commercial world of the West. But they cannot be discounted just because they are different. We are going beyond just saying no. We are creating alternatives by building community seed banks, strengthening farmers' seed supply and searching for sustainable agriculture options suitable for different regions.

The seed has become for us the site and symbol of freedom in the age of manipulation and monopoly. It plays the role of Gandhi's spinning wheel in this period of recolonization through free trade. The *charkha* (the spinning wheel) became an important symbol of freedom not because it was big and powerful, but because it was small and could come alive as a sign of resistance and creativity in the smallest of huts and the poorest of families. In smallness lay its power. The seed too is small. It embodies diversity. It embodies the freedom to stay alive. And seed is still the common property of small farmers in India. In the seed, cultural diversity converges with biological diversity. Ecological issues combine with social justice, peace and democracy.

In spite of the brutal violence of globalization, we have hope because we build alternatives in partnership with nature and people. As a Palestinian poem called the 'Seed Keepers' has stated:

> Burn our land,
> burn our dreams, pour acid on to our songs,
> cover with sawdust
> the blood of our massacred people,
> muffle with your technology
> the screams of all that is free,
> wild and indigenous.
> Destroy,
> Destroy
> our grass and soil, raze to the ground
> every farm and every village
> our ancestors had built,
> every tree, every home,
> every book, every law
> and all the equity and harmony.
> Flatten with your bombs
> every valley, erase with your edits
> our past,
> our literature, our metaphor.
> Denude the forests
> and the earth
> till no insect,
> no bird,
> no word,
> can find a place to hide. Do that and more.

I do not fear your tyranny,
I do not despair ever
for I guard one seed,
a little live seed
that I shall safeguard
and plant again.

References

ABARE (Australian Bureau of Agricultural and Resource Economics) (1998) Australian Commodity Statistics, Canberra: ABARE.

Alexander, T. (1996) *Unravelling Global Apartheid*, Cambridge: Polity Press.

Altvater, E. (1993) *The Future of the Market*, London: Verso.

Alvares, N. (1995) Paper presented at 'Globalisation and Panchayati Raj' Meeting, December (unpublished).

Amnesty International (1997) *The 'Enron Project' in Maharashtra: Protests Suppressed in the Name of Development*, Amnesty International Report, July.

Athanasiou, T. (1998) *Slow Reckoning: The Ecology of a Divided Planet*, London: Vintage Books.

Beck, U. (1995) *Ecological Politics in an Age of Risk*, Cambridge: Polity Press.

Bija Newsletter (1995) 'Patents: Violating Dignity of Life', Issue No. 14, December, New Delhi: Research Foundation for Science, Technology and Natural Resource Policy.

—— (1996) 'Hands Off: Forbidden Seeds', Issue No. 17–18, November, New Delhi: Research Foundation for Science, Technology and Natural Resource Policy.

Edwards, R. (1995a) 'Dirty Tricks in a Dirty Business', *New Scientist*, 18 February: 12–13.

—— (1995b) 'Leaks Expose Plan to Sabotage Waste Treaty', *New Scientist*, 18 February: 4.

Khor, M. (1993) 'Free Trade and the Third World' in R. Nader *et al.* (eds) *The Case against 'Free Trade' and the Globalization of Corporate Power*, Berkeley, CA: Earth Island Press, pp. 97–107.

Kohler, G. (1982) 'Global Apartheid', in R. Falk, S. Kim and S. Mendlovitz (eds) *Towards a Just World Order*, Boulder, CO: Westview Press.

Low, N. P. and Gleeson, B. J. (1998) *Justice, Society and Nature: An Exploration of Political Ecology*, London: Routledge.

Makhijani, A. (1992) *From Global Capitalism to Economic Justice*, New York: Apex.

Martin, H.-P. and Schumann, H. (1996) *The Global Trap: Globalization and the Assault on Democracy*, London and New York: Zed Books.

Raghavan, Chakravarthi (1995) 'A Global Strategy for the New World Order', Report on G77, *Third World Economics*, Issue No. 81182, January.

Research Foundation for Science, Technology and Ecology (1997a) *The Enclosure and Recovery of the Commons*, New Delhi.

—— (1997b) *Towards Sustainable Aquaculture*, RFSTE: New Delhi.

Shiva, V. (1993) 'Biodiversity and intellectual property rights', in R. Nader *et al.* (eds) *The Case against 'Free Trade' and the Globalization of Corporate Power*, Berkeley, CA: Earth Island Press, pp. 108–20.

—— (1996) 'Of Spies, Crime and IPR', *Hindustan Times*, 21 November.

Shiva, V., Ramprasad, V., Hegde, P., Krishnan, O. and Holla-Bhar, R. (1995) *The Seed Keepers*, New Delhi: Research Foundation for Science, Technology and Ecology.

Smith, D. and Blowers, A. (1992) 'Here Today, There Tomorrow: The Politics of Hazardous Waste Transfer and Disposal', in M. Clark, D. Smith and A. Blowers (eds) *Waste Location: Spatial Aspects of Waste Management, Hazards, and Disposal*, London: Routledge, pp. 208–26.

Times of India (1997) 'SC bans import of hazardous wastes', 6 March.

United Nations Development Programme (1998) *Human Development Report*, New York: United Nations (Summary of the Report).

Wackernagel, M. and Rees, W. E. (1996) *Our Ecological Footprint: Reducing Human Impact on the Earth*, Gabriola Island, BC: New Society Publishers.

5 Chernobyl, global environmental injustice and mutagenic threats

Kristin Shrader-Frechette

Introduction

The 1986 Chernobyl nuclear accident, according to one United Nations (UN) volume, was 'the greatest technological catastrophe in human history' (Savchenko, 1995: xv; Shrader-Frechette, 1999, 2000). Nearly 7 tons of irradiated reactor fuel was released into the environment – approximately 340 million curies. Included in the release were radioactive elements with a half-life of 16 million years. Yet we humans cannot protect ourselves from such radiation because we are biologically not equipped to do so. We are unable to taste, touch, smell, hear or see radiation. Its effects are silent but deadly. Less than six years after the accident, already there had been a hundredfold increase in thyroid cancers in Belarus, Russia and Ukraine. Most of these will be fatal (for verification of these comments see Henshaw, 1996: 1052; Rytömaa, 1996; Poiarkov, 1995; NEA and OECD, 1996: 28; Makhijani *et al.*, 1995: 98).

On the one hand, the Soviets, the French, UN agencies and many proponents of nuclear power have tended to claim that the consequences of the Chernobyl reactor explosion and fire were minimal. They say Chernobyl caused only 28 casualties (MacLachlan, 1994c: 11ff.). The IAEA (International Atomic Energy Agency), a UN agency dominated by the nuclear industry, places the number of Chernobyl fatalities at 31 (IAEA, 1991b: 4). On the other hand, many health experts, scientists and environmentalists, especially in developed nations, have argued that the effects were catastrophic. They say that the accident has caused 32,000 deaths so far (Shcherbak, 1996: 46; Konoplev *et al.*, 1996). Others put fatalities at 125,000 (Campbell, 1996), or half a million if one counts future premature cancer deaths induced by germline mutations (Gofman, 1995).

Why has the international scientific and political community allowed the global environmental injustice of Chernobyl – the half million premature cancer deaths and the permanent contamination of millions of acres of land? Why has there been so much cover-up and denial? Apart from outright ethical corruption, one reason may be that we, who could make a difference, have remained largely silent, despite the fact that half of the half-million premature cancer deaths caused by Chernobyl will occur outside the former Soviet Union. Another reason for the cover-up may be that the entire international nuclear establishment relies on the ICRP International Commission on Radiological Protection (ICRP) recommendations for radiation protection. The keystone of these recommendations is the Justification Principle, the standard for justifying or allowing radiation exposures. The principle is controversial because it allows either egalitarian or utilitarian benefits (benefits to society) to offset (or 'justify') the risks of radiation exposure to particular people. In other words, the dominant radiation-protection principle sanctions environmental injustice. The obvious question is whether the Justification Principle, despite its being enshrined in international recommendations about radiation protection, is ethically defensible. Does it satisfy equity? Does it satisfy utility? The remarks which follow argue for negative answers to all three questions. They also call for regulatory reform at national and global levels.

Global environmental injustice

Chernobyl fatalities are disturbing not only because of their sheer magnitude, but also because they exhibit environmental injustice on a global scale. Russians and Ukrainians were responsible, in large part, for the accident, but Belarussians have suffered the worst damage, and more than half the global deaths and injuries from Chernobyl will occur outside the former Soviet Union – many in Europe. Moreover, proponents of nuclear power have controlled the databases on Chernobyl's effects, and they have grossly underestimated and undercounted deaths and injuries (Gofman, 1995: 4–9; Yaroshinskaya, 1995: 100–101). In 1994, for example, the clinical department of the Institute of Biophysics (IBF) in Moscow set up a database of those who have died following acute radiation exposure. 'The official number of [Chernobyl] casualties is still 28' (MacLachlan, 1994c: 11ff.). Ukrainian officials and US scientists, at the same time, placed Chernobyl-caused deaths at 125,000 and argued that the accident's effects would cause fatalities for centuries (Campbell, 1996; Gofman, 1995; Brall and MacLachlan, 1995: 17ff.). One difficulty with the IBF database is that it does not include the people who participated in the Chernobyl recovery and clean-up – especially the 800,000 'liquidators'. Even though these workers wore no dosimeters (MacLachlan, 1994a: 10ff.; Franklin, 1990b: 1ff.; MacLachlan, 1990: 1ff.; Brall and MacLachlan, 1995: 17ff.), the IBF officials said their exposures were insufficient for the acute-injury register (MacLachlan, 1994c: 11ff.).

In 1995 a UNESCO volume warned that the radioactivity released by Chernobyl would never disappear completely from the biosphere (Savchenko, 1995: 5). The World Health Organization (WHO) estimated that the total amount of radioactivity released from Chernobyl was at least two hundred times that released by the atomic bombs dropped on Hiroshima and Nagasaki (Edwards, 1995: 14), or about seven times as much as the Soviets claimed (NEA and OECD, 1996: 28), and about seven times as much as the pro-nuclear UN agency, the IAEA, claimed (Yaroshinskaya, 1995: 130–33). These releases caused background radiation, in many European countries, to rise to a hundred times the natural levels (Savchenko, 1995: 9). Although Chernobyl-induced deaths will not peak until well after the year 2000, medical doctor and molecular biologist John Gofman, from the University of California at Berkeley, says Chernobyl will cause 475,000 premature cancer deaths and 475,000 premature non-fatal cancers (Gofman, 1995: 1–2).

Just the exposure levels, two hundred times that at Hiroshima and Nagasaki, make it inconceivable that only 29 or 30 people died, as official nuclear agencies claim. After all, more than fifty years after Hiroshima and Nagasaki, radiation-induced, premature, fatal cancers in Japan are still rising in statistically increasing numbers. Now, even in districts and villages a hundred kilometres from Chernobyl, such as Lugini, doses of radiation are more than 10 rems a year – or approximately 100 times the allowable limit set by international authorities, and more than 1,000 times what an average Westerner receives annually. (A rem is a standard unit for measuring radiation dose, in terms of its destructive biological effects; 100 rems = 1 sievert.) People in these villages are often ill, and they have no interest in life. Nearly everyone has cancer, blood diseases or other ailments. The children are the worst affected (Yaroshinskaya, 1995: 106ff.).

Apart from the cancers, blood disorders and immune-system problems, Chernobyl survivors face a high number of mental disorders. Effects on them are similar to those that have occurred in Hiroshima and Nagasaki. Japanese researchers, studying the consequences of the two atomic bombings, discovered that patients' distance from the hypocentre of radiation and the clinical symptoms they exhibited at the time of the bombings could predict their psychological state, even forty years later. Psychologists are discovering similar results after Chernobyl (Coryn and MacLachlan, 1995: 1ff.).

In addition to the 800,000 contaminated liquidators, a total of 9 million people are still living in contaminated areas, eating contaminated food, facing premature cancer and death, and receiving a significant extra radiation dose because of Chernobyl. Most of the Chernobyl deaths, over the long term, arise from water and food-chain contamination, not direct external exposure. Approximately 825,000 people are living on land with more than 5 Ci/km^2 (curies per square kilometre) of radiation, and approximately 270,000 people are living in locations with contamination exceeding more than 15 Ci/km^2. In an area with an initial contamination of only 5 Ci/km^2, the average dose of radiation is approximately 1.5 mSv (millisieverts) per year of external radiation, plus anywhere from 5 to 25 mSv per year of inhaled radiation, in addition to the radiation received from eating contaminated food. Even if the food satisfies the current radiation norms for Belarus, the radiation dose to a Belarus citizen from food alone would amount to approximately 1.25 mSv per year. Even in areas of the lowest contamination for these 9 million people, the annual dose of radiation for approximately 8 million people from external, internal and food sources would amount to 7.75–27.74 mSv per year, or approximately 8 to

28 times the ICRP maximum recommended annual dose of radiation, or approximately a hundred times what a Westerner would receive (Hill *et al.*, 1995; Likhtariov *et al.*, 1996; Balonov *et al.*, 1996; Edwards, 1994: 115–16; and see Savchenko, 1995: 5–15, 70–77).

If one uses the UN and IAEA assumption that each 1 mSv of radiation exposure causes an additional cancer per 10,000 people, then (on average) the 8 million people – living in areas of least contamination and receiving approximately 8 to 28 mSv of additional annual radiation exposure – can expect 6,400–22,400 additional, Chernobyl-induced, premature cancers per year because of the contamination (UNSCEAR, 1994; González, 1994; and see Savchenko 1995: 5–15, 70–77). Note that these statistics do not include global deaths annually induced by Chernobyl, and that the predictions are conservative because they arise from pro-nuclear databases, like those of the IAEA.

Rather than civilians, 800,000 liquidators received the highest doses of radiation; most of them were in their teens or early twenties and were military personnel when they helped clean up the accident (Savchenko, 1995: 78–81; Edwards, 1994; UN, 1995; Oganesian, 1995: 69–70; Ministry for Emergencies and Population Protection from the Chernobyl NPP Catastrophic Consequences, 1995: Part 5.3). Although there are no records for most of them, among the Ukrainian liquidators (for whom some records exist) more than two-thirds have become seriously ill with radiation-induced cancers and blood diseases, and nearly two-thirds of their children are unhealthy (MacLachlan, 1992: 1ff.). At about 35 years of age, many of them have become invalids and 20,000 of them already have died (Nussbaum and Köhnlein, 1995: 198–213; Franklin, 1990b, pp. 1ff.; MacLachlan, 1992, pp. 1ff.), 8,000 of whom are Ukrainian (Marples, 1995: xv). Their disorders are proportional to their radiation exposure, and they develop afflictions appropriate to much older people, because radiation causes premature ageing (MacLachlan, 1994a: 10ff.). As many as 80–90 per cent of the liquidators have serious mental problems such as depression and anxiety (MacLachlan, 1994a, pp. 10ff.), and many have radiation-induced immune disorders and reduced mental capacity, typically only 60 per cent of average mental capacity (Coryn and MacLachlan, 1995, pp. 1ff.). Unable to cope with the magnitude of health problems of the liquidators, the government has accused them of 'faking' their illnesses, in spite of their high death rates (Franklin, 1990a: 9ff.; Yaroshinskaya, 1995: 70-73; Ministry for Emergencies and Population Protection from the Chernobyl NPP Catastrophic Consequences, 1995: Part 5.3; MacLachlan, 1992: 1ff.; Kesminie *et al.*, 1995; MacLachlan, 1994a: 10ff.; Coryn and MacLachlan, 1995: 1ff.; Ivanov, 1996).

Besides the 800,000 liquidators, an important high-risk group because of Chernobyl is the children, many of whom have symptoms similar to those of the liquidators. In some cities near Chernobyl, 75 per cent of the children have mental health deviations such as mental retardation or borderline retardation (Coryn and MacLachlan, 1995, pp. 1ff.). Medical experts from the former Soviet Union are worried because they have already confirmed radiation-induced increases in the number of births of retarded children as well as higher rates of mental illness such as schizophrenia (*ibid.*). In all, 1,600,000 children have received very dangerous doses of radiation, and many thousands of children received 100 rems (Yaroshinskaya, 1995: 83). Birth defects, thyroid disorders, leukaemias and cancers are soaring among children hundreds of kilometres from Chernobyl. In Narodichi district, allegedly outside the zone of fallout, 50 percent of children are ill and absent from school on any given day, and 80 percent of all children have enlarged thyroids (Marples, 1995: x–xi; Yaroshinskaya, 1995: 41). In some areas of Belarus only 10 percent of the children are not chronically ill, and at least 100,000 abortions were carried out because of Chernobyl. In Belarus since Chernobyl, birth defects have risen by 161 percent, malignant tumours in children by 39 percent, diabetes by 28 percent and breast cancer by 45 percent (UN, 1995; Prisyazhniuk *et al.*, 1996; Souchkevitch, 1996; and see Edwards, 1994; Savchenko, 1995: 81–84).

International agencies and global environmental injustice

Chernobyl represents a classic case of global environmental injustice not only because millions of innocent and vulnerable people have been killed or injured, but also because the international nuclear industry helped cause these injustices, and the global human community has remained largely silent about them. Two and a half years before the disaster, experts writing in a German nuclear-industry journal said that the Chernobyl-type reactor, the RBMK, was very reliable, in part because the plants are equipped with three parallel, and therefore independent, safety systems, because they are resistant to hurricanes, earthquakes and aeroplane crashes, and because Soviet plants were built a considerable distance from densely populated areas (Born, 1983: 645–49). Less than three years prior to the accident, writing about the Chernobyl-type plant, B. Semenov, Head of the Nuclear Energy and Safety Department of the IAEA, wrote that 'a serious accident in which coolant is lost is virtually impossible' (Semenov, 1983: 47–60). Yet for their own plants, in 1947 the British rejected the Chernobyl-type design on safety grounds, and in 1958 the Germans criticized it as inherently dangerous (see MacLachlan, 1994b: 1ff.). Nevertheless, the IAEA was giving reassurances about the very types of plants that Western nations had rejected.

When the IAEA began its large-scale investigation, the International Chernobyl Project (ICP), former Chernobyl workers and liquidators said its effort was doomed to failure because the pro-nuclear IAEA was an ally of the corrupt Soviet nuclear establishment (MacLachlan, 1990: 1ff.). Because the IAEA relied on data furnished only by the Soviet government, the ICP underestimated the effects of the accident. It also failed to verify these central government data against medical records at the regional clinics (which told a more catastrophic tale). In addition, the IAEA used control populations from areas also contaminated by the radiation. The international nuclear-industry group considered only two villages, and it did not take account of the health of evacuees and liquidators – who had moved (IAEA, 1991b: 6). Moreover, the

IAEA committee ignored all long-term effects of the accident. It ended its study only three years after the accident, not long enough for most of the Chernobyl cancers, blood disorders and immune-system defects to manifest themselves (Nussbaum and Köhnlein, 1995: 209; Marples, 1995: xii). The IAEA even claimed in 1991 that it was not possible to attribute thyroid cancers and leukaemias to Chernobyl (IAEA, 1991b: 509), and that there were no serious consequences of the accident (*ibid.*: 510). It concluded that there were 'no health disorders that could be attributed directly to radiation exposure. The accident had substantial negative psychological consequences in terms of anxiety and stress' (*ibid.*: 508). Its ICP report claimed that there were only 31 deaths (IAEA, 1991b: 4) and no perceived increase in mortality in the population; the agency said the main impact was psychological (IAEA, 1991a: 61).

The IAEA stated that:

> The official data that were examined did not indicate a marked increase in the incidence of leukemia or cancers. . . . Reported absorbed thyroid dose estimates in children are such that there may be a statistically detectable increase in the incidence of thyroid tumours in the future.
>
> (IAEA, 1991b: 508)

But how could there be an increase in cancers if the IAEA scientists ended their study after only three years, examined only two villages, ignored the 800,000 highest-exposed liquidators, and failed to assess long-term effects? The UN, the WHO and many other groups and scientists have contradicted the IAEA claims (MacLachlan, 1992: 1ff.; WHO, 1995: 22–24; Nikiforov and Gnepp, 1994; Kazakov *et al.*, 1992; Baverstock *et al.*, 1992; Scholz, 1994; Stsjazhko *et al.*, 1995; Ito *et al.*, 1994: 259; and see Williams, 1994: 556; Likhtarev *et al.*, 1994; Nussbaum and Köhnlein, 1995). After its short study which ignored the most serious victims of Chernobyl, the IAEA then attributed many alleged Chernobyl problems to radiophobia (IAEA, 1991b: 508; 1991a; and see Marples, 1995: xii), a conclusion that later researchers, including scientists from the Red Cross, refuted (WHO EURO Working Group on the Effects of Nuclear Accidents, 1990; Yaroshinskaya, 1995; Nussbaum and Köhnlein, 1995: 198–213; Lazyuk et al., 1995; Ministry for Emergencies and Population Protection from the Chernobyl NPP Catastrophic Consequences, 1995: 35–37). The IAEA, in its International Chernobyl Project, even concluded, 'The children who were examined were found to be generally healthy' (IAEA, 1991b: 508). Despite falsification of the ICP claims, the IAEA has never withdrawn its erroneous conclusions. Such misinformation and covering up not only has harmed the Chernobyl victims physically but also has induced feelings of despair. They believe – correctly – that they cannot rely even on the international scientific and political community for unbiased information about their plight. Certainly the IAEA has been little help.

Environmental justice: lessons from Chernobyl

Why has the international scientific and political community allowed the global environmental injustice of Chernobyl? One reason may be that the entire international nuclear establishment relies on the ICRP recommendations for radiation protection. The keystone of these recommendations is the Justification Principle, the standard for justifying or allowing radiation exposures. The principle is controversial because it allows either egalitarian or utilitarian benefits to society to justify the harms or risks of radiation exposure to particular people. In other words, the dominant radiation protection principle sanctions environmental injustice. Does the Justification Principle satisfy equity? Does it satisfy utility? Subsequent sections argue for negative answers to both questions. They also call for regulatory reform at national and global levels.

The Justification Principle

The main international agency responsible for radiation protection, the ICRP, stipulates that unless additional radiation exposures provide sufficient benefit either to the exposed individuals or to society, they are not acceptable: 'No practice involving exposures to radiation should be adopted unless it produces sufficient benefit to the exposed individuals or to society to offset the radiation detriment it causes' (ICRP, 1991: 28, 71; IAEA, 1995: 12). The ICRP argues that, in cases of intervention to reduce exposure, 'The proposed intervention should do more good than harm' (ICRP, 1991: 28, 71). Because the Justification Principle allows *societal* benefits to be exchanged for *individual* harms (radiation exposures), it appears to be sanctioned by one version of utilitarian ethics. This utilitarian stance, however, is offset by some protections recommended for all individuals. The ICRP recommends that all radiation exposures 'should be constrained by restrictions on the doses to individuals' (*ibid.*).

In introducing discussion of individual radiation doses, the ICRP emphasizes the notion of equity in exposure reduction. It warns that when a population that receives benefits from the use of radiation does not bear the associated risks or detriments, there can be environmental injustice: inequities within the present generation or among current and future generations (ICRP, 1991: 26; and see IAEA, 1996: 7; NEA and OECD, 1995; Lochard and Grenery-Boehler, 1993: 11). For example, pregnant women may receive benefits from an X-ray, but if any foetus is exposed to radiation *in utero*, between 8 and 15 weeks after conception, it will lose, on average, 30 IQ points for each sievert of exposure (UNSCEAR, 1993; NRC, 1990; NCRP, 1993; Sinclair, 1995: 391; Sinclair, 1992: 319). Nevertheless, following the Justification Principle, the ICRP clearly makes an alleged utilitarian claim that a system of radiological protection should do more good than harm and should 'maximize the net benefit' (ICRP, 1991: 70).

Equity versus utility

The obvious question is whether the Justification Principle, despite its being enshrined in the dominant international recommendations about radiation protection, is ethically defensible, especially in terms of environmental justice. Egalitarians typically argue that utilitarian benefits to a *group* can almost never, if ever, offset detriments to *individuals*. They would say that the exposure 'losers' ought to be the same as the 'gainers', or else a particular involuntary and uncompensated radiation risk is discriminatory and therefore inequitable (Illich, 1974; Shrader-Frechette, 1991; MacLean, 1986). For example, the World Health Organization in its 1993 guidelines took a fundamentally egalitarian stand:

> Differences in distribution of burdens and benefits are justifiable only if they are based on morally relevant distinctions between persons . . . the equitable distribution of the burdens and benefits . . . raises no serious problems when the intended subjects do not include vulnerable individuals or communities . . . [such as] people receiving welfare benefits . . . poor people . . . unemployed . . . and racial minority groups.
> (CIOMS and WHO, 1993: 10–11, 29–31)

Utilitarian ethicists, however, typically allow trade-offs or discrimination – unequal protection from risks such as radiation – if they believe that these inequities promote the common good. They might say that the unequal radiation risks imposed on particular groups are offset by the benefits of allowing higher radiation exposures for some individuals. For example, the Society for Nuclear Medicine makes a utilitarian argument when it points out that stricter radiation standards would mean higher costs for the users and beneficiaries of nuclear-related activities (see Hiam, 1991). Because zero risk is unachievable, and because radiation-protection dollars are not infinite, utilitarians also argue that some allegedly inequitable trade-offs between risks and benefits are, in reality, ethical and desirable. In Sweden, for example, the government has determined that if the marginal cost of preventing a serious case of radiation injury is US$1 million or less, then preventing the individual exposure is 'strongly justified'. If the marginal cost of prevention (to industry) exceeds US$5 million, then the Swedes say that very strong reasons are required to implement the prevention (Bengtsson and Moberg, 1993). To determine whether the utilitarians are right in tending to support the Justification Principle, or whether the egalitarians are wrong in tending to reject it, consider cases of radiation-induced temporal and geographical inequities. The following discussion shows that, notwithstanding the ethical difficulties of utilitarianism, the Justification Principle appears to violate both utilitarian *and* egalitarian ethics.

Radiation and temporal equity

Some philosophers believe that risk–benefit trade-offs ought not to be allowed in cases like radiation exposure, particularly those in which one set of people gain economically by imposing higher risks on another set of people spatially or temporally distant from them who do not gain substantially from the exposure. Derek Parfit argues that temporal differences are not a morally relevant basis for discounting future costs and thus for discriminating against members of later generations. He and other egalitarians, as well as many proponents of an ethics of care (see Goodin, 1985), would maintain that it is unacceptable to impose most of the risks and costs of an activity on future people but to award most of its benefits to people now living. Economists have shown that commercial nuclear fission, for instance, probably benefits mainly present generations, whereas the risks and costs of its long-lived wastes (for a million years or more) will be borne primarily by members of future generations (Kneese *et al.*, 1983: 219; and see Raloff, 1996; Parfit, 1983a: 31–37; Parfit, 1983b: 166–79; Shrader-Frechette, 1993: 160–212).

Apart from radioactive wastes, commercial nuclear fission imposes inequitable risks on future people because of the genetic injuries induced by radiation-related accidents. In the contaminated Mogilev district of Belarus, approximately 300 km from Chernobyl, children born eight years after the accident showed double the normal number of germline mutations (Dubrova *et al.*, 1996). These Chernobyl-induced mutations will be passed on from generation to generation, and they will cause 500,000 premature fatal cancers and hundreds of non-fatal cancers and other defects, until natural selection removes the mutations (Dubrova *et al.*, 1996; see Baker *et al.*, 1996; Hillis, 1996: 665–66; Campbell, 1996: 653; Abbott and Barker, 1996: 658).

In general, egalitarian ethical theory requires society to do everything possible to minimize temporal and geographical inequities like those associated with the Chernobyl accident. Many egalitarians believe that utilitarian value judgements about acceptable radiation exposure are open to criticism in part because they can sanction using members of vulnerable geographical or temporal minorities to benefit the majority. They believe such utilitarian judgements wrongly condone using people as means to the ends of other people (Silini, 1992: 139–48; Shrader-Frechette, 1991: 117ff.). Egalitarians also argue that all people and all generations ought to receive equal consideration of their interests with respect to societal risks such as radioactive pollution since all humans (future as much as present) have the same capacity for a happy life. Moreover free, informed and rational people would be likely to agree to *prima facie* principles of equal rights or equal protection. *Prima facie* principles are those that apply in the absence of compelling arguments to the contrary. *Prima facie* principles of equal treatment are likewise important because they provide the basic justifications for other central concepts of ethics: all schemes involving consistency, justice, fairness, rights and autonomy. In fact, 'law itself embodies an ideal of equal treatment for persons similarly situated' (Blackstone,

1969: 121; and see Rawls, 1975: 284, 277, 280, 282; Beardsley, 1964: 35–36; Pennock, 1974: 2, 6).

Defending an explicitly egalitarian standard, the IAEA (in its *Safety Fundamentals*) sanctions treating future generations equally with present ones. It calls for using radiation protection standards for future people that would be acceptable if they were imposed in the present (IAEA, 1996: 4, 8; see also RPNSA, 1993). Similarly, the ICRP explicitly requires those who use nuclear technology to restrict doses to all individuals 'so as to limit inequity' (ICRP, 1991: 28, 71). Despite such claims, the ICRP recommendations and the IAEA standards might not actually result in temporal equity, in treating present and future people equally. Consider the case of permanent disposal of nuclear waste.

Most nations of the world have decided to use permanent geological disposal of high-level radioactive waste (see, for example, NEA and OECD, 1995: 14, 21–28; NRC, 1995), so that future generations 'need not take any action to protect themselves against the effects of waste disposal' (RPNSA, 1993: 27). Proponents of permanent burial argue that it is passively safe and places fewer burdens on our descendants (e.g. NEA and OECD, 1995). Opponents, however, argue that such geological disposal will not result in complete containment of the waste. They say that because the canisters will remain intact only for several hundred years, burial will actually place the greatest chemical and radiological risks on members of future generations (Hedelius and Persson, 1992: 159–65; Buchheim and Persson, 1992: 303–15). In other words, they maintain that permanent disposal is not really permanent and may implicitly sanction a risk distribution that violates temporal equity. Moreover, because many aspects of proposed sites (for permanent geological disposal of nuclear waste) 'are not well suited for quantitative risk assessment' and are 'fraught with substantial uncertainties that cannot be quantified using statistical methods', future risks could be significant (Younker *et al.*, 1992: B2). The uncertainties are so massive that a recent committee of the US National Academy of Sciences (NAS) argued that it was impossible to predict repository intrusion over the million-year lifetime of a facility (NRC, 1995). If the NAS is right, then even though nuclear policymakers affirm that they wish to keep future radiation risks below those that are acceptable in the present (see, for example, RPNSA, 1993: 24), permanent disposal may be at odds with the goal of temporal equity.

Radiation and geographical equity

Radiation protection also raises a number of environmental justice questions about the desirability of accepting principles of 'geographical equity' that would require radiation risk to be distributed equally across regions and nations (Shrader-Frechette, 1984: 210–60). In the case of the Chernobyl accident, for example, the Ukrainian government has paid no money to compensate for the 1 billion Swedish crowns that Sweden spent to reimburse its citizens who lost sheep and reindeer because of the Chernobyl radiation exposures. Neither has anyone compensated the Swedes for their own increased health risks caused by Chernobyl. The same is true for Chernobyl's effects on Norway, Denmark, Finland, Greece, Germany, Italy and all the nations facing a quarter of a million premature cancer deaths because of Chernobyl.

On the one hand, geographical equity requires that, in the absence of adequate compensation and consent, the set of beneficiaries of radiation exposures should be the same as the set of losers. Otherwise 'environmental injustice' or 'environmental racism' could result (Westra and Wenz, 1995; Bullard, 1993; Wigley and Shrader-Frechette, 1995: 135–59; Shrader-Frechette, 1993: 182–212). On the other hand, classical economists, who generally tend to follow utilitarian (rather than egalitarian or contractarian) ethical norms, question whether egalitarian distributions of radiation risk are necessary to attain geographical equity. Industry groups, nuclear proponents and many neoclassical economists tend to accept utilitarian ethical theories. They argue that siting hazardous facilities – for example, nuclear power plants, radioactive waste facilities or uranium enrichment centres – in economically and socially disenfranchised areas is justified, because the risks are small. They also say that, overall, society benefits from inequitable risk distributions and is often unable to provide equal protection from risks such as radiation (see Maxey, 1988: 4–5).

Utilitarians frequently claim that communities surrounding risky facilities have voluntarily accepted them, and that they provide employment as well as tax benefits. They say that a bloody loaf of bread is better than no loaf at all (Lewis, 1990: 8, 221–23; Cohen, 1990: 295; Jaeschke, 1981: 49–57; Crandall, 1981). Near the disabled Chernobyl reactor, for example, people have made a choice for the bloody loaf of bread, in part because assuring Ukrainians and Belarussians of equal radiation protection appears virtually impossible. A recent Nuclear Energy Agency report warned that 'the Ukraine cannot afford' to impose a dose limit of 2 rem/yr (0.2 Sv/yr), the current annual occupational dose limit that the ICRP recommends (NEA and OECD, 1994: 75–76). Such a dose limit would be extremely costly and probably would require managers at the disabled unit 4 reactor to perform all activities remotely (*ibid.*).

Cost considerations also won over equity in a prominent US radiation decision. Recently the US Advisory Committee on Human Radiation Experiments concluded that cost–benefit balancing argued against notifying radiation-experiment victims whose experiment-induced risk of premature death was lower than 1 in 1,000 for their lifetimes (Advisory Committee on Human Radiation Experiments, 1994; Faden, 1995; and see Gordon, 1996: 35). Utilitarians worry that such notification and compensation for allegedly very small risks could be costly to society. They also claim that, to the degree that one is unwilling to trade some equal protection (against radiation risks) for societal utility, then one is behaving not ethically but as an uncompromising, unrealistic ideologue (Gross and Levitt, 1994: 160).

Many proponents of egalitarian or contractarian ethical theories argue that it is unfair for economically, educationally or socially disenfranchised people to bear larger burdens of societal and workplace risk, as they typically do. They claim that, just as it was wrong for US Southerners to defend slavery as necessary to the economy of the South, so it is also wrong for risk evaluators to defend inequitable distributions of radiation and other risks as necessary to the common good. Contractarians and egalitarians argue that the end (societal benefit) does not justify the means (inequitable individual exposures to radiation risks). They say that equity and rights to equal protection ought to be inviolate, otherwise the concept of individual rights is meaningless (Bullard, 1993; Shrader-Frechette, 1984: 210–60; Shrader-Frechette, 1993: 182–212). They point out that even a low-probability risk – conforming to ICRP exposure standards – may be unacceptable if it is inequitably distributed (Lindell, 1980: 110).

Psychometric studies likewise reveal that, in their risk evaluations, lay people are often more averse to a small, inequitably distributed societal risk than to a large, equitably distributed one (Shrader-Frechette, 1991: 77–130). One reason may be that there is virtually no compensation for, or direct consent to, societally increased radiation risk. Another reason may be that there is great 'variability in the appearance of stochastic damage among members of a population' and that 'susceptibility to cancer is not uniform' (Silini, 1992: 144–45). As a result, inequitable effects of similar exposures to radiation are highly likely, yet often difficult, to prove in individual cases. Inequity also occurs because most risk, cost and benefit calculations are, to some degree, arbitrary, in so far as they depend on the chosen population and geographical distribution (Baram, 1981: 123–28; Ashford, 1981: 129–37).

Justification and authentic utility

If the preceding analysis is correct – as the Society for Nuclear Medicine, Margaret Maxey, the Nuclear Energy Agency and the US Advisory Committee on Human Radiation Experiments suggest it is (see preceding citations) – then the Justification Principle sacrifices equity for utility, for the greater societal good. But is the radiation case this simple? Do current applications of the Justification Principle really serve utility?

Although the principle aims to satisfy utility (maximizing the general welfare) or efficiency (maximizing the balance of goods over bads), it is not obvious that it does so. Allowing uncompensated radiation exposure inequities – whether of Chernobyl victims, indigenous peoples or others – in the name of efficiency may *not* be truly efficient and may *not* lead overall to a greater balance of good over bad. If one assumes that society needs to take account of future generations and that there will be future generations, then radiation-induced inequities will harm a great many people in the future. And if so, these inequities could harm utility as well as equity (fairness). Not preventing Chernobyl-induced germline mutations, for example, may have been expedient but these mutations may harm overall utility as well as equity. They may not serve the greatest good of the greatest number of people, over all time. John

Stuart Mill, in fact, specifically distinguishes between immorality and mere expediency (Mill, 1910: ch. 5; and see Lyons, 1979: 176ff.). He also points out that even utilitarians have particular obligations:

> The moral rules which forbid mankind to hurt one another (in which we must never forget to include wrongful interference with each other's freedom) are more vital to human well-being than any maxims, however important, which only point out the best mode of managing some department of human affairs.
>
> (Mill, 1910: ch. 5, par. 33)

Mill explains that bodily security is 'the most vital of all interests', 'the most indispensable of all necessaries, after physical nutrition', and 'the very groundwork of our existence' (*ibid.*: ch. 5). He affirms, 'to have a right, then, is, I conceive, to have something which society ought to defend me in the possession of. If the objector goes on to ask, why it ought? I can give him no other reason than general utility' (*ibid.* par. 25; for a discussion of this point see Baier, 1986: 49–74).

These passages suggest that, if Mill is right, then people have something like basic 'rights' to security, 'rights' not to have their liberty constrained (*except to prevent harm to others*). Mill's remarks also suggest that perhaps society ought to recognize these rights to security because they serve utility or the general welfare (see Gewirth, 1982: 157, who makes a similar point). On this view, classical utilitarian doctrine is not 'a hunting license, allowing the infliction of whatever wounds one likes, provided only that one's pleasure in the infliction is greater than the victim's pain' (Shue, 1981: 122; for criticism of the position that Shue rejects see Samuels, 1988). Rather, it is arguable that one is not allowed, under classical utilitarian doctrine, to threaten another's security. Were one allowed to do so, then maximization of net benefits could be said to justify the worst sort of barbarism or sadism (Shrader-Frechette, 1991: 150).

Of course, sometimes people must tolerate inequities because they are unavoidable in some situations (Jaeschke, 1981; Crandall, 1981; Maxey, 1988). However, most of the Chernobyl fatalities were and are avoidable. And it makes sense to avoid them, particularly if the IAEA radiation standards, based on ICRP principles, are intended to 'cover all people who may be exposed to radiation, including those in future generations who could be affected by present practices or interventions' (IAEA, 1996: 11). To the degree that radiation causes genetic defects that may induce, in turn, a slow loss of fitness in the population, then those exposures threaten not only equity but also utility – the good of the majority (see Muller, 1964: 42–50). To the extent that long-lived radioactive isotopes, like iodine-129 (with a half-life of 16 million years), threaten posterity, then particular exposures may harm the good of a majority of people and not just a few members of the public or a few workers. As a result, many preventable, inequitable radiation exposures may violate utilitarian principles as well as egalitarian ones.

If the preceding reasoning is correct, then there may be a false dichotomy between alleged utilitarian and egalitarian responses to radiation protection under the Justification Principle. Utilitarian ethical theory, correctly understood, provides no justification for environmental injustice like that caused by Chernobyl. By framing questions of global nuclear injustice in the rhetoric of an allegedly utilitarian justification, nuclear proponents such as the IAEA may have influenced public responses to the apparent injustice. Those who frame the questions control the answers. Once citizens, scientists and other professionals frame the questions of global environmental justice more correctly, then disinformation and injustice – about Chernobyl, low-level radiation and global destruction from it – may be less possible.

Once both the public and professionals recognize that neither utilitarianism nor the Justification Principle clearly justifies either the nuclear culture that apparently helped cause Chernobyl or the international nuclear response to it, society will be better able to address global environmental injustice. Solving these problems will require the commonsensical recognition that the *only* way the Justification Principle could serve utility, as well as equity, would be for radiation dose limits to become more and more strict in the future. Assuming that society ought to take account of future people and that there will be future people, only by increasing the strictness of radiation exposure standards can one protect the gene pool. But if so, only increasingly strict radiation standards can serve general utility or the greatest number of people. Egalitarians likewise might note that stricter standards are necessary in order to ensure that members of future generations have protection equal to that of members of present generations. Biocentrists, in addition, might note that stricter radiation standards are necessary to ensure genuine utility for all beings, both human and non-human.

Whether or not one argues for increasing the strictness of radiation standards, as some egalitarians in fact do and as utilitarians perhaps ought to do, future genetic damage cannot be justified merely on grounds of expediency (see Snihs, 1994). Expediency is not the same as ethics. And if one does not, then many arguments that attempt to justify radiation-related genetic damage, on purely utilitarian grounds, fail. Nuclear physicist Bernard Cohen argues, for example, that threats to the gene pool are insignificant because natural selection will eventually remove undesirable mutations caused by radiation (Cohen, 1990: 277). He argues that such radiation-induced genetic damage 'can therefore only improve the human race', an assertion which does not in fact follow from the preceding claim. What he forgets, moreover, is that even utilitarianism does not allow humans to be used as a means to some end. People's suffering is not insignificant just because radiation-induced damage may not continue for ever. If one accepted the argument that natural selection would eventually rid the gene pool of some harmful condition, and that some current damage 'can therefore only improve the human race', it could take millions of years for natural selection to end the damage. Besides, this natural selection argument is questionable because it would allow any genetic harm, no matter how heinous.

The natural selection argument fails to take account of basic human rights, fairness, equity and utility.

Conclusion: vehicles for change

If the arguments in this chapter are correct, then global environmental injustice (as exemplified by Chernobyl) may have been possible because citizens, professionals and intellectuals have not demanded the truth and worked to obtain it. Just as there is no 'technological fix' for the complex ethical problems raised by environmental degradation, so also there is no 'ethical fix' for the complex equity problems raised by global environmental injustice. The vehicles for change may lie within us. The vehicles for change may include citizen, professional and intellectual education about radiation and public-interest advocacy. The vehicles for change may be our own demands for information. Without public information and committed activism, any possible solutions – national or international laws, a UN forum, a world environmental council – will fail. Later contributions to this volume, especially those by Oran Young (Chapter 14) and John Dryzek (Chapter 15), suggest some institutional reforms that committed activists might pursue.

References

Abbott, A. and Barker, S. (1996) 'Chernobyl Damage "Underestimated"', *Nature* 380 (6576): 658.

Advisory Committee on Human Radiation Experiments (1994) *Interim Report of the Advisory Committee on Human Radiation Experiments*, Washington, DC: US Government Printing Office.

Ashford, N. A. (1981) 'Alternatives to Cost–Benefit Analysis in Regulatory Decisions', in W. J. Nicholson (ed.) *Management of Assessed Risk for Carcinogens*, New York: New York Academy of Sciences, pp. 129–37.

Baier, A. (1986) 'Poisoning the Wells', in D. MacLean (ed.) *Values at Risk*, Totowa, NJ: Rowman and Allanheld, pp. 49–74.

Baker, R. J., Van Den Bussche, R. A., Wright, A. J., Wiggins, L. E., Reat, E. P., Smith, M. H., Lomakin, M. D. and Chesser, R. K. (1996) 'High Levels of Genetic Change in Rodents of Chernobyl', *Nature* 380 (6576): 707–708.

Balanov, M., Jacob, P., Likhtarev, I. and Minenko, V. (1996) 'Pathways, Levels and Trends of Population Exposure after the Chernobyl Accident', in *Conference on the Radiological Consequences of the Chernobyl Accident, Minsk, 18–22 March (1996)*, Brussels: European Commission.

Baram, M. S. (1981) 'The Use of Cost–Benefit Analysis in Regulatory Decision-Making is Proving Harmful to Public Health', in W. J. Nicholson (ed.) *Management of Assessed Risk for Carcinogens*, New York: New York Academy of Sciences, pp. 123–28.

Baverstock, K. F., Egloff, B., Pinchera, A., Ruchti, C. and Williams, D. (1992) 'Thyroid Cancer after Chernobyl', *Nature* 359 (6390): 21–22.

Beardsley, M. C. (1964) 'Equality and Obedience to Law', in S. Hook (ed.) *Law and Philosophy*. New York: New York University Press, pp. 35–42.

Bengtsson, G. and Moberg, L. (1993) 'What is a Reasonable Cost for Protection Against Radiation and Other Risks?', *Health Physics* 64: 661–66.

Blackstone, W. T. (1969) 'On the Meaning and Justification of the Equality Principle', in W. T. Blackstone (ed.) *The Concept of Equality*, Minneapolis: Burgess, pp. 115–33.

Born, H.-P. (1983) 'Kernenergie in der Sowjetunion', *Atomwirtschaft–Atomtechnik* 28 (12): 645–48.

Brall, A. and MacLachlan, A. (1995) 'Kiev Defines Chernobyl Position, Asks $4-billion for New Plant', *Nucleonics Week* 36 (18): 17ff.

Buchheim, B. and Persson, L. (1992) 'Chemotoxicity of Nuclear Waste Repositories', *Nuclear Technology* 97: 303–15.

Bullard, R. D. (1993) *Confronting Environmental Racism*, Boston: South End Press.

Campbell, P. (1996) 'Chernobyl's Legacy to Science', *Nature* 380 (6576): 653.

Cohen, B. L. (1990) *The Nuclear Energy Option: An Alternative for the 90s*, New York: Plenum Press.

Coryn, P. and MacLachlan, A. (1995) 'Mental Disorders Said to be Spreading among Chernobyl-Affected People', *Nucleonics Week* 36 (27): 1ff.

Council for International Organizations of Medical Sciences (CIOMS) and World Health Organization (WHO) (1993) *International Ethical Guidelines for Biomedical Research Involving Human Subjects*, Geneva: CIOMS.

Crandall, R. W. (1981) 'The Use of Cost–Benefit Analysis in Regulatory Decision-Making', in W. J. Nicholson (ed.) *Management of Assessed Risk for Carcinogens*, New York: New York Academy of Sciences, pp. 99–107.

Dubrova, Y. E., Nesterov, V. N., Krouchinsky, N. G., Ostapenko, V. A., Neumann, R., Neil, D. L. and Jeffreys, A. J. (1996) 'Human Minisatellite Mutation Rate after the Chernobyl Accident', *Nature* 380 (6576): 683–86.

Edwards, M. (1994) 'Chornobyl', *National Geographic* 186 (2): 100–116.

Edwards, R. (1995) 'Will It Get Any Worse?', *New Scientist* 148 (2007): 14–15.

Faden, R. (1995) 'Testimony', in US Senate, Committee on Governmental Affairs, *Human Radiation and Other Scientific Experiments: The Federal Government's Role. Hearing before the Committee on Governmental Affairs, January 25, 1994*, Washington, DC: US Government Printing Office.

Franklin, B. A. (1990a) 'Chernobyl Pilot's Marrow Transplant a Guarded Success', *Nucleonics Week* 31 (18): 9ff.

—— (1990b) 'USSR Marks Chernobyl Date with Closure Promise, Pleas for Help', *Nucleonics Week* 31 (18): 1ff.

Gewirth, A. (1982) *Human Rights: Essays on Justification and Applications*, Chicago: University of Chicago Press.

Gofman, J. W. (1995) Foreword to A. Yaroshinskaya, *Chernobyl: The Forbidden Truth*, Lincoln: University of Nebraska Press, pp. 1–13.

González, A.J. (1994) 'Biological Effects of Low Doses of Ionizing Radiation: A Fuller Picture', *IAEA Bulletin* 4: 37–45.

Goodin, R. E. (1985) *Protecting the Vulnerable: A Reanalysis of our Social Responsibilities*, Chicago: University of Chicago Press.

Gordon, D. (1996) 'The Verdict: No Harm, No Foul', *Bulletin of the Atomic Scientists* 52 (1): 32–40.

Gross, P. R. and Levitt, N. (1994) *Higher Superstition: The Academic Left and Its Quarrels with Science*, Baltimore: Johns Hopkins University Press.

Hedelius, G. and Persson, L. (1992) 'Legal Problems Connected with the Chemotoxicity of Nuclear Waste', in N. Pelzer (ed.) *Stillegung und Beseitigung kerntechnischer Anlagen*, Baden-Baden: Nomos Verlagsgesellschaft, pp. 159–65.

Henshaw, D. L. (1996) 'Chernobyl 10 Years On', *British Medical Journal* 312 (7038): 1052–53.

Hiam, J. (1991) 'NRC and ICRP Lower Radiation Exposure Limits', *Journal of Nuclear Medicine* 32 (5): 29N.

Hill, P., Hille, R. and Heinzelmann, M. (1995) 'Radiation Exposure of the Population in Regions of Belarus, Russia and the Ukraine Affected by the Chernobyl Accident', in World Health Organization, *International Conference on Health Consequences of the Chernobyl and Other Radiological Accidents, 20–23 November*, Geneva: WHO.

Hillis, D.M. (1996) 'Life in the Hot Zone around Chernobyl', *Nature* 380 (6576): 665–66.

Illich, I. (1974) *Energy and Equity*, New York: Harper & Row.

International Atomic Energy Agency (IAEA) (1991a) *International Chernobyl Project, Proceedings of an International Conference, Vienna, 21–24 May 1991. Assessment of Radiological Consequences and Evaluation of Protective Measures*, Vienna: IAEA.

—— (1991b) *International Chernobyl Project: Technical Report*, Vienna: IAEA.

—— (1995) *Organization and Operation of a National Infrastructure Governing Radiation Protection and Safety of Radiation Sources*, Vienna: IAEA.

—— (1996) *International Basic Safety Standards for Protecting against Ionizing Radiation and for the Safety of Radiation Sources*, Vienna: IAEA. Safety Series No. 115.

International Commission on Radiological Protection (ICRP) (1991) *1990 Recommendations of the International Commission on Radiological Protection: Adopted by the Commission in November 1990*, Oxford: Pergamon Press. ICRP Publication 60; *Annals of the ICRP* 21 (1–3).

Ito, T., Seyama, T., Iwamoto, K. S., Mizuno, T., Tronko, N. D., Komissarenko, I. V., Cherstovoy, E. D., Satow, Y., Takeichi, N., and Dohi, K. (1994) 'Activated RET Oncogene in Thyroid Cancers of Children from Areas Contaminated by the Chernobyl Accident', *Lancet* 344 (8917): 259.

Ivanov, V. (1996) 'Health Status and Follow-Up of Liquidators in Russia', in A. Karaoglou (ed.) *Radiological Consequences of the Chernobyl Accident, Conference on the Consequences of the Chernobyl Accident, Minsk, March (1996)*, Luxembourg: Office for Official Publications of the European Communities, pp. 861–70.

Jaeschke, W.C. (1981) 'Anatomy of Unreasonable Risk', in W. J. Nicholson (ed.) *Management of Assessed Risk for Carcinogens*, New York: New York Academy of Sciences, pp. 49–57.

Kazakov, V. S., Demidchik, E. P., and Astakhova, L. N. (1992) 'Thyroid Cancer after Chernobyl', *Nature* 359 (6390): 21–22.

Kesminie, A. *et al.* and Chernobyl Medical Centre, Vilnius (1995) 'Chernobyl Clean-Up Workers from Lithuania: Monitoring of Health Effects', in *11th Symposium on Epidemiology in Occupational Health, 5–8 September (1995), Noordwijkerhout, The Netherlands. Epidemiology* 6 (4): 1044–3983.

Kneese, A. V. *et al.* (1983) 'Economic Issues in the Legacy Problem', in R. Kasperson (ed.) *Equity Issues in Radioactive Waste Management*, Cambridge, MA: Oelgeschlager, Gunn & Hain.

Konoplev, A. V., Bulgakov, A. A., Popov, V. E., Popov, O. F., Scherbak, A. V., Shveiken, Y. V. and Hoffman, F. O. (1996) 'Model Testing Using Chernobyl Data: I. Wash-off of ^{90}Sr and ^{137}Cs from Two Experimental Plots Established in the Vicinity of the Chernobyl Reactor', *Health Physics* 70 (1): 8–12.

Lazyuk, G. I., Kirillova, I. A., Nikolaev, D. L., Novikova, I. V., Fomina, Z. N. and Khmel, R. D. (1995) 'Frequency Changes of Inherited Anomalies in the Republic of Belarus after the Chernobyl Accident', *Radiation Protection Dosimetry* 62 (1–2): 71–74.

Lewis, H. W. (1990) *Technological Risk*, New York: W. W. Norton.

Likhtarev, I. A., Gulko, G. M., Kairo, I. A., Los, I. P., Henrichs, K. and Paretzke, H. G. (1994) 'Thyroid Doses Resulting from the Ukraine Chernobyl Accident, Part I: Dose Estimates for the Population of Kiev', *Health Physics* 66 (2): 137–46.

Likhtariov, I., Kovgan, L., Novak, D., Vavilov, S., Jacob, P. and Paretzke, H. G. (1996) 'Effective Doses Due to External Irradiation from the Chernobyl Accident for Different Population Groups of the Ukraine', *Health Physics* 70 (1): 87–98.

Lindell, B. (1980) 'Ethical and Social Issues in Risk Management', in R. L. Shinn (ed.) *Faith and Science in an Unjust World: Report of the World Council of Churches' Conference on Faith, Science, and the Future*, Vol. 1: *Plenary Presentations*, Geneva: World Council of Churches, pp. 105–15.

Lochard, J. and Grenery-Boehler, M. C. (1993) 'Optimizing Radiation Protection: The Ethical and Legal Bases', *Nuclear Law Bulletin* 52: 9–27.

Lyons, D. (1979) 'Human Rights and the General Welfare', in D. Lyons (ed.) *Rights*, Belmont, CA: Wadsworth, pp. 174–86.

MacLachlan, A. (1990) 'Chernobyl Health Effects Debate Continues in USSR', *Nucleonics Week* 31 (29): 1ff.

—— (1992) 'Rising Children's Thyroid Cancers Indicate Growing Chernobyl Link', *Nucleonics Week* 33 (37): 1ff.

—— (1994a) 'Chernobyl Liquidators Experience More Illnesses, not More Death', *Nucleonics Week* 35 (46): 10ff.

—— (1994b) 'IAEA Sounds Alarm on Chernobyl Safety Level', *Nucleonics Week* 35 (14): 1ff.

—— (1994c) 'Official Russian Register Keeps Chernobyl Death Toll at 28', *Nucleonics Week* 35 (46): 11ff.

MacLean, D. (1986) 'Social Values and the Distribution of Risk', in D. MacLean (ed.) *Values at Risk*, Totowa, NJ: Rowman & Allanheld, pp. 75–93.

Makhijani, A., Hu, H. and Yih, K. (eds) (1995) *Nuclear Wastelands: A Global Guide to Nuclear Weapons Production and its Health and Environmental Effects*, Cambridge, MA: MIT Press.

Marples, D. R. (1995) Introduction to A. Yaroshinskaya, *Chernobyl: The Forbidden Truth*, Lincoln: University of Nebraska Press, pp. ix–xvii.

Maxey, M. N. (1988) 'Radiation Risks: The Ethics of Health Protection', in SRP, *Radiation Protection Practice, Proceedings of the Seventh International Congress of the International Radiation Protection Association, Sydney, 10–17 April (1988)*, vol. 1, Sydney: Pergamon Press, pp. 2–5.

Mill, J. S. (1910) *Utilitarianism, On Liberty, and Representative Government*, New York: Dutton.

Ministry for Emergencies and Population Protection from the Chernobyl NPP Catastrophic Consequences (1995) *Republic of Belarus – 9 Years after Chernobyl: Situation, Problems, Action*, Minsk: Ministry for Emergencies and Population Protection from the Chernobyl NPP Catastrophic Consequences.

Muller, H. J. (1964) 'Radiation and Heredity', *American Journal of Public Health* 54 (1): 42–50.

National Council on Radiation Protection and Measurements (NCRP) (1993) *Risk Estimates for Radiation Protection: Recommendations of the National Council on Radiation Protection and Measurements*, Bethesda, MD: NCRP. NCRP Report No. 115.

National Research Council, Committee on the Biological Effects of Ionizing Radiation, Board on Radiation Effects Research Commission on Life Sciences (1990) *Health Effects of Exposure to Low Levels of Ionizing Radiation*, Washington, DC: National Academy Press. BEIR V.

National Research Council (NRC) (1995) *Technical Bases for Yucca Mountain Standards*, Washington, DC: National Academy Press.

Nikiforov, Y. and Gnepp, D.R. (1994) 'Pediatric Thyroid Cancer after the Chernobyl Disaster: Pathomorphologic Study of 84 Cases (1991–1992) from the Republic of Belarus', *Cancer* 74 (2): 748–66.

Nuclear Energy Agency and OECD (1994) *Nuclear Waste Bulletin: Update on Waste Management Policies and Programmes No. 9*, Paris: NEA.

—— (1995) *The Environmental and Ethical Basis of Geological Disposal*, Paris: NEA.

—— (1996) *Chernobyl Ten Years On: Radiological and Health Impact*, Paris: OECD.

Nussbaum, R. and Köhnlein, W. (1995) 'Health Consequences of Exposures to Ionising Radiation from External and Internal Sources: Challenges to Radiation Protection Standards and Biomedical Research', *Medicine and Global Survival* 2 (4): 198–213.

Oganesian, N. M. (1995) 'The State of Health of Chernobyl NPP Accident Liquidators', *Radiation Protection Dosimetry* 62 (1–2): 69–70.

Parfit, D. (1983a) 'Energy Policy and the Further Future: The Social Discount Rate', in D. MacLean and P. G. Brown (eds) *Energy and the Future*, Totowa, NJ: Rowman & Littlefield, pp. 31–37.

—— (1983b) 'Energy Policy and the Further Future: The Identity Problem', in D. MacLean and P. G. Brown (eds) *Energy and the Future*, Totowa, NJ: Rowman & Littlefield, pp. 166–79.

Pennock, J. R. (1974) Introduction to J. R. Pennock and J. W. Chapman (eds) *The Limits of the Law, Nomos 15, Yearbook of the American Society for Political and Legal Philosophy*, New York: Lieber-Atherton, pp. 2–6.

Poiarkov, V. A. (1995) 'Post-Chernobyl Radiomonitoring of the Forest Ecosystem', *Journal of Radioanalytical and Nuclear Chemistry, Articles* 194 (2): 259–67.

Prisyazhniuk, A., Fedorenko, Z., Okeanov, A. and Ivanov, V. (1996) 'Epidemiology of Cancer in Populations Living in Contaminated Territories of Ukraine, Belarus, Russia after the Chernobyl Accident', in A. Karaoglou (ed.) *Radiological Consequences of the Chernobyl Accident: Conference on the Consequences of the Chernobyl Accident, Minsk, March*, Luxembourg: Office for Official Publications of the European Communities, pp. 909–24.

Radiation Protection and Nuclear Safety Authorities in Denmark, Finland, Iceland, Norway, and Sweden (RPNSA) (1993) *Disposal of High-Level Radioactive Waste: Consideration of Some Basic Criteria*, Stockholm: Swedish Radiation Protection Institute.

Raloff, J. (1996) 'Radiation Damages Chernobyl Children', *Science News* 149: 260.

Rawls, J. (1975) 'Justice as Fairness', in J. Feinberg and H. Gross (eds) *Philosophy of Law*, Encino, CA: Dickenson, pp. 280–84.

Rytömaa, T. (1996) 'Ten Years after Chernobyl', *Annals of Medicine* 28 (2): 83–87.

Samuels, S. (1988) 'The Arrogance of Intellectual Power', in A. Woodhead, M. Bender and R. Leonard (eds) *Phenotypic Variation in Populations*, New York: Plenum, pp. 113–20.

Savchenko, V.K. (1995) *The Ecology of the Chernobyl Catastrophe: Scientific Outlines of an International Programme of Collaborative Research*, London/New York: UNESCO and the Parthenon Publishing Group.

Scholz, R. (1994) 'On the Sensitivity of Children to Radiation', *Medicine and Global Survival* 1: 38–44.

Semenov, B. A. (1983) 'Nuclear Power in the Soviet Union', *IAEA Bulletin* 25 (2): 47–60.

Shcherbak, Y. M. (1996) 'Ten Years of the Chornobyl Era', *Scientific American* 274 (4): 44–49.

Shrader-Frechette, K. (1984) *Science Policy, Ethics, and Economic Methodology*, Boston: Kluwer/Reidel.

—— (1991) *Risk and Rationality: Philosophical Foundations for Populist Reforms*, Berkeley: University of California Press.

—— (1993) *Burying Uncertainty: Risk and the Case Against Geological Disposal of Nuclear Waste*, Berkeley: University of California Press.

—— (1999) *Taking Action, Reclaiming Democracy: Environmental Advocacy and Public Ethics*, forthcoming.

—— (2000) *Radiation Protection and Ethics*, forthcoming.

Shue, H. (1981) 'Exporting Hazards', in P. Brown and H. Shue (eds) *Boundaries: National Autonomy and its Limits*, Totowa, NJ: Rowman & Littlefield, pp. 107–45.

Silini, G. (1992) 1992 Sievert Lecture: 'Ethical Issues in Radiation Protection', *Health Physics* 63: 139–48.

Sinclair, W. K. (1992) 'Radiation Protection: Recent Recommendations of the ICRP and the NCRP and their Biological Basis', in O. F. Nygaard, W. K. Sinclair and J. T. Lett (eds) *Advances in Radiation Biology*, vol. 16, San Diego, CA: Academic Press, pp. 303–24.

—— (1995) 'Radiation Protection Recommendations on Dose Limits: The Role of the NCRP and the ICRP and Future Developments', *International Journal of Radiation Oncology, Biology, Physics* 31 (2): 387–92.

Snihs, J. O. (1994) *The Approach to Individual and Collective Risk in Regard to Radiation and Its Application to Disposal of High Level Waste*, Stockholm: Swedish Radiation Protection Institute. SSI Report 18.

Souchkevitch, G. (1996) 'WHO-IPHECA: Epidemiological Aspects', in A. Karaoglou (ed.) *Radiological Consequences of the Chernobyl Accident, Conference on the Consequences of the Chernobyl Accident, Minsk, March (1996)*, Luxembourg: Office for Official Publications of the European Communities, pp. 899–908.

Stsjazhko, V. A., Tsyb, A. F., Tronko, N. D., Souchkevitch, G. and Baverstock, K. F. (1995) 'Childhood Thyroid Cancer since Accident at Chernobyl', *British Medical Journal* 310 (6982): 801.

United Nations (1995) 'Strengthening of International Cooperation and Coordination of Efforts to Study, Mitigate and Minimize the Consequences of the Chernobyl Disaster', in *Strengthening of the Coordination of Humanitarian and Disaster Relief Assistance of the United Nations, Including Special Economic Assistance: Report of the Secretary-General, Fiftieth Session, Item 20 (d) of the Provisional Agenda, 8 September*, New York: UN.

United Nations Scientific Committee on the Effects of Atomic Radiation (UNSCEAR) (1993) *Report of the United Nations Scientific Committee on the Effects of Atomic Radiation*, New York: United Nations.

—— (1994) *Sources and Effects of Ionizing Radiation: UNSCEAR (1994) Report to the General Assembly, with Scientific Annexes*, New York: United Nations.

Westra, L. and Wenz, P. (eds) (1995) *Faces of Environmental Racism: Confronting Issues of Global Justice*, Lanham, MD: Rowman & Littlefield.

Wigley, D. and Shrader-Frechette, K. (1995) 'Consent, Equity, and Environmental Justice', in L. Westra and P. Wenz (eds) *Faces of Environmental Racism Confronting Issues of Global Justice*, Lanham, MD: Rowman & Littlefield, pp. 135–59.

Williams, D. (1994) 'Chernobyl', *Nature* 371: 556.

World Health Organization (1995) *Health Consequences of the Chernobyl Accident: Results of the IPHECA Pilot Projects and Related National Programmes, Summary Report*, Geneva: WHO.

World Health Organization EURO Working Group on the Effects of Nuclear Accidents (1990) 'Psychological Dimensions of the Chernobyl Nuclear Accident', Copenhagen: unpublished document No. EUR/ICP/CEH 093.

Yaroshinskaya, A. (1995) *Chernobyl: The Forbidden Truth*, Lincoln: University of Nebraska Press.

Younker, J. L. et al. (1992) *Report of the Peer Review Panel on the Early Site Suitability Evaluation of the Potential Repository Site at Yucca Mountain, Nevada*, Washington, DC: US Department of Energy. SAIC-91/8001.

6 Justice, the market and climate change

Clive Hamilton

Introduction: facts and fairness

Climate change is possibly the most serious environmental threat ever faced by the world community and unsurprisingly the causes, impacts and the solutions to it are fraught with ethical issues. Indeed, ethical undercurrents swirl beneath almost every aspect of the international negotiations under the Framework Convention on Climate Change (FCCC), especially in the process begun by the Berlin Mandate in 1995 that reached a conclusion in Kyoto in December 1997. The principal concerns are distributive: who is responsible for the problem, who will suffer most from climate change, and who will bear the costs of abatement measures. There are in addition deeper philosophical concerns about the way in which decision-makers conceive of the problem and the solutions as fundamentally economic. This raises profound issues about the relationship of human beings to the natural environment.

This chapter explores the most significant of these issues of environmental justice. A persistent theme will be the role of facts, especially numerical data, in forming perceptions of fairness. In international negotiations it is not enough for a nation to claim that it is being treated unfairly; it must substantiate its claim with credible scientific and economic evidence. In the United States and Australia, economic modelling results have assumed particular importance in domestic debates between environmentalists and industry lobbyists and, especially in the Australian case, in government attempts to convince the world community that it would be especially disadvantaged by uniform greenhouse gas reduction targets.

First the question is considered of the distribution of the causes and impacts of climate change between developed and developing countries. The second section comments on the distribution of the impacts of climate change and abatement policies between social classes within developed countries. The third section discusses some of the vital issues of the distribution of the costs of reducing emissions among developed countries, since this has been fundamental to international negotiations to find agreement on mandatory emission reduction targets. The fourth section considers some of the philosophical underpinnings of the debate, and in particular the way in which the problem of climate change has been captured by the economic way of thinking. The final section considers the outcome of the Kyoto Protocol and some of the issues of justice that it raises. Throughout, I illustrate the arguments with reference to the Australian government's position on climate change.

At the outset, there are several key facts about global climate change that provide the essential background to the international debate (see IPCC, 1995a).

1   The problem of climate change is one of global commons. The greenhouse gas emissions of one country contribute to climate change in all countries, and each country will benefit from the efforts of every other country to reduce its emissions. Thus progress is possible only with international cooperation.
2   The increased concentrations of greenhouse gases in the atmosphere are due overwhelmingly to the activities of developed countries, and their emissions have been an essential component of the accumulation of industrial wealth.
3   Poor countries – especially those in the tropics – will suffer the most significant impacts of climate change.
4   Developing countries will become the dominant emitters of greenhouse gases in twenty-to-thirty years' time, with China, India and Brazil expected to become the largest emitters.
5   While there is some evidence that anthropogenic emissions have already changed the Earth's climate, the major impacts of climate change will not be felt until the middle of the next century and beyond. Thus climate change is a problem that requires significant action now in order to avoid damage well into the future.
6   There remain major uncertainties about the type, extent and locations of changes in climate and the impacts on humans.

It is not possible here to go into detail on these facts, but they clearly have fundamental implications for notions of fairness and justice. Questions of distribution of the benefits of fossil fuel use, the costs of abatement and the impacts of climate change lie at the heart of the climate change negotiations. Indeed, although the problem of climate change has been and continues to be driven by the science, the international debate about how to tackle the problem is fundamentally about principles of fairness.

The distribution of impacts among countries

The economic benefits from fossil fuel combustion since the start of the Industrial Revolution, and the expected impacts of climate change in the twenty-first century, are each distributed very unequally around the world.

Industrialized countries are responsible for more than 80 per cent of increased greenhouse gas concentrations in the atmosphere, and their industrial wealth has in large measure been possible because of consumption of the fossil fuels from which most greenhouse gases derive. On the other hand, climate scientists believe that the worst impacts of climate change – including sea-level rise, increased frequency of severe weather events, crop damage and the spread of diseases – will occur near the tropics and will therefore affect developing countries more severely (see IPCC, 1995b). These basic facts define the boundaries of the roles assumed by rich countries and developing countries and are reflected in the 1995 Berlin Mandate of the Framework Convention on Climate Change, the agreement that set in train the process leading to mandatory emission reduction targets agreed in Kyoto in December 1997.

Although developed countries are responsible for more than 80 per cent of anthropogenic greenhouse gases in the atmosphere, developing countries currently contribute about 32 per cent of annual global CO_2 emissions from fossil fuel combustion, a share that is expected to increase to around 44 per cent by 2010 (Hare, 1996: table 1). It is thus widely accepted that if a doubling of CO_2 concentrations in the atmosphere is to be avoided, then developing countries must take action to reduce emissions soon after industrialized countries begin to cut their emissions.

However, it is also widely accepted that it would be unfair to ask developing countries to reduce their emissions at the same time and at the same rate as developed countries. Since developed countries largely created the problem and, due to their wealth, are in a much better position to meet the costs of emissions reductions, it is fair to expect them to lead the way. It is envisaged that leadership by developed countries will also provide the technologies that will reduce the costs for developing countries of cutting their emissions. These principles are acknowledged explicitly in the Framework Convention on Climate Change.

While the OPEC countries have attempted to delay international agreement (and have even asked for financial compensation as demand for oil declines in response to emissions reduction measures), developing countries as a whole have pressed for strong measures to reduce emissions. The strongest emission reduction proposal has been put forward by the Alliance of Small Island States (AOSIS): a protocol requiring Annex 1 countries (i.e. industrialized countries) to reduce their emissions by 20 per cent below 1990 levels by the year 2005. Representing the group of countries with the most to lose from climate change, the small island states have gained considerable moral purchase in the international negotiations despite their tiny size. Their position has a major influence on the Group of 77 (G77) – the major grouping of developing countries – and they have received strong backing from the members of ASEAN (the Association of South East Asian Nations). Their influence is tempered somewhat by the voices of the OPEC countries, and at times by the caution of the biggest developing countries (China, India and Brazil), which, mindful of the fact that attention will shift to their emissions after Annex 1 countries agree to targets, stress the need for developing countries to meet basic needs and the primary role of developed countries in mitigating climate change.

Ultimately, however, the influence of G77 at Kyoto was not strong. Developing countries were put on the defensive as a result of late demands by the United States and Australia that they too be required to adopt mandatory targets. The belief that by insisting on targets for developing countries sooner rather than later the rich countries are trying to deprive poor countries of the opportunity of following their path to wealth is nicely summarized in a cartoon that appeared in *Tiempo*. It shows a besuited Western bureaucrat leading the G77 children away from a smoke-bellowing carousel named 'Fossil Fuel Fun'. The little boy from G77 is saying, 'But Dad, you always played on it, why can't WE?' (*Tiempo* Issue 23, March 1997).

The influence of the AOSIS nations, and their precarious position, was graphically illustrated at the South Pacific Forum in September 1997. The annual forum brings together the Pacific island states and Australia and New Zealand. The latter have played a major role historically in protecting and supporting the Pacific nations. While the Australian government sees Australia as the country with the most to lose from mandatory emission cuts at Kyoto, the AOSIS nations see themselves as having the most to gain. The Intergovernmental Panel on Climate Change (IPCC) predicts sea-level rise of up to 95 cm by the end of the twenty-first century, and this would seriously jeopardize low-lying islands in the Pacific. The Hadley Centre model of the UK Meteorological Office predicts a sea-level rise of 21 cm by 2050, which would put an additional 46 million people at risk of flooding, most of them in South Asia (Meteorological Office, 1998: 9). In addition, an increase in extreme weather events may result in more cyclones devastating some of these islands (IPCC, 1995a,b).

The undisguised self-interest of the Australian position attracted severe criticism from some Pacific leaders and Australian commentators. Even the conservative business daily, the *Australian Financial Review*, editorialized against the government's 'ham-fisted diplomacy' and 'intransigent stand'. The Australian Prime Minister dismissed the concerns of Pacific island states as 'exaggerated' and 'apocalyptic', apparently insensitive to the Pacific view that if 2,500 of the world's top climate scientists produce a comprehensive report predicting that your country may well disappear under rising seas, this does indeed look like an apocalypse. Perceptions of the moral bankruptcy of the Australian position were reinforced by comments from the government's chief economic adviser on climate change, who said that it might be more efficient to evacuate small island states subject to inundation than to require industrialized countries like Australia to reduce their emissions.

Distributional issues within developed countries

Climate change policies in the United States and Australia are heavily influenced by beliefs about the distribution of the costs of reducing emissions. This is the classic dilemma of policy change in which a small but influential sector bears the costs while the beneficiaries are spread out and are less aware of the impacts. Thus in the United States and Australia, the fossil fuel industry has had a major influence on policy, with the effect of diluting the desire to take serious measures to reduce emissions.

Trade unions representing workers in the fossil fuel industries have generally supported the arguments of their employers. This position has been reinforced by economic modelling studies that predict significant job losses, particularly among blue-collar workers, as a result of emissions abatement policies and price rises for petrol and electricity that would disadvantage poorer households. This has led to the view that abatement measures may have an inequitable impact, and has made alliances between environmental groups and social justice organizations difficult.

Typically, the debate over the distributional consequences of climate change has been very partial, with almost no attention paid to the distributional consequences of *failing* to pursue measures to reduce climate change. There are good grounds for suspecting that the costs of climate change in industrialized countries would be distributed very inequitably so that future generations of the poor will suffer the consequences of policy failure now.

In the case of households, the principal costs of climate change are likely to be the costs of protection measures such as climate-proofing homes, insurance costs and medical expenses (see Hamilton, 1997a). Insurance costs are likely to increase very substantially, although it may be difficult to obtain insurance against cyclones and storms in some areas.

What can be said about the distribution of these costs? In general, the effects of environmental degradation fall more heavily on the poor. Wealthier people can afford to live in areas with low air pollution, better water quality and superior amenities. They are also better able to protect their environments through exertion of political power. Brajer (1992) provides evidence for the United States that air quality is worse in areas where low-income households are located. A recent Australian study of lead levels in children showed that children from families with annual income levels below Aus\$20,000 had substantially higher levels of lead in their blood than children from families with incomes higher than Aus\$20,000 (Donovan, 1996).

These studies focus on pollution problems that are localized. Climate change is a global phenomenon and there is no *prima facie* reason to suppose that its effects in the form of storms, cyclones, extreme temperatures and so on will affect poor areas in developed countries more than wealthy ones. There may be an exception in the case of flooding if poorer households tend to be concentrated in low-lying areas. However, there does not appear to be any evidence for this. There may also be an exception in the case of diseases, although again there do not appear to be any studies that examine the distribution of diseases such as malaria by household income (though malaria is expected to spread further into densely populated regions of the developed world, e.g the north-eastern United States, south-eastern Australia and the South of France –

Meteorological Office, 1998: 10). Perhaps also a further differential impact may arise in the case of warming. Suburbs that are heavily treed are cooler, so temperatures in the western suburbs of Sydney, for example, are likely to rise by more than those on the north shore where wealthy households are concentrated. But peripheral areas of Australian cities already subject to bush fire threat will be more at risk as the number of days of extreme temperature increases.

The key distributional issue governing private financial costs relates to the different ability to pay for climate-proofing of homes. Clearly, wealthier households will be in a better position to protect themselves from the effects of warming and extreme weather events. Poorer households will tend to invest in less effective measures or simply suffer the consequences. The question of the distributional implications of failing to tackle climate change requires much more research. Moreover, there is a severe imbalance in the debate. While the costs of climate change are likely to be very large, the costs of emissions abatement are likely to be very small (see Hamilton, 1997b, and Interagency Analytical Team, 1997, for estimates of economic impacts on various Annex 1 countries) and can be relatively easily offset.

John Rawls has captured a great deal of interest with the notion of a 'veil of ignorance' (or of 'uncertainty'; see Young, this volume, Chapter 14) for determining just outcomes. If a person must decide on a just distribution of resources but does not know which social class he or she will be born into, then this veil of ignorance is likely to result in a more even distribution. The same idea can be applied across generations. If today's decision-makers did not know whether they would be born into the present generation or one two hundred years into the future, it is probable that they would devote a great deal more attention to considering what life will be like in two hundred years' time and err on the side of caution in making decisions now that will affect the quality of the natural environment.

Distribution of costs of abatement among developed countries

Issues of fairness and justice underlie the negotiations between Annex 1 (industrialized) countries about the size and distribution of the emissions reductions to be mandated. Just as in the case of domestic systems of taxation, the public position of all parties to the FCCC is that they are willing to do their fair share as long as all other parties do the same. As a general principle this is powerful; the problem, of course, lies in differing perceptions of what is fair. In the lead-up to the Kyoto conference, the Australian government took a very strong stance, arguing that a requirement for Annex 1 countries to cut their emissions by a uniform percentage would be unfair for a country like Australia that is heavily dependent on fossil fuels. While plausible on the face of it, this position has turned out to be impossible to defend, and the Australian proposal for differentiated targets was viewed by the rest of the world as one designed to give Australia an unfair advantage.

We can approach the general principles through an examination of the Australian argument for differentiated targets. The basic position of the Australian government was as follows. Since Australia is heavily dependent on fossil fuels for export revenue and relies on fossil fuels as the chief source of domestic energy, uniform emissions reductions targets would be very costly and would impose a disproportionate economic burden on Australia compared to other Annex 1 countries. It advocated 'differentiation'; that is, allocation of different targets for Annex 1 countries on the basis of equal economic cost per capita for each Annex 1 country. Australia would, under the this proposal, have more lenient targets than most other countries.

Table 6.1 provides some vital information for understanding how the Annex 1 parties to the Convention view the question of emission controls. It shows which countries are the biggest polluters overall and which have the highest levels of emissions per capita.

It should be noted that having high per capita emissions is widely believed to give a country a greater obligation to reduce its emissions, a view broadly known as the 'polluter pays' principle. However, having high per capita emissions does not mean that to reduce emissions will be more difficult. Indeed, as a rule the reverse is true. The costs of reducing emissions depend principally on the opportunities for energy efficiency and fuel switching. A low

Table 6.1 CO_2 emissions of selected Annex 1 countries, 1990

Country	Total CO_2 emissions (GgC)	CO_2 emissions per capita (t C/cap/yr)
Australia	78,809	4.61
Canada	126,175	4.74
Denmark	15,894	3.09
France	99,964	1.76
Germany	276,588	3.48
Greece	22,391	2.21
Italy	116,984	2.03
Japan	315,000	2.55
Netherlands	47,455	3.18
New Zealand	6,948	2.05
Norway	9,686	2.28
Russian Federation	651,469	4.39
Spain	61,997	1.59
Sweden	16,706	1.95
United Kingdom	157,367	2.74
United States	1,351,915	5.41

Source: ECO NGO Newsletter, Climate Change Negotiations, Bonn, July–August 1997, issue no. 5, based on FCCC data.
Note: These figures exclude emissions from forestry and land-use change.

level of per capita emissions may be the result of previous efforts, as in the case of Japan in response to the oil shocks of the 1970s. For those countries that have already 'picked the low fruit', equal percentage reductions in emissions may be viewed as unfair. This is why Japan advocated emissions cuts for high per capita emitters, but the option to expand emissions for low per capita emitters.

However, the European Union (EU), which has relatively low emissions and has pursued energy efficiency with some vigour (much more so than the United States, Canada and Australia) was easily the most determined of the major powers to see big cuts made at Kyoto. The EU advocated a uniform 15 percent cut in emissions below 1990 levels by 2010 for all Annex 1 Parties. The United States and Australia, the countries that could cut emissions most cheaply and easily, were the most resistant to a strong protocol. So, despite the fact that high per capita emissions provide both the obligation and opportunity to reduce emissions by more, the per capita emissions levels in Table 6.1 are in practice a poor guide to a country's expressed willingness to pursue emissions reductions.

Within the European Union, agreement has been reached for a variety of targets, with the deepest cuts allocated to Germany and Britain and increases in emissions permitted in Spain, Portugal and Greece. These differing targets have been allocated roughly on the principle that countries that are less wealthy and have lower emissions per capita should receive more lenient targets. As we will see, this is an extremely important point. Since the level of cooperation within the EU cannot be translated on to the international stage, anything other than uniform reduction targets for Annex 1 countries would be too difficult to negotiate on the basis of an agreed set of principles.

Fairness has thus been interpreted in different ways. The dominant view was put by a Norwegian delegate to the FCCC negotiations:

> Parties should take their share of the burden in proportion to their relative contribution to the climate change problem. Those who currently emit more than their fair share should thus contribute more. Also, Parties that have greater capacity, economic or otherwise, to deal with the problem, should in principle do more than other Parties to reduce emissions.
>
> (Dovland, 1997)

Dovland's last point is also particularly important in understanding the position of developing countries. The rich countries have not only caused the problem of climate change but are seen to be in a much stronger position to begin to solve it.

The broad principles of 'polluter pays' and 'ability to pay' inform the differential targets agreed within the EU. Australia, on the other hand, developed and pursued a view of fairness based on the 'principle' of equal economic cost per capita, which rejects both the doctrine of ability to pay and the simple 'polluter pays' principle. Indeed, the criteria Australia put forward turned the 'polluter pays' principle on its head by arguing that countries with high emissions (due to high dependence on fossil fuels for domestic production and exports) should be given more lenient targets.

But perhaps a bigger difficulty with the Australian position was the way in which equal economic cost was to be determined. The Australian government position was supported by economic analysis that purported to show that Australia would suffer a large and disproportionate burden as a result of uniform binding targets. However, the rest of the world did not accept the arguments or the evidence provided by Australia and in fact believed that Australia would find it easier than most other Annex 1 countries to reduce emissions. There is substantial evidence that the latter point is true, including a recent review of the Australian energy economy by the OECD's respected International Energy Agency (IEA) which noted that, over the previous ten years, energy consumption in Australia had increased at the rate of 2.1 per cent per annum, compared with the IEA average of 1.1 per cent, and that the average ratio of energy use to output is significantly higher than in Europe, and about comparable with that in the United States. The review said that there were substantial opportunities to reduce emissions at zero net cost including higher fuel prices and mandatory energy efficiency standards (IEA, 1997).

In addition, Australian economic modelling results that showed high and disproportionate economic costs to Australia were flatly contradicted by modelling carried out by the US government. The US study estimated that Australian GDP would fall by 0.5 per cent at its peak in 2010 as a result of measures to stabilize greenhouse gas emissions at 1990 levels (Interagency Analytical Team, 1997). This contrasts sharply with the Australian government's estimate that GDP in Australia would fall by 1.5 per cent in 2010, a fall three times greater than the US estimate (ABARE – Australian Bureau of Agricultural and Resource Economics – 1997). (The credibility of the ABARE modelling results was severely tarnished when it was revealed that the work was overseen by a steering committee mainly comprising representatives of fossil fuel-based companies. The companies paid $50,000 each for the privilege of joining the committee. An inquiry by the Commonwealth Ombudsman was highly critical of the apparent bias introduced into the policy process.) Some of the results of the US study are summarized in Table 6.2. While few would suggest that these modelling results have any precision, they do indicate the broad assessments of negotiators from other countries about the relative national impacts of emissions reduction targets.

The US government study also found that the impact on Australia would be less than the impact on Western Europe (GDP down by 0.7 per cent in 2010), Japan (0.6 percent) and Canada (1.1 percent). Only the United States is expected to experience a lower economic impact (0.2 percent). By contrast, modelling by the Australian government was used to claim that the per capita economic costs for Australia would be *twenty-two times higher* than those for Europe (ABARE, 1997).

Table 6.2 Impacts on GDP of stabilization of emissions at 1990 levels by 2010 using the SGM model (without emissions trading)

Country	Year 2005	Year 2010
Australia	−0.2%	−0.5%
Canada	−0.4%	−1.1%
Eastern Europe	0.0%	0.0%
Former Soviet Union	0.0%	0.0%
Japan	−0.2%	−0.6%
Western Europe	−0.2%	−0.7%
United States	−0.1%	−0.2%

Source: Interagency Analytical Team, 1997, table 7b.

The German government commissioned a study specifically to examine the implications of a range of possible formulas for assigning differential targets to Annex 1 countries (Walz *et al.*, 1997). It considered various principles and criteria that have been suggested as the basis of a 'fair' allocation of national targets. The criteria were selected from the international literature and included emissions per capita (CO_2 per capita), level of wealth (GDP per capita), emissions intensity of output (CO_2/GDP), dependence on primary energy (TPES/GDP), national climate characteristics (heating degree days) and dependence on fossil fuels (share of fossil energy sources in TPES, where TPES is total primary energy supply). Five variants combined these criteria in different ways. The study then asked how each Annex 1 country would fare if each variant were used to assign differential targets so that overall emission reductions in all Annex 1 countries reached 15 percent. The results appear in Figure 6.1.

It is apparent that under any feasible differentiation proposal, far from it being given more lenient targets, Australia would in most cases have more stringent targets. To the extent that studies like these inform opinion, governments of other Annex 1 countries believed that Australia would find it relatively *easy* to meet mandatory emissions reduction targets.

Philosophical foundations and the role of economics

More generally, the debate over climate change policies represents one of the most disturbing examples of the conceptual imperialism of economics. The policy debate over how to respond to climate change has been dominated by economists who come to the issue with the standard framework provided by utilitarian philosophy, instrumentalist attitudes and cost–benefit tools. Using this approach, a thorough assessment of the costs and benefits of global warming (including the benefits of improved agricultural productivity, the costs of building dykes, the costs of transmigration to avoid famine and so on) might well show that the benefits exceed the costs, especially when compared to the alternative scenario, namely a dramatic reduction in dependence on fossil fuels. It would be entirely consistent with this line of argument to advocate measures

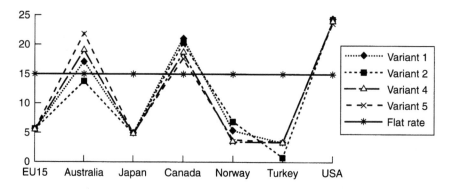

Figure 6.1 Country targets under various differentiation formulae. Variant 1: CO_2 per capita; TPES/GDP. Variant 2: CO_2 per capita; TPES/GDP; GDP per capita. Variant 3: not calculated. Variant 4: CO_2 per capita; TPES/GDP; fossil share in TPES. Variant 5: CO_2 per capita; CO_2/GDP
Source: Walz *et al.*, 1997, table 3-2

to speed up global warming so that we can enjoy the benefits of a hotter planet sooner.

The most influential economist advocating this viewpoint is William Nord-haus, who, in a string of papers, has analysed the economic impact of the enhanced greenhouse effect on the United States (see in particular Nordhaus, 1990a, b). Nordhaus begins with the most fundamental assumption of economics, that a course of action is desirable according to whether or not the sum of the economic benefits outweighs the sum of the economic costs: 'Whether preventive action should be taken depends on the costs of preventing GHG [greenhouse gas] emissions relative to the damages that the GHGs would cause if they continued unchecked' (Nordhaus, 1990a). Fortunately, Nordhaus reassures us,

> Most of the US economy has little *direct* interaction with climate, and the impacts of climate change are likely to be very small in these sectors. For example, cardiovascular surgery and microprocessor fabrication are under-taken in carefully controlled environments and are unlikely to be directly affected by climate change.
>
> (Nordhaus, 1990b)

Nordhaus calculates that only 3 per cent of national output in the United States is produced in 'climate-sensitive sectors' – mainly agriculture – while another 10 per cent is generated by 'moderately impacted sectors', such as construction, water transportation and energy utilities.

However, he argues, it is by no means obvious that these sectors will be affected negatively. In agriculture, for instance, while rising temperatures may reduce yields, the fertilization effect of higher levels of carbon dioxide will tend to raise yields. As another example, 'investments in water skiing will appreciate while those in snow skiing will depreciate'. Nordhaus comes to the soothing conclusion that when the money values of all of these effects are added up and suitably discounted, US citizens at least have little to worry about: 'In sum, the economic impact upon the U.S. economy of the climatic changes induced by a doubling of CO_2 concentrations is likely to be small . . . around one-fourth of 1 per cent of national income.'

There are many responses to the simple-minded economists' view of the relationship between humans and the biosphere that underlies Nordhaus's analysis, not the least of which draw on the principles of sustainability, including the precautionary principle. For example, Daily *et al.* (1991) take up Nordhaus's argument that the best guess of an increase of between 1.5 degrees C and 4.5 degrees C in the Earth's mean temperature is trivial because most people experience greater temperature fluctuations daily:

> At the peak of the last ice age, the mean surface temperature of the Earth was 'only' 5 degrees C cooler than it is now. But then, an ice cap a mile or more thick covered most of Canada and parts of both Eurasia and the northern United Sates (including all of Manhattan).
>
> (*ibid.*: 3)

But one attitude stands out, and that is the belief that faced with a potentially drastic change to the entire global ecosystem we can add up the costs and benefits to humans and decide the extent to which we want to modify the Earth's atmosphere. The clear implication of the idea of an optimal level of greenhouse gas emissions is that economic rationality requires human beings to transform the Earth's climate optimally, and any other aspect of the natural environment, in order to maximize our welfare.

Nordhaus's conclusions have been influential in slowing down international action to tackle the greenhouse problem. His results show that any but the easiest actions to slow greenhouse warming would be too costly compared to the calculated small net costs of global warming in the longer term. Nordhaus concedes that his analysis fails to take account of 'goods and services' that are difficult to value monetarily. If only he could do it in practice, Nordhaus would reduce the ecological and social justice effects of global warming to commodities to be bought and sold in the marketplace, for then, according to economic theory, they would be valued precisely according to their contribution to human welfare.

Many valuable goods and services escape the net of the national income accounts and might affect the calculations of the economic effects of climate change in the United States. Among the areas of importance are human health, biological diversity, amenity values of everyday life and leisure, and environmental quality. Some people will place a high moral, aesthetic or environmental value on preventing climate change, but I know of no serious estimates of what people are willing to pay to stop greenhouse warming.

Nordhaus even goes so far as to concede that 'many values cannot be incorporated into a quantitative cost–benefit analysis. . . . While being unable to put a price tag on Venice, we might decide that it is unacceptable to take actions that threaten Venice's existence'. Even though Nordhaus admits that within his own framework there are important effects that are excluded, this does not prevent him from reaching the conclusion that global warming is not a very serious problem and that, therefore, policies to reduce carbon emissions substantially are unnecessary. Within his own framework of cost–benefit analysis, he attributes zero values to human health, biological diversity, amenity values of everyday life and leisure, and environmental quality – every effect other than those which the national accounts, and thus private markets, permit him to value monetarily. Given the use to which this sort of analysis is inevitably put by the fossil fuel lobby and by governments swayed by it, Nordhaus's analysis can only be described as irresponsible; it pretends to draw conclusions based on scientific analysis when in fact they reflect a peculiar philosophical and methodological approach.

The inescapable logic of Nordhaus's method of analysis could just as easily lead to the conclusion that welfare could be maximized by increasing the rate at which greenhouse gases are pumped into the atmosphere so that humanity does not have to wait so long to enjoy the benefits of global warming. Indeed, Nordhaus finally admits that 'the greenhouse effect might on balance actually be economically advantageous'.

Quite apart from the unsustainable assumption that there is a close relationship between economic growth and improvements in human well-being, this whole line of argument is breath-taking for its faith in humanity's ability to control the natural world (for an Australian version of an alternative index of well-being to GDP see Hamilton, 1997c). With global climate change, we are talking about change to a complete system of unimaginable complexity – a system whose intricacies we have barely begun to understand. Who can possibly predict the consequences of global warming except at the crudest level? And what does it reflect about our respect for Gaia to imagine that we can regulate the whole to suit our own needs? The belief that twentieth-century humans can manipulate a global ecosystem that has evolved to its present delicate equilibrium over millions of years is perhaps the lowest point of the contempt for nature that has so characterized the scientific vision of the past three centuries, a vision to which most economists still naively cling.

The Kyoto Protocol and the Australian precedent

The outlines of the Kyoto Protocol are well known. The Annex B countries – that is, the OECD and the Economies in Transition – agreed to limit greenhouse gas emissions to an average of 5.2 per cent below 1990 levels in the commitment period 2008–12. The European Union agreed to cut emissions by 8 percent, the United States by 7 per cent and Japan by 6 percent. The average was dragged up to 5.2 per cent largely by the fact that Russia and

Ukraine agreed to stabilize emissions at 1990 levels. Australia was granted an 8 per cent increase.

As we have seen, developing countries did not, and were never expected to, agree to mandatory targets for their emissions. At the meeting of the parties in Bonn in June 1998 an intriguing discussion took place over the use of the words 'rights' and 'entitlements' in debates over emission trading. The United States opposed the use of these words, arguing that the Protocol does not refer to these concepts but simply to assigned amounts that may be traded. The use of the notion of rights clearly implies an allocation of control over a common property resource, namely the Earth's atmosphere. As a wealthy and powerful nation, the United States feels uneasy in the face of the assertion of rights over the atmosphere by some very poor people. The next step to flow from the assertion of rights is the principle of equal per capita entitlements for every citizen of the world. This proposal for 'global justice', known as 'contraction and convergence', is already forcing itself on to the international agenda. It has been endorsed in principle by the European Parliament. It would mean that if a rich country wanted to pollute at higher than average levels then it would need to purchase the right to do so from poor countries that own them. We can anticipate some convoluted arguments in an attempt to discredit this proposition.

Subsequent to Kyoto, the US government has come under intense pressure from a recalcitrant Congress determined to obtain the acquiescence of developing countries to mandatory targets before it agrees to ratify the Protocol. This creates a serious problem for the success of the Protocol. The agreement also contained anomalies that will generate difficulties for future rounds of negotiation.

Australia was granted extraordinary concessions at Kyoto; not only was it given a target of an 8 per cent increase above 1990 levels, but it was allowed to include emissions from land clearing as part of its base-year emissions. Because land clearing is declining for commercial reasons, the effect of this clause will be to permit Australia's fossil emissions to increase by at least 25 per cent, higher than the government's own forecast.

The idea of differentiated targets based on equal economic cost was the foundation-stone of the Australian position before Kyoto. But before Kyoto and in the Protocol itself, differentiation was never accepted by the international community as a basic concept or major influence on targets in the way advocated by Australia. Almost every target agreed at Kyoto was within the narrow range of 1 per cent increase to 8 per cent reduction, and thirty-two out of thirty-eight countries (with widely differing characteristics according to Australia's proposed indicators) accepted cuts between 5 per cent and 8 per cent. Under any feasible differentiation criteria (including those put forward by Australia), Japan and the United States would have been given markedly different targets, whereas in practice they differed by only 1 per cent. The Kyoto outcome was therefore very close to uniform reductions for almost all countries, with a few deviations of a few percentage points. The Australian concession was agreed because Australia threatened to break the consensus and withdraw, something that would have jeopardized the entire agreement.

Nevertheless, the precedents established to keep Australia in the Kyoto negotiations will bedevil future negotiations. An 8 per cent increase for a country that is wealthy and the world's highest per capita polluter is unfair by any standard and will make it difficult to gain the agreement of developing countries to begin cutting their emissions.

In the longer term, the shape of global climate change controls beyond 2012 has become clearer. The Kyoto conference foreshadowed a move towards equal per capita emission rights and the institutionalization of the polluter pays principle. These are powerful instruments of 'global justice' and would represent a remarkable breakthrough in international relations.

References

Australian Bureau of Agricultural and Resource Economics (ABARE) (1997) *The Economic Impact of International Climate Change Policy*, Research Report 97.4, Canberra: ABARE.

Brajer, V. (1992) 'Recent Evidence on the Distribution of Air Effects', *Contemporary Policy Issues* 10 (2): 63–71.

Daily, G., Ehrlich, P., Mooney, H. and Ehrlich, A. (1991) 'Greenhouse economics: learn before you leap', *Ecological Economics* 4 (1): October.

Donovan, J. (1996) *Lead in Children: Report on the National Survey of Lead in Children*, Australian Institute of Health and Welfare, Canberra.

Dovland, H. (1997) 'Climate Change: Some Views from a Norwegian Perspective', in *Commodity Markets and Resource Management*, papers from the Outlook 97 conference, ABARE, Canberra.

Hamilton, C. (1997a) 'The Distributional Aspects of Climate Change', National Academies Forum, The Challenge for Australia on Global Climate Change, Australian National University, 30 April 1997, Canberra.

—— (1997b) 'Climate Change Policies in Australia: A Briefing to a Meeting of the Ad Hoc Group on the Berlin Mandate, Bonn', Background Paper No. 9, The Australia Institute, Canberra.

—— (1997c) 'The Genuine Progress Indicator for Australia', Discussion Paper No. 14, The Australia Institute, Canberra.

Hare, B. (1996) 'Prospects for International Negotiations on a CO_2 Emissions Reduction Protocol', Amsterdam: Greenpeace International, November

Interagency Analytical Team (1997) *Economic Effects of Global Climate Change Policies* (draft) US Department of Energy and the Environmental Protection Agency, June.

Intergovernmental Panel on Climate Change (IPCC) (1995a) Second Assessment Report.

—— (1995b) *Summary for Policy Makers*, Working Group II.

International Energy Agency (1997) *Energy Policies of IEA Countries: Australia 1997 Review*, Paris: IEA.

Meteorological Office (1998) *Climate Change and Its Impacts: Some Highlights from the Ongoing UK Research Programme, a First Look at Results from the Hadley Centre's New Climate Model*, London: Department of Environment, Transport and the Regions.

Nordhaus, W. (1990a) 'Economic Approaches to Greenhouse Warming', mimeo, Yale University, New Haven, CT.

—— (1990b) 'To Slow or Not to Slow: The Economics of the Greenhouse Effect', mimeo, Yale University, New Haven, CT.

Walz, R. *et al.* (1997) *Derivation of Internationally Comparable Reduction Targets for the Annex-1 Countries of the Climate Convention*, Study commissioned by the Federal Ministry for the Environment, Nature Conservation and Nuclear Safety, Fraunhofer Institute for Systems and Innovation Research, Karlsruhe, May.

Part II

Environmental justice: issues of principle

7 Considerations on the environment of justice

David Harvey

Introduction

I begin with a couple of paradoxes that may be relevant. Some recent journalistic accounts in the United States have suggested that the environmental justice movement there is 'sputtering', that its hoped-for political potential is not being realized and that it is fading as a political force from the scene (Braile, 1997). On the other hand, rapidly proliferating conferences and writings indicate a booming theoretical and intellectual interest in the topic – particularly within the groves of academia. I am not sure this paradox really exists (and I certainly have no conclusive evidence for it), but the mere suggestion of it worries me. It has obvious implications for the question of an adequate environmental ethics for the twenty-first century. Will all the talking and all the conferences contribute to or detract from the empowered capacity for political action on the part of those who suffer most from environmental injustice?

The second paradox arises out of my own gut response to the question: 'Is a universal environmental ethic possible or desirable?' – a question posed to keynote speakers at the conference 'Environmental Justice: Global Ethics for the 21st Century', at the University of Melbourne, in October 1997. I found myself saying, 'Of course it is impossible – of course it is desirable.' In this, I heard echoes of one of the slogans of the movement of '68: 'Be realistic, demand the impossible!' I then wondered how this contradictory response might fold into the first, more general paradox.

The impossibility of a universal environmental ethic

It is not hard to show why a universal environmental ethic is such an impossible idea. First enumerate some of the major axes of differentiation to be found within what is loosely called the environmental movement. Then consider how each axis produces its own polarized sense of how to formulate an environmental ethic. The aggregate result, no matter how we cut it, is a plethora of confusing and chaotic conceptions of environmental ethics and justice that muddy rather than clarify political choices. To hammer that point home, just consider some of the major axes of difference:

Ecocentric versus anthropocentric views

Do justice and ethics lie in nature? Are they natural virtues, and if so how do we know them and, just as important, who is privileged to know or interpret natural virtues and ethics for us? Or are they social constructs that relate solely to relations within society? In which case, which institutions have the power to define them and which social relations and struggles enter into, or should enter into, their definition?

Individualistic versus communitarian (collectivist) views

Is justice (or ethics) a concept that can be attached only to clearly identified individuals – in which case, are all individuals equal, and what happens when, for example, corporations, non-human animals or trees, or even whole eco-systems are given the status of individual? Or is there some collectivity (social group, territorially bounded state, super-aware environmentalists, capitalist entrepreneurs, etc.) that defines a sense of justice over and above that which can be defined for the individuals contained within or comprised by it?

There are, of course, ecocentric versions of this argument – for example, rights of individual animals or sentient beings (e.g. Regan, 1983) versus species and habitats (Leopold, 1968) – and a multitude of equally contentious arguments within the anthropocentric tradition over whether methodological individualism is the only proper foundation for all theories of justice and political action or, if not, what constitutes a valid social group.

Culturally and historically-geographically embedded views

The rights and claims of so-called indigenous peoples to practise their own ways of life and patterns of resource use lead the way here (see Chapter 13 by Fourmile in this volume), but there is a more general argument that definitions of justice and of ethics are a product of uneven historical-geographical development (economic, cultural, legal) and that the particularistic grounding (ecological and social) this gives to claims about justice is antagonistic to any and all universal principles, the application of which would inevitably be unjust and oppressive to some cultural configuration somewhere.

This view of the culturally embedded definition of ethics and justice is widely shared in practice and increasingly so in theory, particularly through the recognition of the Eurocentrism of most universality claims, to say nothing of the inroads of post-modern thinking. This issue was most clearly posed in the call for papers issued for the Melbourne Conference on Environmental Justice:

> If local cultural conceptions of justice are to determine environmental questions, what principles can be applied to ensure that these locally derived conceptions are not oppressive to humanity and (unsustainably) exploitative and destructive of nature? Is there a universal principle which

underwrites the right to cultural self-determination, and is such a principle in conflict with the demands of ecological sustainability?

Materialist versus spiritual concerns

Broadly materialist and economistic bases for claims in which access to life chances and material possibilities (whether it be of species, individuals, social groups or habitats) are the measure of justice, are frequently opposed to aesthetic, spiritual and religious readings (including those that are redemptive and apocalyptic) as bases for formulating ethical positions. If ethics and justice are simply about the distribution of economic goods, then we get a set of definitions completely different from those which acknowledge cultural and religious beliefs, identities and personal or group sensitivities as fundamental.

Hubristic versus humble attitudes

Emotive oppositions often play subterranean havoc with our thinking. On one side, there is the fear so palpably felt by many that we are far too hubristic, arrogant, Promethean and domineering for our own good and that, like Icarus, we court environmental disaster by our insatiable desire to fly too close to the sun. We should be more humble, self-effacing, prostrate ourselves even, before the mighty and wondrous powers of nature – sometimes construed as vengeful when provoked – tiptoe upon the land and tread lightly upon the face of the earth. On the other side lies the aggressive desire to dominate and control, to master, conquer and subdue the powers of nature through scientific, engineering and political-economic enterprises that shape the earth into a dwelling place fit for civilized and superior beings (sometimes felt also to be proximate to God in the Great Chain of Being). While this second positionality is sometimes denigrated as crass and purely materialistic, it possesses its own sense of heroism, nobility and messianic (even quasi-religious) fervour in which humans, often in the figure of scientist and engineer, pit themselves David-like against the overwhelming Goliath powers of nature, refusing to be cowed and to the end (death) remaining unbowed in their grand struggle to tame natural forces to human desires.

On villains and heroes

Different designations as to who is the villain of the piece give rise to radically different explanations of both what is wrong, who is responsible, and what to do about it. Enlightenment reason, speciesism, modernity and modernization, scientific/technical rationality, materialism (in both the narrow and broader sense), technological change (progress), multinationals (particularly oil), the World Bank, patriarchy, capitalism, the free market, private property, consumerism (usually of the supposedly mindless sort), state power, imperialism, state socialism, meddling and bumbling bureaucrats, military industrial complexes, human ignorance, indifference, arrogance, myopia and stupidity, and the like all jostle (singly or in some particular combination) for the position of arch-enemy of environmental ethics and justice. They become the 'other' – the sources of injustice – to which resistance must be mounted. They conversely define the oppositional role of hero in saving us from environmental ills, as well as a specific ideology and definition of justice and of ethics necessary to political resistance and revolutionary change.

Processes versus outcomes and means versus ends

The long-standing debate over ends versus means has plenty of echoes in environmental politics. There is much to consider here: how, for example, authoritarian concepts and exclusionary measures are frequently appealed to within various segments of the environmental movement as means to arrive at a more democratic world of sustainable environmental justice. But beyond that there is a broader question as to whether justice and ethics are about processes and procedures internalized within democracy, the market, self-governance or whatever without regard to the endstate (if there is one), or whether they should attach to outcomes. If the latter, then are there well-defined end-states, specifiable in advance, which define the goals of a truly just and ethical condition of human life in and with nature? Some (e.g. both Nozick and Derrida) insist that any definition of justice or of ethics that specifies an end-state is unjust and that a justice of process ('doing justice') is all that matters.

I know these are caricatures of some of the binary positions to be found within the environmental/ecological movement. But the central point remains that these sorts of oppositions – even if we concede that most opinion lies on a spectrum rather than at the poles of each axis – make for innumerable confusions, particularly when taken in combination. But throw into the pot a few other disturbing ideas like:

- that claims worked out and agreed upon at one geographical scale (the local, the bioregion, the nation) do not necessarily make sense when aggregated up to some other scale (e.g. the globe);
- that the formative institutional frameworks and settings for discourses and practices of justice are multilayered, multiscalar and often either inconsistent or in overt conflict;
- that what makes sense for one generation will not necessarily be helpful to subsequent generations;
- that nobody knows exactly where an 'ecosystem' or a 'community' begins and ends (even if either concept can be taken as anything more than a convenient fiction that hides as much as it reveals – see O'Neil *et al.*, 1986);
- that it is hard to give a sharp enough definition to 'the environment' to know when a problem is distinctively environmental as opposed to urban, economic, political, psychological, social, etc., and it is hard to know how

serious or threatening some so-called environmental problems are or will be. Also, the range of even conventionally defined environmental issues is so vast – including, for example, questions of biodiversity, global warming, ozone holes, acidification, bad water and air quality, resource depletion, food insecurity, loss of cultural identity, desertification and deforestation, the resurgence of major diseases (e.g. malaria), contaminants and carcinogens in the food supply, aesthetic degradation, occupational hazards in the workplace, cruelty to animals in the food chain, lead paint on the walls and radon in the basement – that it seems impossible to collapse all these diverse questions requiring their own specific forms of political action into one environmental ethic or one programme of environmental justice;

- that there are a variety of definable principles and discourses of justice (contractarian, egalitarian, utilitarian, intuitive, positive law, etc.) available to us and innumerable shades of received opinion as to what an appropriate ethical system might be to regulate human behaviour with respect to other humans, let alone to the natural world (see, for example, Wenz 1988);

- that every political movement under the sun – from Nazis to free-market liberals, from feminists to social ecologists, from capitalists to socialists, from religious fundamentalists to atheistic scientists – necessarily considers it has the exclusive and correct response to environmental issues because it needs must claim the mantle of 'the natural' as a basis for its political programme.

Put all this together and we have a witches' brew of concepts and difficulties that can conveniently be the basis of endless academic, intellectual, theoretical and philosophical debate. There is enough grist here to engage participants at learned conferences until kingdom come – which, in a way, is what makes the whole topic so intellectually engaging and attractive. But the effect will surely be that we (let alone that mythic force called 'the general public') will remain hopelessly confused, if not lost, while the articulation of so many divergent principles can only divide all activists one from the other (and thus rule) in a veritable soup of hopelessly differentiated discourses and particularities. This will help ensure either:

1   that the very concept of justice gets called into question (in much the same way that Marx mocked Proudhon's appeal to some concept of universal and eternal justice as a basis for political action), that nothing of real import gets done in spite of all the chaos and ferment of divergent and non-complementary actions engaged in, at all manner of scales in political practice, unless, that is, one believes – as I do not – in the Foucauldian/ Unger thesis that aggregate social change builds up organically from micro-oppositions pursuing divergent goals;

or:

2   that we surrender our capacity for thought and reflection as well as our political autonomy and power to some demagogue (or demagogic group) who cuts the crap, derides the endless discussion and pulls us into the immediacy of political protest and action in the name of justice while usually crying wolf about the imminence of some environmental Armageddon ('we are running out of time . . .').

The starkness of those unpalatable alternatives lies at the root of my desire for some kind of universal statement – however tentative and incomplete – to which we might adhere in our struggles for a more socially just (as well as saner) approach to environmental issues. But there are more general arguments that make a more universalistic approach to the environmental ethic both necessary and desirable as part and parcel of political struggle.

The desirability of a universal environmental ethic

If Armageddon is just around the corner and the survival of the species (in this case ours) is truly threatened, then the case for coming up with a universal environmental ethic fast is overwhelming. As King Charles I of England reputedly commented on his way to the scaffold, 'the prospect concentrates the mind wonderfully'. In practice, this idea of imminent collapse plays an overwhelmingly powerful role in shaping most varieties of environmental discourse. But I don't think the evidence points that way. More of us are collectively living longer and with higher material standards on average (though the average conceals spiralling inequalities); there are few real signs of resource exhaustion; the paths of technological change are proliferating (providing ever more choices of resource base as well as lifestyle options that can be more environmentally friendly); governments, international organizations, the media, and even segments of business are far more responsive to 'green issues' than they were twenty years ago and now even see it as 'good business' for capitalism (see, for example, von Weizsäcker *et al.*, 1998). This does not mean there are no problems or that we are necessarily behaving prudently or well, but I do think it important to resist the idea that the world is on the verge of some catastrophic environmental collapse, or that the very existence of a fragile planet Earth is threatened.

Coming up with a universal environmental ethic on the grounds that catastrophe is imminent is, I believe, a sign of weakness. To begin with, if the collapse does not materialize in the near term, or the grounds for such expectations are seriously disputed (with strong appeals to both scientific theory and evidence), then environmentalism gets discredited, and there is now a whole genre of writing along those lines. And looking for signs of catastrophe (always popular with the media) may divert our attention from some of the longer-term, more gradual changes that ought also to command our attention. Besides, I am by no means as sanguine as many that a rhetoric of imminent catastrophe will sharpen our minds in the direction of cooperative, collective and democratic responses as opposed to a 'lifeboat ethic' in which the powerful pitch the rest overboard.

It is primarily for this reason also that the invocation of 'limits' and 'eco-scarcity' as a means to focus our attention upon environmental issues makes me as politically nervous as it makes me theoretically uncomfortable. While there are versions of this argument that thoroughly accept the proposition that 'limits' and 'ecoscarcities' are socially evaluated and produced, it is hard to stop this line of thinking slipping back into some version of naturalism or, worse still, of Malthusianism (even to the point where many radical environmentalists would now claim that Malthus was right rather than wrong). If, however, the focus remains on processes of social construction, then the whole emphasis ultimately shifts away from imperatives located in nature, and the basis from which we are urged to construct an appropriate environmental ethic becomes as shifting sand. There are, of course, plenty of intermediate positions, but in that case the invocation of 'natural' limits becomes a redundant issue.

We have a choice of background metaphor for our deliberations. Against the idea that we are headed over the cliff into some irrecuperable abyss (collapse) or that we are about to run into a solid, impermeable and immovable brick wall (limits), I think it far more consistent with the general thrust of environmental thinking to construe ourselves as embedded within an ongoing flow of living processes that we can individually and collectively affect through our actions at the same time as we are profoundly affected by all manner of events within the world we inhabit.

To construe ourselves as active agents caught within the web of life is a much more useful metaphor than the linear thinking that entails heading off a cliff or crashing into a brick wall (see, for example, Capra, 1996). But it is then necessary to find a way to construct environmentalist perspectives within the general 'web of life' metaphor. And here I think it useful to consider the 'negative' and 'positive' consequences of our actions, for ourselves as well as for others (including non-human species and whole habitats), as these actions are filtered through the web of interconnections that makes up the living world around us. We can all too easily foul our own nest within the web of life, and that seems to me to be a better conceptualization of the ecological problem than painting apocalyptic scenarios of 'the end of nature' or the 'destruction of planet Earth' (see the debate between Foster, 1998, and Harvey, 1998).

Here, I think, a strong case can be made that the environmental transformations collectively under way in these times are larger-scale, riskier and more far-reaching and complex in their implications (materially, spiritually, aesthetically) than has ever been the case before in human history. The quantitative shifts that have occurred in the last half of the twentieth century in, for example, scientific knowledge and engineering capacities, industrial output, waste generation, urbanization, population growth, international trade, fossil fuel consumption, resource extraction – just to name some of the most important features – imply a qualitative shift in environmental impacts, and potential unintended consequences that require a comparable qualitative shift in our responses and our thinking.

The environmental movement (broadly understood) has pioneered in alerting us to some of the risks and uncertainties and in charting responses. As a result, we now see far more clearly than half a century ago that there is far more to the environmental issue than the conventional Malthusian view that population growth might outstrip resources and generate crises of subsistence (up until as late as the 1970s this was the dominant form environmentalism took). Furthermore, the evidence for widespread unintended consequences (some distinctly harmful to us and others unnecessarily harmful to other species) of such massive environmental changes, though not uncontested, is far more persuasive than the idea that we are reaching some limit, that environmental catastrophe is just around the corner or, even more bizarrely, that we are about to destroy the planet Earth itself.

Prudence in the face of such risks, then, appears a perfectly reasonable posture. This provides a more likely basis for forging some collective sense of how to approach a range of environmental issues, particularly because many risks and uncertainties can strike anywhere, even among the rich and the powerful. While Rawls, for example, asked us to think of justice theoretically behind a veil of ignorance (always hard to do), the veil of ignorance created by risk and uncertainty of environmental impacts (many of which, like smoke from the forest fires that raged in Indonesia in 1998, do not respect national or class boundaries) has a more powerful prospect of forcing us to think of justice under Rawlsian-type conditions – which, I hasten to add, does not entail an endorsement of Rawls's specific theory, though it is of interest, as De-Shalit (1995) shows (and see Chapter 14 by Oran Young in this volume). The fact that cholera knew no class boundaries under conditions of urbanization in the nineteenth century thus provoked a universal, rather than specifically class-based, approach to public health.

But there is another rationale for strong universality claims, and here I do, in a way, have to identify the role of 'villain of the piece' (even if in a rather qualified way). For what is almost beyond dispute is that the risk and uncertainty we now experience acquires its scale, complexity and far-reaching implications by virtue of a certain set of hegemonic processes of the sort that have produced the massive industrial, technological, urban, demographic, lifestyle and intellectual transformations that we have witnessed in the latter half of the twentieth century, and that a relatively small number of key institutions, such as the modern state and its adjuncts, multinational firms and finance capital, and 'big' science and technology, have played a dominant and guiding role in this process.

What is also apparent is that for all the inner diversity, some sort of hegemonic economistic-engineering discourse has come to prevail which, unless we watch it, has the effect of making us 'puppets of the institutional and imaginary worlds we inhabit' (to use Roberto Unger's, 1987: 37, words) by commodifying everything and subjecting almost all transactions (including those connected to the production of knowledge) to the singular logic of commercial profitability and the cost–benefit calculus. If, as seems very much to be the case, the scalars and dimensions of our environmental problems (and so-called 'solutions') are set basically by such singular parameters, then presumably we desperately need a rather single-minded environmental ethic (replete with strong universalistic claims) to attach to or rebut that singular process. In so doing it is important to recognize a danger: hegemonic processes have the habit of perpetuating themselves by defining hegemonic forms of opposition. To avoid that danger requires challenging, as I shall hope to do, the prevailing definitions of what universality is all about.

But what might prudence in the face of risk and uncertainty mean here? This is not easy to define. And what criteria of prudence are internalized within the environmental movement itself? On this point there are strong grounds for some self-criticism and reflection. For one of the paradoxes within the environmental movement is the certainty (even absolutism) with which views are often advanced about the unintended consequences of the actions of others and the lack of any attention to the unintended consequences of wrong opinions within our own ranks. While, for example, we may all admire the tenacity of Greenpeace in preventing Shell from sinking an oil rig in the North Sea, the environmental consequences of land-based disposal may have been far worse. And while I admire the certainty with which Murray Bookchin (1990) depicts a decentralized municipal federalism as an answer to the problems of both democracy and environmental sanity, I am not at all convinced that such a programme would be any more friendly with respect to, say, problems of biodiversity than is currently the case. It is in general a strange aspect of the environmental movement that, on the one hand, it recognizes the incredible complexity of interactions, while on the other it exhibits a penchant for extraordinarily simplistic answers ('if only we cared . . .' or 'if only we changed our values . . .').

This difficulty is compounded by the ways in which some groups raise a clamour against the lack of humility, the Prometheanism and the arrogance of others. It might be better to recognize that all truth claims depend on certainty of opinion and of results, and that a certain degree of hubris and of arrogance is implicated in the pursuit of any kind of political action. For example, while the thesis of the domination of nature has been thoroughly trashed in recent times, are those who claim to know that the Earth is fragile and in danger and who prescribe drastic remedies to put it right any less hubristic? Are the restoration ecologists any less concerned with domination and mastery than the western rancher who supports the Wise Use movement? Are the social ecologists who insist, with Élisée Reclus (1982), that human beings are 'nature becoming conscious of itself' and that the only form of social organization that can deliver on environmental sanity and justice is a decentralized municipal socialism, any less arrogant in principle than the World Bank, which believes the market can do it best?

I am not attacking any of these particular positionalities *per se* as being bastions of a hidden arrogance (though there are many signs of that in certain instances). All I wish to record here is the simple fact that to propose a line of political action is always presumptive concerning the possession of certain inalienable truths. This is always at the heart of political choices which are 'either–or' rather than 'both–and'. When we make those choices we do one thing rather than another (we support the clear-cutting of forests or we don't), and when we connect those choices to principles such as those of environmental justice and of ethics, we invoke a truth claim that is not open to challenge at the moment of political choice even though we may worry afterwards whether we were right.

It would, I think, be helpful if we would all recognize that this is what we do. And it might also be helpful for all groups to recognize that life consists in making such choices even though we may reserve judgement and be ultimately self-reflexive with respect to the principles invoked. Much of the dogmatism to be found within the environmental movement could then be attenuated, and since I regard that dogmatism as one of the main barriers to progress through broad-based discussions and alliance formation, I need at the outset to raise the important question of the risk (both to others as well as to the environment) of wrong opinions within the environmental movement itself.

This in turn raises the question of the conditions and practices of dialogue across a fractious movement where deeply held beliefs are common. And it turns out that this is also one of the hardest of all questions to answer if we seek some universal ethic out of the multilayered, multiscalar and highly geographically differentiated discourses and practices as these currently exist across the face of the earth, pitting *oeconomy* (in the classical sense), community, nation, cultural configuration, ecological condition, belief system and opinion each against the other.

As a beginning point here I take up the idea of 'justice as translation' in the work of James Boyd White. Translation is

> the art of facing the impossible, of confronting unbridgeable discontinuities between texts, between languages, and between people. As such it has an ethical as well as an intellectual dimension. It recognises the other – the composer of the original text – as a center of meaning apart from oneself. It requires one to discover both the value of the other's language and the limits of one's own. Good translation thus proceeds not by the motives of dominance and acquisition, but by respect. It is a word for a set of practices by which we learn to live with difference, with the fluidity of culture and with the instability of the self. It is not simply an operation of mind on material, but a way of being oneself in relation to another being.
>
> This is the sense in which translation can be an image of thought, a model of social life, including in the law, for all of our life together requires the constant reading of one another's texts and the creation of texts in response to them. The activity of translation in fact offers an

education in what is required for this interactive life, for, as I have suggested, to attempt to 'translate' is to experience a failure at once radical
and felicitous: radical, for it throws into question our sense of ourselves,
our languages, of others; felicitous, for it releases us momentarily from the
prison of our own ways of thinking and being.

(White, 1990: 257–59)

We must, as we search for a strong and consistent environmental ethic,
acknowledge the risks that attach to getting the ethic wrong. We must also
avoid what White calls 'the radical intellectual vice of our day', which is 'to
insist that everything can be translated into one's own terms'. If a universal
environmental ethic is desirable, it must be approached through translations of
this kind.

The inevitability of a universal environmental ethic

There are many ways to think of some universal environmental ethic. The
problem of conflict between universality and particularity that arises out of
uneven geographical development and cultural differentiation is obviously
important, but I think it equally important to think about the parallel problem
of arbitrating between radically different environmental discourses. In what
follows, I shall consider the two problems as broadly homologous and inter-
related.

I start out, quite unashamedly, with my own language with all of its possibilities and shortcomings, for it is into this language that I hope to translate, as
faithfully as I can, the opinions of others. That language is one of dialectics and
of historical-geographical materialism. I assert it not as some 'universal language of authority into which all others can be translated and in which the
truth can be spoken plainly and clearly' – that would be some kind of tyranny –
but because respect for the other in no way 'obliges us to erase ourselves, or
our culture, as if all value lay out there and none here' (White, 1990: 264). My
own tradition is entitled to as much respect as any other, and if I aspire to be a
good translator then I must first be faithful to my own language and traditions.

I begin with one feature of Marxian dialectics which may seem rather abstract
but which is important to my own thinking on the topic at hand (Harvey,
1996: chapter 2). According to this argument, universality always exists in relation to particularity: neither can be separated from the other even though they
are distinctive moments within our conceptual operations and practical engage-
ments. The notion of justice, for example, acquires universality through a process of abstraction from particular instances and circumstances, but becomes
particular again as it is actualized in the real world through socio-ecological
practices. The actualization and administration of justice crucially depend, however, upon the mediating institutions (those, for example, of law and custom)
that help to translate between particularities and universals. It is here that some

of the work done by political scientists has a great deal to tell us, for the exact manner in which institutions mediate can have wide ranging differential effects.

Consider how this has worked in the case of the environmental justice movement in the United States. Particular local struggles emerged around the location of environmental bads (e.g. incinerators, toxic waste dumps, lead paint poisoning). Such particular struggles all internalized universalistic claims (e.g. 'we, all of us, have a right to live in a decent and healthy living environment'). Such an argument appealed to an already-established universalism (e.g. that of civil rights) that made differential distributions of environmental burdens potentially discriminatory with respect to race, class and/or gender. The movement focused on mediating institutions, both constructing its own (such as the Citizens Clearing House on Hazardous Wastes) and confronting others (such as federal and local governments, the courts, the Environmental Protection Agency) to force an adjustment to (a) the conception of universality (to include environmental damage as a locus of civil rights discrimination for example); (b) to change the behaviour and mandates of mediating institutions (subsequently forced to take environmental justice into account in their decisions); and (c) to alter outcomes in particular communities (the actual modification of distribution of environmental bads in particular places with impacts on particular individuals and communities). The discourse on environmental justice flowed from grassroots specifications of universalism to mediating institutions and back again.

I construe each moment (that of particularity, universality and mediating institution) as interrelated but identifiably separate. Particular struggles could not occur without making universal claims, universal concepts of environmental ethics and of justice could not be formulated outside of the particular struggles that highlighted them. I make much of this point because I am distressed at the way in which particularity and universality are so often posed as mutually exclusive determinations. I would want to treat of them not in 'either–or' terms but as 'both–and', with the 'and' focusing in particular upon the mediating institutions, which can be weak or strong, partial or comprehensive, etc. Mediating institutions (which become centres of both discursive and political power in their own right) derive their powers from the capacity to translate between particularities and universalities and give the latter some permanence of interpretation and meaning. But it then follows that the moments are potentially contradictory or antagonistic to each other as well as internally related.

The implication is that the multiple particular struggles that we are now witnessing around environmental justice and environmental ethics inevitably internalize their own claims to universality and project them outwards as more general propositions. Attention then switches to the nature of the mediating institutions which give shape, form and some sense of permanence to beliefs, discourses and practices of environmental ethics and justice by translating from concrete claims to more abstract formulations of principles and laws. How such mediating institutions are to be constructed to deliver on environmental justice at local, regional, national and global levels then becomes a pressing problem for institutional reform.

Such mediating institutions can become alienated from the processes within which they are embedded. They no longer mediate but become centres of power through which a given abstract universal language is articulated and imposed on others as a tyrannical form of governance. They no longer translate freely between particularity and universality, but fix meanings, arbitrarily arbitrate, freeze the dialectic and internalize powers of dominion and domination over the socio-ecological process. At this point the mediating institutions and the dominant discourses they project become objects of critique (in the fashion that Foucault demanded) and may need to be liberated from their authoritarian and tyrannical role. But such struggles are not between particularity of, for example, grassroots and revolutionary movements on the one hand and the universal principles articulated within mediating institutions on the other. On the contrary, particular struggles gain much of their strength and power from the universalisms they articulate. The struggle is to liberate the dialectic and unfreeze the mediating institutions by either reforming or revolutionizing them.

In part, the story of the environmental movement these past few years has been the mobilization of a new sense of universality out of particular struggles that have crashed through the barriers posed by institutions (like the World Bank) that had become totally alienated from the dialectic which originally nourished them. The Bank, while still dominated by the language of economics (probably the most singular example of that 'radical intellectual vice of believing that everything can be translated into its own terms'), has been forced to listen and to respect opinions it formerly dismissed out of hand. Not that that battle is entirely won – indeed, there are many signs of backsliding from what was, in any case, a minimalist position – but the point here is to see that attacks upon mediating institutions like the World Bank are not attacks upon universality *per se*, but on the particular kind of universality that poses as unbending and unquestioned authority.

But there are multiple warning signals to be heeded here. Mediating institutions even within the environmental movement can and do take on alien forms. The rise of the environmental justice movement within the United States exposed the lack of engagement of much of the mainstream movement with questions that stretched across class, race and gender and, when challenged, the principles that guided the mainstream movement appeared just as authoritarian and single-minded as anything that could be found in the World Bank. On an international scale, issues of food security and access to life-supporting resources are among the most pressing items on the environmental agenda, yet they are not a consideration (let alone a priority) for organizations like the World Wide Fund for Nature.

So where does this leave us? If, following White (1990: 257), we agree that 'each of us is a distinct center of meaning and experience that cannot be reduced to the language of another', then the only grand universal we are left with is 'that we all speak languages none of which can become a universal language'. We must, then, 'give up the dream of an Objective or Universal language of authority, one into which all others can be translated'. But this does not lead to 'the kind of relativism that asserts that nothing can be known, but is itself a way of knowing: a way of seeing one thing in terms of another', because our 'universal question is how to relate to one another' across the fact that each of us is a distinct centre of meaning (White, 1990: 263–64).

To this, I would add that if we also accept that each of us exists primarily as a relation to 'significant others' (understood not solely as human beings but also as the natural world we inhabit), then we can no more avoid this relating than we can abstract ourselves as individuals from the socio-ecological conditions in which our existence is embedded. And in that world we do make judgements and choices (both individual and collective) all the time, and through those judgements and choices we choose a mode of relating to others and express our own universal ideals, however tentative and contingent, as to what an environmental ethic and what environmental justice might amount to.

Translations

So how, then, would I set about translating the various elements of environ-mental rhetoric and the grand diversity of environmental discourses into the language of someone who has been deeply immersed in the Marxist political-intellectual tradition of dialectical and historical materialism? There are plenty of attempts in recent years to confront this question, some of which have pro-duced awkward compromises rather than translations. Translation requires that I be as faithful as I can to ideas expressed in other languages while in no way abandoning the powers of my own. As I hope to show, translation properly conducted can reveal hidden powers within my own language and so alter the balance of its meaning. For example, the shift from a purely historical to a geo-historical materialism begins to take on firmer shape not as a challenge to the pre-existing Marxist tradition but as a deepening of its own forms of under-standing.

The basic formulation in my own language goes roughly like this. We are a species on Earth like any other, endowed, like any other, with specific capacities and powers that are put to use to modify environments in ways that are condu-cive to our own sustenance and reproduction. In this we are no different from all other species (from termites to beavers) that modify their environments while adapting further to the environments they themselves help construct. This is the fundamental conception I would offer of the dialectics of social and ecological change. It is, as Marx (1973: 290) put it, 'the nature imposed con-dition of our existence' that we are in a metabolic relation to the world around us, that we modify it at the same time as we modify ourselves through our activities and labours. But we, like all other species, have some very species-specific capacities and powers, arguably the most important of which in our case is our ability to alter and adapt our forms of social organization, to build a long historical memory through language, to accumulate knowledge and understandings that are collectively available to us as guides to future action, to reflect on what we have done and do in ways that permit learning from experience, and, by virtue of our particular dexterities, to build all kinds of adjuncts (e.g. tools and communications systems) to enhance our capacities and powers. The effect is to make the speed and scale of adaptation to and transformation of our species environment highly sensitive to the pace and direction of cultural/economic changes. And in recent times, those cultural/economic changes have increasingly become captive of capitalistic modes of behaviour, organization, social relations and ways of thought.

This conception is species-centred and thereby commits me resolutely to a particular form of anthropocentrism (or speciesism in Singer's terms). I simply cannot see that we can ever avoid asserting our own identity, being expressive of who we are and what we can become, and asserting our species' capacities and powers in the world we inhabit. To construe the matter any other way is, in my view, to fool ourselves (alienate ourselves) as to who and what we are. In this sense the Marxian concept of 'species-being' continues to resonate. But if our task is, as White (1990) puts it, 'to be distinctively ourselves in a world of others', this does not mean that we cannot, if we wish, 'create a frame that includes both self and other, neither dominant, in an image of fundamental equality'. We can strive to think like a mountain, like the Ebola virus or like the spotted owl, and construct our actions in response to such imaginaries, but it is still we who do the thinking and we who choose to use our capacities and powers that way. And that principle applies cross-culturally too. I can strive to think like an Aborigine, like a Chipko peasant, like Rupert Murdoch (for he inhabits a cultural world I find hard to comprehend). In these cases, however, my capacity to empathize and put myself in the other's shoes is further aided by the possibility to translate across languages and to study activities through careful observation. But it is still an 'I' who does the imagining and the translation, and it is always in the end through my language that the thinking gets expressed. The ethical thrust here lies, of course, in the choice to try to think like the other, the choice of who or what I try to think like (why a mountain and not the Ebola virus?) and the effort to build frames of thought and action that relate across self and others in particular ways.

I here learn a great deal from trying to understand the ecocentric lines of thought and the works, for example, of deep ecologists, land ethicists and animal rights theorists. I may not accept their views but I do respect them and try as faithfully as I can to transcribe and translate their thoughts into my own resolutely anthropocentric and Marxian framework. They help concentrate my mind wonderfully on the qualitative as well as the quantitative conditions of our metabolic relation to the world and raise important issues about the manner of relating across species and ecological boundaries that have traditionally been left on one side in many Marxist accounts.

I am aided in this by a striking parallel between a relational version of dialectics (which has always been central to my own interpretation of the Marxian tradition, though I have to recognize that this may be a minority view) and many other forms of environmental discourse. From deep ecology and other 'green' critiques of Enlightenment and Cartesian instrumentality (including those developed in ecofeminism) I find sustenance for a more nuanced dialectical and process-based argument concerning our positionality in the natural world. Writers as diverse as Whitehead and Cobb, Naess and Plumwood, have something important to say on this and I do not find it impossible to translate at least some of what they say into the language of a relational Marxism. This does not lead me to accept some of the more strident rejections of Enlightenment thought (indeed, I think there was much that was positive and liberatory about it), but it reinforces a rejection of mechanistic and positivist accounts that have often infected Marxism as well as conventional bourgeois forms of analysis.

There is a more specific way to frame this connection. The relational conception of self (which is acceptable within my Marxist language) recognizes the porosity of the supposed boundary between our selves and our cultures and the environmental transformations upon which we engage. Ecocentric arguments frequently elaborate upon this relational way of thinking and help expand understandings when translated into the Marxian frame. This does not mean that I accept the idea that values, ethics or justice reside in nature, that these are or ever can be attributes of the natural world. Nor does it mean that the Self, defined in that non-egotistical sense that Arne Naess (1989) proposes, has superior political legitimacy to act in the world and to define policies and outcomes that might seriously affect the lives of the millions of other 'selves' who eke out a difficult livelihood on planet Earth. But while it might be helpful for deep ecologists to worry more about the misanthropic strain that can so easily attach to the elitist and exclusionary aspect of their thinking, the inherent porosity they envisage between self and other, between body and world, between culture and nature, is vital to recognize. Only in such terms, for example, can Marxism more powerfully explore the implications of its own rhetoric about dialectics, relationality, metabolism and transformative activity. If we are relational beings in the world, then modes of thought that reflect that fact (rather than the isolated and alienated individualism that everywhere confronts us) must be nurtured at every turn. It is that capacity that the deep ecologists and the land ethicists have quite properly struggled to enhance.

From Marxism I draw the recognition that the role of transformative activity – human labour – is fundamental to our species being and that the evolution of human societies through organization of the labour process is an integral part of the evolutionary process in general. Marx was, I believe, broadly right to see his studies as being on a continuum with those of Darwin, though of course it is the evolution of our specific species-powers that form the main focus of Marx's attention. This focus on the labour process as the active point at which we as a species appropriate the grand other of the natural world we inhabit is vital to my own conception and I wish that more environmentalists would focus on it (rather than drifting off, as so many do, into more mystical, contemplative or 'consumerist' – e.g. nature as a positional good – ways of thinking). From this perspective it becomes plain that we can never ignore the conditions (social, political, economic) under which we appropriate and transform the world around us in accordance with our needs, wants and desires. How we 'make a living' and organize our life chances is always, therefore, a fundamental materialist reference point to my own mode of thinking, and to abandon that reference point is, in my language, to fall into alienated modes of thought.

But here too there is room for expansion of the Marxian argument, in part through encounters with various environmentalisms. For if, as Marx put it, we change ourselves through changing the world and if, at the end of every labour process, we get a result that existed in our imagination before being converted into a material fact through labour, then there is a distinctive role for the imaginary both in defining the nature of labour processes and, even more importantly, in defining who and what we might or will become. The implication is that no socialist project to transform social relations can afford to ignore the experiential qualities (including aesthetic and emotive responses and meanings) of metabolic relations; the imaginary of socialist transformation must focus as much upon its relational embeddedness in the natural world and upon its metabolic conditions as upon social relations and power structures.

Struggles for emancipation and self-realization are multi- rather than uni-dimensional. Here, the general lines of the 'green' critique of Marxism have been helpful in forcing thinkers like myself back to reassess the powers of our own linguistic tradition. But a series of reverse judgements also holds: from my perspective there is nothing 'unnatural' or 'inauthentic' in what we do (even New York City is to be construed as an ecosystem), all projects to transform ecological relations are simultaneously projects to transform social relations, and transformative activity (labour) lies at the heart of the whole dialectics of social and environmental change.

But which social relations need to be transformed? The traditional Marxist focus has, of course, been on those of class, but environmentalists as well as many others have insisted that there is much more to it than that. Emancipation from conditions of dependence coupled with self-realization are both noble Enlightenment aims deserving of unashamed reaffirmation and extension across the whole spectrum of sociality. Issues of gender, of reproduction activities, of what happens in the living space as well as in the workspace, of group difference and of cultural diversity and divergence deserve careful consideration. A more nuanced view of the interplay between environmental transformations and sociality is seriously called for. But it is also important to recognize that systems of domination are not built purely upon essentialist or 'natural' qualities of being white, male, European (North American). 'Moral subjects' are relationally constructed, and the dominant institutions in our society are so organized as to entail hierarchical forms of class domination that reduce much of the population to a condition of repressive dependence. This is the hallmark of a capitalistic mode of production, and it is only in these terms that we can understand how opening these institutions to traditionally excluded others so often has the effect of moulding those others as agents of hierarchical domination (cf. Margaret Thatcher and Clarence Thomas) rather than liberating such institutions of class domination from their performative role.

The grassroots and democratizing impulse that grounds the environmental justice movement here offers some sort of alternative from which inspiration can be drawn. It also connects in my own case with long-standing concerns for justice in urban settings, and consideration of how environmental issues are embedded in substantive questions of the search for a better quality of life across the boundaries of class, race and gender. But in the United States, at least, the principles of justice advanced within that movement too often become purely defensive (oriented solely to the issue of equal protection under the law). Of course, strong defences of this sort, when mounted across the board, force a rethinking of how problems of, for example, toxic or nuclear waste disposal arise in the first place, but the environmental justice movement in the United States has a rather weak presence in the proactive debate. I continue to believe it must take on the rhetoric of ecological modernization in a positive sense, radicalize it, and articulate plans and ideas for the transformation of production/consumption systems and distributive networks, for it is only in that way that it can make good on its promise to shift from a politics of Not In My Back Yard to Not In Anyone's Back Yard. Much of what is said in this movement translates with ease into my own language but through that act of translation I find myself yearning for a more inclusive and programmatic politics.

Given my own tradition, I scarcely need any urging to see that nothing short of a radical replacement of the capitalist mode of production will suffice to instantiate a new and saner regime of socio-ecological relations. Commodification and market processes cannot provide answers to most of the environmental problems we encounter (indeed, they are a central part of the problem). The importance of a radical agenda, clearly recognized throughout much of the non-managerial wing of environmentalism, arises for the very simple reason that the options open within the hegemonic powers of capitalistic institutions and processes are far too limited in relation to the risks we face. If, for example, the strategy of the automobile industry is to bring China up to the levels of the United States in car ownership, then it hardly needs emphasizing that the environmental risks are huge. And even that wing of capitalism that acknowledges that something called 'sustainability' has importance judges sustainability as much in terms of continuous capital accumulation as it does in terms of socio-ecological well-being. From this standpoint, it seems that there is no option except to engage in massive confrontations with many of the central institutions of a capitalistic world order.

But the content and spirit of any such 'revolutionary' movement is a very much more open question. I am certainly prepared to listen, for example, to some of the more radical decentralizing and communitarian, even bioregionalist proposals that circulate within green politics. But as I seek to translate such potentially fecund ideas into my own language, I find myself wanting to transform them through that dialectical conception of the relations between universality and particularity with which I began. For example, mediating institutions play a vital role in giving shape and permanence, solidity and consistency to how we relate with others and with the natural world we inhabit. To give up on some version of the central state apparatus, for example, is to surrender an extraordinarily powerful instrument (with all of its warts and wrinkles) for guiding future socio-ecological transformations. To opt out of considering global forms of governance is similarly to abandon not only hope about but also real concern for a wide range of global environmental issues (or to presume, without evidence, that acting purely locally will have the desired global effect). We cannot cease to transform the world any more than we can give up breathing, and it seems to me neither feasible nor desirable that we try to stop such a process now, though there is much to be said about what directions such transformations might take and with what risks and socio-political effects.

I also find myself questioning the ways in which many environmental groups imagine the future geography of the world to be. Of course, we have to recognize that socio-environmental relations vary geographically, that structures of feeling and cultural understandings understood as whole ways of life exist both in a state of uneven geographical development (a somewhat unilinear conception of a singular developmental process) and in a world differentiated historically and geographically by radical differences in what the language of 'development' might mean. But, being a geographer, I did not need the environmentalists to tell me this (it has been a focus of my own attempts to create a more geographically aware historical materialism for many years now). The bioregionalists and social ecologists, as well as many of those drawn to more communitarian aspects of environmentalism, certainly reinforce the important idea that place-based or regionally based politics are more often than not a significant seed-bed for radical socio-ecological politics as well as key sites for radical change, but in so doing they do not really provide an adequate framework for thinking about how future geographies of production, distribution, consumption and exchange might be produced.

If, as many ecologists maintain in principle, everything relates to everything else, then how can we possibly carve out distinctive entities called bioregions, communities, ecosystems even, from the web of interdependent and integrating processes, without doing an incredible disservice – even perhaps irreparable damage – to them? This interconnectivity is as much a human construct as it is an ecological condition (with, for example, money flows leading the way). While there is much to criticize in the modalities of such connections (financial markets being a prime focus of critical attention), there are many positive aspects to that interconnectivity (including, for example, reduction of local vulnerabilities to ecological risks) that ought not to be given up without serious pause for thought.

The geographical imagination on show here appears remarkably wooden, undynamic and as arbitrary and old-fashioned as regional geography ever was at its worst. To be sure, the activity of bounding and securing, of defining and classifying, has always been a crucial aspect to human affairs. But walls and barriers to restrict movement do not make sense without doors and bridges to facilitate movement and interchange. A better appreciation of that dialectic might provide the environmental movement with a more powerful set of metaphors to guide its geographical vision than is currently the case. It is precisely at this point also that so many communitarian approaches to environmental issues seem to get it so drastically wrong. For not only do they tend to bury conflict as a positive shaping force in human affairs but they often circumscribe their geographical vision in exclusionary and isolationist ways.

At this point the concept of ecofascism – a difficult, unfortunate and pejorative term in many ways – has to be confronted with all of its echoes (expulsion of exotic and alien species of all sorts in the name of ecological and social purity) as more than just a minor worry. But the danger is rather more widespread than can be captured by overtly neo-fascist movements that focus on relations to nature. For there are many exclusionary, populist and authoritarian ideologies and practices at play within the environmental movement that hide behind respect for something deemed 'natural' and therefore 'authentic' in relations to the land. And these ideologies typically invoke notions of national identity, regionalism, authenticity and socio-ecological purity as badges of a certain kind of privilege.

While we may, therefore, insist (along with many in the environmental movement) upon the greatest level of local or regional autonomy to shape places and ways of life in innovative and particularist ways according to beliefs, imaginaries, institutions and interests and needs, we must also recognize that such transformative projects (the production of diversity) cannot be abstracted from broader processes of cooperation and competition over the multiscalar aspects of socio-environmental change. Geographical diversification and decentralization have to be rendered consistent with a hierarchical multiscalar concept of global political, social and ecological relations. For this reason, for example, I find it hard to automatically accede to the view that the claims of indigenous populations to resources, ways of life, territories are somehow more authentic than our own. There is something inherently unjust about starving urban populations being held back from seeking land and resources on lightly populated frontiers because of the rights of indigenous groups long ensconced there. But there is something equally unjust about such indigenous groups being blasted unwillingly and sometimes even unwittingly out of the way in the name of commercial development by multinationals or preservation movements backed by multinational environmental organizations.

It is important in such instances to recognize the possible production of antinomies in which both sides can with equal justice claim the mantle of right. 'Between equal rights,' Marx (1973: 344) long ago observed, 'force decides', and I think it best to recognize here that there will always be a terrain of struggle and open conflict over environmental justice in which force (preferably in the sense of mobilized political power) will be decisive because outcomes can never be fixed by appeal to some higher-order set of concepts and institutions. Conflict is not necessarily to be feared or shied away from, though much will depend on the forms such conflict takes. But if political struggles and even wars can be viewed as 'diplomacy by other means', then there is surely abundant opportunity here for the hard work of translation to be undertaken with all the honesty that it demands before overt conflict erupts. Then at least we will know what the conflict is truly about and why it cannot be resolved any other way except by main force.

I will not pretend that it is easy to translate and negotiate across cultural and linguistic realms. But I do have the impression that honesty of translation – between cultural traditions and languages as well as between the different environmental discourses we have available to us – is not necessarily upheld as much as it could or should be within the environmental movement as a guiding virtue. Yet without such honesty of translation no environmental ethic and no agreed-upon concept of environmental justice can be procured.

Disputes over interpretations and meanings that are both understood and honestly held have a very different quality from disputes over fantasies, fabrications, histrionic declarations and radical misrepresentations. The environmental issue is far too serious a question and the attendant risks far too great to be allowed to dissipate in mangled meanings, and there is here a lot of work to be done to extricate ourselves from that muddle of our own making. There is, I insist, no reason to 'erase ourselves' individually or collectively as centres of meaning. But there is a crying need to prepare ourselves for that kind of interactive life that allows us to live in dialogue with difference and pursue our goals. It is in and around such processes that a conflictual but nevertheless common sense of an appropriate environmental ethic for the twenty-first century can arise.

The possibilities of theoretical reflection

There are, of course, no privileged philosophical/academic answers to the many questions posed here. But it is useful for us as academic underlabourers to try to separate out the wheat from the chaff, the gold from the dross, in all the political arguments that surround us. If the humpty dumpty of environmental justice and an adequate environmental ethic has been thoroughly pushed off the wall (even presuming it ever sat there in the first place), no collection of the finest minds of academia will ever be in a position to put it back together again.

But that does not mean that academics and intellectuals exist only to deconstruct what activists do. One of the more intriguing and constructive tasks we academics and activists can jointly undertake is to see where hidden commonalities might lie between ostensibly highly differentiated positions and what translations might be possible between the multiplicities of environmental discourses. In so doing, we may find surprising bases for unlikely alliances for political action.

References

Bookchin, M. (1990) *Remaking Society: Pathways to a Green Future*, Boston: South End Press.

Braile, R. (1997) 'Environmental Justice: The Movement Is Sputtering', *Boston Globe*, 27 July 1997.

Capra, F. (1996) *The Web of Life*, New York: Anchor Books.

De-Shalit, A. (1995) *Why Posterity Matters: Environmental Policies and Future Generations*, London: Routledge.

Foster, J. B. (1998) 'The Scale of Our Ecological Crisis' and 'Rejoinder to Harvey', *Monthly Review* April: 5–16; 31–36.

Harvey, D. (1996) *Justice, Nature and the Geography of Difference*, Oxford: Blackwell.

Harvey, D. (1998) 'Marxism, Metaphors, and Ecological Politics', *Monthly Review* April: 17–31.

Leopold, A. (1968) *A Sand County Almanac*, Oxford: Oxford University Press.

Marx, K. (1973) *Capital*, vol.1, New York: Vintage Press.

Naess, A. (1989) *Ecology, Community, Lifestyle*, Cambridge: Cambridge University Press.

O'Neil, R., De Angelis, D., Waides, J. and Allen, T. (1986) *A Hierarchical Concept of Ecosystems*, Princeton, NJ: Princeton University Press.

Reclus, É. (1982) *L'Homme et la terre*, Paris: La Découverte.

Regan, T. (1983) *The Case for Animal Rights*, Berkeley: University of California Press.

Unger, R. (1987) *False Necessity: Anti-necessitarian Social Theory in the Service of Radical Democracy*, Cambridge: Cambridge University Press.

von Weizsäcker, E., Lovins, A. and Lovins, L. (1998) *Factor Four: Doubling Wealth, Halving Resource Use*, London: Earthscan.

Wenz, P. (1988) *Environmental Justice*, Albany: State University of New York Press.

White, J. B. (1990) *Justice as Translation: An Essay in Cultural and Legal Criticism*, Chicago: University of Chicago Press.

8 Care-sensitive ethics and situated universalism

Karen Warren

Introduction

Is a universal ethic possible or desirable? Current philosophical debate over 'universalism versus contextualism' – over abstract, impartial, absolute, universal perspectives versus concrete, local, historically specific, contextual perspectives – suggests radically different answers to this question. From the perspective of canonical Western philosophy, the answer is yes. Various competing consequentialist theories (e.g. Ethical Egoism and Utilitarianism) and deontological theories (e.g. Divine Command Theory, Kantianism, Rights-Based theories, perhaps Virtue Ethics) have been offered to fill this role. From the perspective of some feminisms, as well as what has come to be called post-modernism, the answer seems to be no. Various alternative theories (e.g. a feminist 'ethic of care', radical deconstructionist positions) have been proposed. Navigating these waters is no easy task, especially if one finds oneself, as I do, wanting to defend some version of both universalism and contextualism.

In this chapter I suggest a way through the horns of the dilemma of 'universalism versus contextualism' (and absolutism versus relativism). I do so by defending three claims. First, we need to think of the universality of 'universal ethical principles' not in the canonical sense of ahistorical, transcendent, abstract principles but, instead, as guidelines, strategies or 'useful heuristic devices' (Kellert, 1995). As guidelines, ethical principles remind us to pay close attention to local, historically situated, contextually dependent practices. In this sense, universal principles are 'situated'. I call this position about universals '*situated universalism*'. Second, regardless of which particular ethical principle one adopts in a given situation, a moral requirement of ethics, in general, and any candidate ethic, in particular, is that it must be '*care sensitive*'. Third, which particular ethic or situated ethical principle ought to be adopted as most appropriate in a given situation will be determined by the extent to which application of that ethic or ethical principle reflects, creates or results in care practices. I call this the '*care practices condition*'.

Taken together, the three claims that ethics must be care sensitive, that universal principles are 'situated' and that the appropriate ethic or ethical principle for a given context must reflect or result in 'care practices' presuppose a commitment to ethical pluralism (rather than to ethical monism). On the pluralistic account I am proposing, the various values instantiated by such canonical, monist, ethical theories as utilitarianism, rights-based theories or Kantian ethics are still morally relevant; each may be appropriate for some contexts (but not others). But which value and which associated principle (e.g. a principle of utility, a Kantian principle of duty or a principle of rights), now understood as situated universal principles, is the best principle for a given situation will be determined by the condition of care practices. I conclude that by reconceiving ethics in terms of the three conditions of situated universalism, care sensitivity and care practices, one has both a way of understanding how and why a universal ethic is both possible and desirable, and a way of centralizing the moral significance of care to morality.

Consider the image of a fruit bowl full of lots of kinds of fruit, for example apples, oranges, bananas, mangoes, pineapples, tangerines, blueberries. Which particular fruit is best suited for a particular situation depends on that situation (e.g. whether one is baking an apple or a blueberry pie, or selecting a fruit to carry on a backpacking trip). Different ethical principles (e.g. principles of utility, duty, rights, even an 'ethic of care') are like particular fruits in the fruit bowl. As particular fruits, each of them is an appropriate candidate for the fruit bowl. But which ethical principle one chooses as applicable in a particular situation will depend on the situation. The care-sensitivity requirement is a description of the fruit bowl – of ethics and ethical decision-making. Any particular ethic or ethical principle is a particular fruit in the fruit bowl, where these various ethical principles function as situated universals (not as absolute, monist principles). Which fruit (ethical principle) is selected as best suited in a particular situation will be determined, in part, by the care practices condition. At least, this is what I will attempt to show.

Situated universalism

Why is it that one can watch a theatrical production of *Oedipus Rex*, *Hamlet*, or *Death of a Salesman*, plays written in historical time periods spanning more than two thousand years, and yet in some sense 'identify' with the play? Or listen to Australian Aborigine music played on a didgeridoo, probably the oldest human musical instrument, a Celtic harp sonata and a Louis Armstrong jazz composition and find personal meaning in each? Or be moved to care about rain forests in Brazil, deforestation in India or endangered condors in North America, if one lives in Australia?

I think the pre-philosophical answer is obvious and correct: there is something we bring to the situation which we have in common with others and something in the situation to which we, across cultures, resonate. This is sometimes put in ordinary language by saying that there are '*universal*' truths (themes, emotions) expressed by these culturally specific practices.

Struck by the profound simplicity and truth of this pre-philosophical answer, I have come to believe that there are indeed 'universal truths' but that this universality lies in particularity – the guiding principle of 'situated universalism'.

The common themes or emotions any of us experiences as audience participants to a Greek tragedy, Shakespearean drama or Arthur Miller 'common man' play arise out of and are possible because of the context of our own particular circumstances. It is in particular, concrete, felt, lived experiences or understandings of tragedy that 'universal' themes (truths) are located.

If situated universalism is true, then either there are no 'universals' in the Western canonical sense of ahistorical, transcendental, abstract principles, or, if there are, no contemporary person located in and socially constructed from particular historical contexts currently could know what they are. According to situated universalism, there is no, or no currently knowable, set of necessary and sufficient conditions that express a single, supreme, abstract, ahistorical, transcendent principle of morality. Furthermore, even if such an ultimate, necessary and sufficient condition principle were logically possible, it would be undesirable: it would presuppose both an undesirable notion of the epistemological and moral self as a detached, impersonal, objective, rational agent and an undesirable view of ethics and ethical reasoning as abstract, impartial and objective – what Peta Bowden calls 'grand theorizing' (Bowden, 1997: 3). If situated universalism is true, then some of the basic assumptions of modern liberal moral theory (e.g. Kantian ethics) are false.

But it does not follow from the claim that there are no (or no knowable) universal principles in the canonical sense (and it is not true) that there are no 'universals' in a different sense. These 'universals' may be understood as guidelines, rules of thumb or useful heuristic devices which are always morally relevant, even if, on examination, not actually morally appropriate, to a given context. These guideline principles provide latitude in how one understands or justifies a moral practice. Universals in this sense provide only the necessary, and not also the sufficient, conditions for ethics and ethical decision-making. Couched in terms of the 'universalism versus contextualism' (or 'absolutism versus relativism') debate, these universals are 'contextual' in that they grow out of and reflect historically situated, real-life experiences and practices; they are 'universal' in that they express generalizations common to and reflective of lives of diverse peoples situated in different historical circumstances. That is, they are *'situated universals'*.

If universal principles of ethics are situated, how is an ethic possible? An ethic in the canonical sense of a set of necessary and sufficient conditions for right conduct or a good life is not, or not now, possible. But this is as it should be. Real moral life situations, even the favoured canonical situations involving truth-telling, promising and contracts, are seldom, if ever, clear-cut. They are ambiguous and conflicting, and moral decision-making often involves equally strong, competing and compelling values and reasons for acting one way or another. If a philosophical ethic is to be useful to, and reflective of, real-life decision-making, it must be flexible enough to account for the ethical ambiguities of real moral life, while providing guideline principles for resolving real moral conflicts. This is the underlying good sense of situated universalism.

Situated universalism captures the position of many feminist ethicists, including myself, who reject conceptions of ethical theory in terms of a set of necessary and sufficient conditions. Instead, we claim that ethical theory provides only necessary (and not also sufficient) conditions. It is also the motivating intuition for those, like myself, who accept ethical pluralism, rather than ethical monism, as the proper way to conceive philosophical ethics.

In rejecting 'universal' principles in the canonical sense, is one thereby rejecting principles of justice, or duty, or utility, or rights? No, not at all. What one is rejecting is ethical monism and its assumption of one and only one absolute (in the sense of non-overridable) moral principle. Acceptance of situated universalism requires that one reconceive Western canonical ethical appeals to principles of justice, duty, utility or rights as *situated* appeals and responses to the values named by those principles. Values of justice, duty, utility or rights are important in many moral contexts and do deserve a key place in moral deliberations. But these values find expression in principles understood as 'situated universals' – principles that are always relevant, even if not overriding, to understanding and resolving ethical disputes.

Feminist philosophers are among those who have had a lot to say about the 'universalism versus contextualism' debate in ethics. Perhaps its clearest expression is found in the literature on 'justice versus care', where 'justice' cryptically refers to canonical Western ethics of rights and rules (understood as absolute, non-overridable principles) and 'care' refers to the diversity of contemporary positions advocated by many feminists interested in articulating and defending the importance of care in morality. Feminists defending 'an ethic of care' tend to stress four basic claims: that values of care are morally significant (and are as important as values of justice, equality, or fairness); that emotions or feelings (and not, or not simply, dispassionate reason) have moral significance; that the self is socially constructed and relational (rather than atomistic and separate); and that ethical decision-making and conflict resolution are (or are often) contextual or 'web-like' (rather than abstract and hierarchical). Disagreement among 'care ethicists' turns largely on how to understand these four claims and whether an 'ethic of care' is compatible with, distinct and independent from, or more basic than, 'an ethic of justice'.[1] What all feminist care ethicists agree upon is the gender sensitivity of care and the importance of care in ethical deliberation.

As fruitful as this 'justice versus care' debate has been, I do not discuss that debate here. Instead, I focus on a claim endorsed by all care theorists, namely that care is an important moral value, one which is necessary both to the maintenance of society in general and to any adequate conception of ethics or ethical decision-making in particular. I intend to show wherein the moral significance of care is located.[2] I begin my argument by appealing to recent scientific research which sheds light on the central relevance and significance of care to ethics and ethical decision-making.

Emotional intelligence

In his 1995 book *Emotional Intelligence*, psychologist Daniel Goleman presents impressive scientific research to support a view of intelligence 'in terms of what it takes to lead life successfully' (Goleman, 1995: 43). According to Goleman, there are two 'minds' or 'brains', 'one that thinks and one that feels' (*ibid*.: 8): rational minds ('reason') and emotional minds ('emotion'). Goleman's thesis is not Cartesian dualism. Rather, Goleman is claiming that the anatomical centres for reason and emotion are located in different, though connected, parts of the brain. The neocortex is the seat of logic, reflection and what is traditionally referred to as reason or rationality. The limbic system, comprising the hippo-campus and amygdala, is the seat of memory and emotions. According to Goleman, the 'rational mind' and 'emotional mind' provide very different, though intertwined, ways of knowing; they provide two different kinds of intel-ligences (*ibid*.: 9).

Goleman claims that 'emotional intelligence' – both the sense in which there is intelligence *in* the emotions and the sense in which intelligence can be brought *to* the emotions (*ibid*.: 40) – is what maintains an appropriate balance between the two 'minds'. 'Emotional intelligence' recognizes that emotions matter for rationality. According to Goleman's research, 'The intellect (rational mind) simply cannot work effectively without emotional intelligence'; what we do and ought to do in life is determined by both (*ibid*.: 28).

The implications for ethics of this research into emotional intelligence are stunning. For one thing, the research shows that the separation of reason from emotion, the elevation of reason to a higher status than emotion, and the predicating of ethics, ethical knowledge and ethical action on dictates of reason unencumbered by emotion – mainstays of much canonical philosophical ethics – are all mistaken. Goleman himself makes this implication explicit:

> This [the interaction of rational and emotional intelligence] turns the old understanding of the tension between reason and feeling on its head: it is not that we want to do away with emotion and put reason in its place . . . but instead find the intelligent balance of the two. The old paradigm held an ideal of reason freed of the pull of emotion. The new paradigm urges us to harmonize head and heart.
>
> (*ibid*.: 28–29)

Both 'emotional intelligence' and 'rational intelligence' are needed for moral reasoning and decision-making.

Second, Goleman's data provide welcome scientific support for many of the claims of feminist care ethicists who insist on the importance of emotions, particularly care, to ethics. Care is no longer a 'dumb emotion' (Jaggar, 1989: 148–49). It is a 'moral emotion' – an emotion crucially relevant to ethical prac-tice and decision-making. According to Goleman's research, the ability to empathize through care is part of what it *means* to have 'emotional intelligence' (Goleman, 1995: 44).

Third, Goleman's work on 'emotional intelligence' substantiates the claims of feminist care ethicists that an ethic must be based in human psychology.[3] As Rita Manning argues,

> If an ethic is seen as providing guidance for living a good human life, then we must recognize important features of human psychology – the attachments of humans to other humans. . . . An ethic of care takes [human psychology] as central.
>
> (Manning, 1992: 84)

Goleman's scientific case for the importance of 'emotional intelligence' to ethics and ethical decision-making is illustrated clearly by the case of Elliot. Elliot underwent surgery for the removal of a brain tumour. The surgery was declared a success. However, afterwards, those who knew him said he was no longer Elliot, that a drastic personality change had occurred. To use Goleman's language, Elliot's 'rational intelligence' was as sharp as ever, but his 'emotional intelligence' was seriously impaired. Elliot's neurologist, Dr Antonio Damasio, wondered whether the severing of ties between the lower centres of Elliot's 'emotional brain', especially the amygdala and related circuits, during the surgery could account for this personality change. Dr Damasio was struck by one element missing from Elliot's mental repertoire: though nothing was wrong with his logic, memory, attention or any other cognitive ability, Elliot was virtually oblivious to his feelings about what had happened to him.

> His thinking had become computerlike, able to make every step in the calculus of a decision, but unable to assign *values* to differing possibilities. Every option was neutral. And that overtly dispassionate reasoning . . . was the core of Elliot's problem: too little awareness of his own feeling about things made Elliot's reasoning faulty.
>
> (Goleman, 1995: 52–53)

Dr Damasio found that Elliot remained capable of 'dispassionate reasoning', since his 'rational mind' was fully intact. That is, Elliot's 'rational intelligence' was fine. In fact, on many versions of classical liberal moral theory (e.g. Kantian), Elliot-after-surgery seems to instantiate the (Kantian) ideal of the rational, impartial, detached observer! But Elliot's 'emotional intelligence' was seriously debilitated. Without the ability to feel or to care about any of the options available to him, Elliot was incapable of moral reasoning. Elliot's case suggests what Goleman explicitly argues for, namely that the absence of emotional intelligence does not just produce bad or faulty moral reasoning, it produces no moral reasoning at all.

So, if the research of Goleman and others is correct or even plausible, then research on human brain functioning suggests that reason without emotion –

rational intelligence without emotional intelligence – is inadequate for ethics, ethical decision-making and ethical practice. What is necessary is 'emotional intelligence', of which the ability to care is one of the basic skills. So understood, the ability to care is not simply an 'add-on' feature of ethical deliberation; it is an element of emotional intelligence presupposed by it. Without this ability – as in the case of Elliot – one cannot engage in moral reasoning at all.

My basic argument for care-sensitive ethics can now be stated formally as follows:

1   The ability to care about oneself and others is (where the limbic system is intact) physically possible, causally necessary, and in practice desirable for moral reasoning and action. But

2   If having a certain ability is physically possible, causally necessary, and in practice desirable for moral reasoning and action, then ethics (ethical principles, ethical decision-making, ethical conduct) must be care sensitive. Thus

3   Ethics (ethical principles, ethical decision-making, ethical conduct) must be care sensitive.

The desirability of care-sensitive ethics

Goleman's book compiles an impressive scientific basis in support of several claims. First, emotions matter for rationality: what we do and morally ought to do in life is determined by emotional and rational intelligence in concert. The case of Elliot is important because it shows what happens when emotional intelligence is lacking. So, one reason care is morally desirable for ethics is that ethical reasoning is not possible without it. The ability to care – about oneself and others – is a necessary (but not sufficient) condition of moral reasoning.[4] Minimally, it is necessary to genuine ethical motivation. Second, Goleman's work on 'emotional intelligence' shows that empathy and care are crucial, at least psychologically, to effective moral reasoning. The requirement that any ethic be care sensitive both acknowledges and ensures that an appropriate ethic is based in human psychology. And third, Goleman's research establishes that the ability to care about oneself and others is both physically possible (for those, unlike Elliot, with a functional limbic system) and causally necessary (since it is an element of 'emotional intelligence' presupposed by moral deliberation) for ethical deliberation. It is something humans can and must do to engage in any ethical reasoning or deliberation.

One philosophical issue which arises at this point is how one derives the claim that 'one ought to care' from the empirical claims that one *can*, in fact, care. The 'is–ought' controversy seems to loom large here. I think the answer requires a switching of gears. Defence of care sensitivity as a necessary condition of any viable ethic is not a defence of some supreme, absolute moral principle about care. I have already argued against the desirability of such principles and against doing ethics in the canonical style of 'grand theory'. In this respect, I am not a 'care ethicist'. Attempts to capture the moral significance of care by defending a separate 'ethic of care', one which is more basic than and in competition with traditional canonical ethics ('ethics of justice'), is, I am arguing, the wrong way to proceed. The main questions are not (as feminist care ethicists involved in the 'justice versus care' debate have often assumed): 'What are the absolute principles of care?' 'Is an ethic of care a foundational ethic with primacy over other ethics?' 'Is an ethic of justice presupposed by an ethic of care?' The main issues are not about the logical derivation of some moral 'ought' of care from factual premises about care. Rather, the main questions are: 'What is the ethical significance of care and care practices?' 'Why and how do considerations of care impose moral requirements on any ethical theory?' 'What is it about care – caring about oneself and others – that makes care a necessary condition of ethics?'

Properly speaking, then, there are two distinct issues here. One issue is about the importance of care as a necessary condition for ethics, ethical reasoning and ethical practice (the issue of care sensitivity); that is the 'is' issue. The other issue is about care as providing a criterion for assessing the appropriateness of any given ethical principle in a given context (the issue of care practices); that is the 'ought' issue. But the 'ought' claim is not logically derived from the 'is' claim. It is asserted separately, and it is asserted as not only compatible with but honouring the moral significance of the 'is' claim (the care sensitivity claim).

This leads to a fourth point about the significance of the research Goleman documents: it suggests a unique answer to the question 'Why care?' As a (mistaken) request for some more basic, foundational principle in which to ground care, the question 'Why care?' is somewhat like the question 'Why be moral?' One can appeal to traditional ethical theories for any number of answers to both questions: one ought to care (or be moral) because self-interest requires it (ethical egoism), or utility requires it (utilitarianism), or duty (Kantianism), rights, justice, virtue or the Golden Rule requires it. But, as A. I. Melden (1948) pointed out with regard to the question 'Why should I be moral?', none of these answers will satisfy one who does not accept moral reasons as *bona fide* reasons. So, if the question 'Why care?' is understood as like the question 'Why be moral?' – that is, as seeking some foundational principle in the canonical sense – no answer will satisfy one who does not accept moral reasons (including 'a moral obligation to care') as *bona fide* reasons.

But there is something else going on with appeals to care – something quite different from the traditional 'Why should I be moral?' question and something which suggests that one should not treat 'Why care?' as analogous to 'Why be moral?' If Goleman is correct, one cannot engage in moral reasoning at all without emotional intelligence, of which care is a part. One would be like Elliot. So, unlike possible answers to the question 'Why be moral?', the question 'Why care?' can be answered in a way which does not beg the question. One should care because one cannot reason morally, be motivated to act morally, choose to act morally, or value certain practices as moral and others as

immoral or amoral unless one has emotional intelligence – or, as I am express-
ing it, *unless one cares*. The ability and necessity of care seems to be part of
what it is or what it means to be a moral actor or agent.

As a fifth and last point, Goleman's research substantiates the claims of many
feminist 'care ethicists' that care practices are 'gender sensitive' (historically
associated with females). The exclusion of the significance of care and care
practices from traditional, canonical Western philosophy, then, is a serious
omission. As Peta Bowden states,

> Currently the dominant tradition is focused primarily on the obligations
> owed universally and impartially in the kinds of relations that are typically
> associated with men. Given this focus, gender sensitivity requires an equal
> stress on the ethical implications of the special and 'partial' relations in
> which women are characteristically involved.
>
> (Bowden, 1997: 5)

Historically, practices of care have been associated with mothering, nursing
and friendship. Taking care and caring practices seriously for morality, then,
suggests 'some promise of providing a gender-sensitive corrective to conventional moral theories' (*ibid.*: 9). Since all feminist ethicists (and not just 'care
ethicists') want to expose male-gender bias in ethical theorizing and to offer, in
their place, theories or positions which are not male-gender biased, taking
seriously care as a moral value promises to provide such a corrective.

The place of ethical principles in moral conflict

Before we discuss the place of ethical principles in moral conflict, it might help
to clarify the distinction between the condition of 'care sensitivity' and the
ontological and epistemological status of ethical principles as 'situated universals'. Just as principles of justice, duty, utility or rights are properly expressed
as 'situated universals', so too are any principles of care which might be articulated by an 'ethic of care'. A feminist 'ethic of care', as a particular fruit with
other fruit in the fruit bowl, has no *prima facie* priority over any other ethic
(e.g. utilitarianism, Kantian ethics, rights-based ethics). Any universal principles
of care which are generated by a distinct 'ethic of care' (i.e. principles which
compete with other ethical principles, or fruits in the fruit bowl) must also be
'situated'. In contrast, care sensitivity is not an ethical principle – situated or
otherwise. Care sensitivity is a necessary condition for ethics, and the possibility
of ethical reasoning and practice. Care sensitivity is a characteristic of the fruit
bowl itself. No particular fruit (an ethic or ethical principle) belongs in the fruit
bowl unless this condition is satisfied. The requirement of care sensitivity
cannot be reduced to some more basic value.

How, then, does one choose among competing ethical principles? The
answer, I propose, is by determining which principle(s), when applied,
reflects or creates 'care practices'. 'Care practices' are practices which maintain,

promote or enhance the well-being of relevant parties, or do not cause unnecessary harm to the well-being of those parties. For my purposes here, what constitutes the 'well-being' of selves and 'others' is left open.[5] But the following are some examples of the sorts of concerns about well-being I have in mind: those practices which oppress, torture or exploit selves or morally relevant 'others' are not genuine care practices. Those practices which violate the civil rights of selves and 'others' are not care practices. Those practices which cause unnecessary and avoidable harm to selves and 'others' (e.g. destruction of the stability, diversity and sustainability of 'first peoples' cultures or natural ecosystems such as rain forests, oak savannahs and fragile deserts) are not genuine care practices. Whether a principle of rights, utility, duty or some other principle is the best or most appropriate principle for unpacking relevant moral prohibitions, obligations or responsibilities in each of these cases will turn on the extent to which implementation of these situated universal principles promotes or reflects care practices.

To summarize, canonical principles of rights, utility or duty are really 'situated' principles; how one understands and applies the values reflected by these principles must grow out of and reflect specific historical contexts. The 'universality' of such principles is that the values of rights, utility or duty they honour are always morally relevant, even if, in a given case, these values are not taken to be the most morally significant values in a given situation, or the principles meant to privilege them are overridden in a given situation. A requirement for choosing among these values or principles is to determine the extent to which they reflect or result in 'care practices'. And all of this reflects the morally basic requirement that in moral theory, decision-making and action, we can and ought to care – the 'care sensitivity' requirement.

Moral conflict resolution

In addition to providing a necessary condition criterion for choosing among competing ethical values and the ethical principles intended to capture them (e.g. values of utility, rights and fairness captured by principles of utilitarianism, rights, justice), another feature of the 'care practices' condition is that it helps clarify and resolve six noteworthy and often overlooked characteristics of moral conflict.

First, some moral conflicts, while real, are 'isolated instances running contrary to general patterns' (Clement, 1996: 46). Acting on the basis of care (the care practices condition) may not always be just. Saving one's own child from drowning when another child is also drowning (and one cannot save both) may conflict with a justice of fairness or equal rights to life position. Nonetheless, it may be the right thing to do. The 'care practices' condition can accommodate both 'general patterns' and specific exceptions (e.g. giving preference to one's own child rather than acting as a disinterested observer) without privileging either the general or specific to the exclusion of the other.

Unlike canonical ethical positions, the 'care practices' condition does not pre-determine which values are, in a given context, most basic or relevant.

Second, some conflicts can be understood and resolved by appeal to canonical principles *recast* as situated universals. A 'care practices' condition captures this. However, other conflicts cannot, and ought not to, be so resolved. In cases where inclusive solutions cannot be found, methods of compromise, negotiation and consensus must be used. The 'care practices' condition provides a way of deciding which method of conflict resolution to pursue in a given case. The most fitting method is one which either maintains, promotes, enhances or does not cause unnecessary harm to the well-being of selves and 'others'. Sometimes the most suitable conflict resolution method will be found by weighing alternative situated universal principles in a hier-archical, adversarial, win–lose way; sometimes it will not.

Third, some conflicts are the result of symbolic and institutional structures (e.g. structures which dichotomize reason and emotion, care and justice, or which privilege the point of view of those at the top of various value hier-archies; Clement, 1996: 46). As Grace Clement states, such conflicts are '*defined* in opposition to one another, and thus *necessarily* conflict, such as lying and truthtelling' (*ibid.*: 49). Such structures may work well when, in fact, moral conflicts *are* genuinely oppositional. But when this is not the case, one may need to challenge the relevant 'symbolic and institutional structures' them-selves, and not simply the decisions and portrayals of moral conflict within those structures. One reason canonical monist principles may fail in a given case, or fail to capture the morally significant aspects of a particular situation, is that they are ill-suited to moral conflicts that are the result of defective institu-tions and structures, and moral conflicts that are not best expressed in terms of exclusivist, oppositional dualisms (e.g. representing all morally relevant aspects of abortion simplistically as an adversarial issue of a woman's right to choose what shall happen in and to her body and a foetus's right to life). The 'care practices' condition not only helps us to understand and resolve both those conflicts that do, and those that do not, fit the oppositional, dualistic model, but also challenges those very models when it is 'symbolic and structural' features of the model that are at issue (see Warren, forthcoming).

Fourth, some conflicts are 'gender sensitive'. Since care practices historically have been associated with women, females and 'feminine' traits, the 'care practices' condition has the potential, in any given case, to honour both the contexts in which care is a primary value (e.g. friendship, parenting, nursing) and the lives, labour and voices of women care-practitioners.

Fifth, not all conflict is rationally resolvable without loss of value. Even those conflicts canonical philosophers focus on (e.g. truth-telling, promise-keeping, conflicts of rights) involve loss of important values (e.g. values involved in betraying a friend or grieving a loss). Emotional intelligence is crucial to under-standing the importance of these lost values. Hence care practices are crucial to understanding and resolving moral conflict, and to understanding such conflict in a way that 'rational intelligence' alone is not.

Lastly, 'it is a mistake to limit morality to conflict resolution' (Clement, 1996: 82). The 'care practices condition' can be useful in preventing conflicts by necessitating that one look at local, historical realities relevant to a moral situation. Monistic canonical ethical principles, in effect, predetermine the morally basic considerations before actual moral situations arise (e.g. issues of utility or rights or fairness, as always), and necessitate lexically ordered, hierarchical ways of resolving ethical disputes after moral conflicts arise. Such principles are often unhelpful in unpacking the moral dimensions of diverse moral situations, especially situations involving contextual considerations of care, *before* the situation or conflict arises. The 'care practices condition' permits flexibility about what are the morally relevant variables in a potential conflict situation, and requires familiarity with local situations in resolving moral conflict.

Conclusion

I have argued in this chapter for three basic claims: 'situated universalism' establishes the ontological and epistemological status of ethical principles as guides to theorizing; 'care sensitivity' provides a necessary condition for morality and moral reasoning; and the criterion of 'care practices' provides a necessary condition for choosing among competing moral values and ethical principles. Taken together, these claims establish that a universal ethic, characterized by the three features of care sensitivity, situated universal ethical principles, and a condition of care practices for determining the applicability of any given ethical principle in a specific context, is both possible and desirable.

The possibility and desirability of such a universal ethic was defended by appeal to a variety of concerns: Goleman's scientific research on 'emotional intelligence'; a view of moral life as ambiguous and not a straightforward matter of either–or moral dilemmas which yield right and wrong answers according to some overarching, ahistorical, transcendent principle; gender concerns about the importance of care, including paradigmatic examples of care work by women; worries about problematic symbolic and structural features of traditional canonical moral theorizing; a conception of moral theory in terms of necessary conditions only (not as a set of necessary and sufficient conditions); and a relational concept of the moral self as a *carer* – one capable of caring about selves and 'others'.

The last point, that humans are and must be *carers*, was brought home to me most profoundly during the summer of 1997, when my daughter Cortney and I spent nearly a week swimming with wild spotted bottlenose dolphins off the coast of Key West, Florida. What I learned through relating to these incredibly beautiful, sensitive, responsive, intelligent creatures changed for ever how I think about myself, others and morality. It also is what enabled me to *see* the answer to the question, 'Is a universal ethic possible or desirable?' which I have defended in this chapter.

There were many occasions during the week when I would encounter the dolphins underwater, swimming in their vicinity or around them for ten or fifteen seconds at a time. But, unlike Cortney, I had not yet experienced the joy of being with them – as part of their community, at their invitation, on their terms. On the last day, that changed. Rather than aggressively swimming towards them (as I had on the previous days), on this day when I entered the water I just stayed still. Although I couldn't see the dolphins, I began speaking quietly to them, telling them that I would like to swim with them and asking permission to do so. Before I knew what was happening, several adult dolphins came and 'took me' to their pod (where I found my daughter already swimming in the pod). That began what was an incredible hour. It was profoundly clear to me that the dolphins had *chosen* to swim with me – to swim slowly so that I could be with them, to include me as they engaged in their ordinary activities of feeding, playing and touching each other, to take turns 'baby-sitting' me. (This was Cortney's term for the phenomenon of a few dolphins branching off from the pod to take each of us on short, separate side swims before returning us to the pod.)[6] I felt a part of their community for that hour. And while swimming with them I found myself literally voicing what was for me a profound realization:

> Even if you are sentient, capable of language and communication, rational, rights-holders, deserving of respect and protection – even if all this is true, which I believe it is – that's not what is morally basic. What is morally basic is that *I care about you and you deserve to be cared about.* Without the emotion and motivation of care, there is no moral basis for us humans to ponder whether rights, duties, utility or God's commands are the best avenues to pursue to secure your protection and preservation.[7]

That swim was transformative for me. I realized that without care as a primary value and motivation, I could not experience or understand what, for me, was the most compelling reason to protect and honour these dolphins: that they deserve our care and concern. Traditional principles of rights, duty, justice and utility may provide various philosophical avenues by which to gain protection and preservation of dolphins. But they do not, by themselves, get at what is, and ultimately must be, the basis and motivation for human moral concerns; they do not, by themselves, give expression to what is morally fundamental to human interaction with selves and 'others'.

One simply cannot fit the whole moral story about selves and 'others' (humans and non-humans) into the shoe of rights, or duty, or justice, or utility without loss of crucial moral value. To tell the proper moral story of the matter, attention to and cultivation of human capacities to care and to engage in care practices is needed. That was so clear to me, swimming with wild dolphins. Providing that missing moral piece is what care-sensitive ethics and care practices do.

Notes

I thank my former student Kirsten Nystrom and my colleagues Martin Gunderson, Carla Johnson, Nicholas Low and Henry West for comments on earlier drafts of this chapter.

1 As I hope will become clear, I think feminist attempts to articulate and defend an 'ethic of care' often fare no better than, and commit the same mistakes as, canonical attempts to defend an ethics of rights, duty, utility or justice. This is because proponents of an 'ethic of care' often present and discuss that ethic as if it were a universal ethic (in the canonical sense) in competition with or more basic than a justice ethic.
2 For the purposes of this chapter, I do not discuss the meaning of 'care'. I do use the expression 'care about', rather than 'care for', since, as is clear in the discussion of Goleman's research, I endorse a version of a 'cognitive' (rather than a 'feeling') view of emotions whereby emotions have a cognitive component and are not simply twinges, twanges or physiological occurrences. Accordingly, to 'care about' someone or something is to have certain beliefs or attitudes about that person or thing, whether or not one also has 'feelings' about either.
3 In this respect, ethics must be based on an accurate notion of human capabilities. This is consistent with the position of ethical naturalism and moral realism.
4 For a discussion of the 'boundary conditions' of any ethic, including an environmental ethic, from an ecofeminist point of view see Warren (1990).
5 I put 'others' in quotes because I leave open the referent for 'others'. 'Others' does not mean the problematic 'other' (e.g. subject–object dualistic concepts of 'the other' in which 'the other' is not itself also a subject – see Chapter 12 by Plumwood in this volume); nor does it presume that 'the other' must be a self, since I hold that non-human 'others' (e.g. animals, ecosystems) are deserving of moral consideration.
6 It was also clear to me that the dolphins had not changed; I had. Until that last day, my intentions were to swim with the dolphins, and I single-mindedly pursued that end. On the last day, I entered the water differently. I took deep breaths and settled into the moment. I asked their permission to join them. I had made an internal shift from caring about my having the experience of swimming with them, to caring about them and hoping they would permit me to swim with them. Although it is only anecdotal evidence, I believe that this motivational shift made possible what I subsequently experienced with the dolphins.
7 Does 'emotional intelligence' apply beyond the human species? I don't know. But, in one sense, it does not matter ethically whether dolphins (or other non-humans) are incapable of caring about themselves or others. What matters is what human moral agents are capable of caring about; human relationships to non-human animals and nature are thereby morally assessable, whether or not the 'other' is a moral agent or carer.

References

Bowden, P. (1997) *Caring: Gender Sensitive Ethics*, London: Routledge.
Clement, G. (1996) *Care, Autonomy, and Justice: Feminism and the Ethic of Care*, Boulder, CO: Westview Press.
Goleman, D. (1995) *Emotional Intelligence: Why It Can Matter More than IQ*, New York: Bantam Books.
Jaggar, A. M. (1989) 'Love and Knowledge', in A. M. Jaggar and S. Bordo (eds) *Gender/Body/Knowledge: Feminist Reconstructions of Being and Knowing*, Princeton, NJ: Rutgers University Press.

Kellert, S. H. (1995) 'Never Coming Home: Positivism, Ecology, and Rootlessness', unpublished manuscript, Philosophy Department, Hamline University, St Paul, MN.
Manning, R. (1992) *Speaking from the Heart: A Feminist Perspective on Ethics*, Lanham, MD: Rowman & Littlefield.
Melden, A. I. (1948) 'Why Be Moral?', *Journal of Philosophy* 45 (17): 449–56.
Warren, K. J. (1990) 'The Power and the Promise of Ecological Feminism', *Environmental Ethics* 12: 125–46.
—— (forthcoming) 'Environmental Justice: Some Ecofeminist Worries about a Distributive Model', *Environmental Ethics*.

9 Ethics across the species boundary

Peter Singer

Introduction: choosing among non-speciesist ethics

Here is a very brief summary of a position that I have defended on many occasions, most fully in my book *Animal Liberation* (see Singer, 1990): our present treatment of animals is based on speciesism; that is, a bias or prejudice towards members of our own species, and against members of other species. Speciesism is an ethically indefensible form of discrimination against beings on the basis of their membership of a species other than our own. All sentient beings have interests, and we should give equal consideration to their interests, irrespective of whether they are members of our species or of another species.

My aim in this chapter is to defend this position and explain why I hold to it, despite criticisms both from those who seek to defend a speciesist ethic and from those who think that the kind of ethic I hold does not go far enough. It is the latter criticism, in particular, that I address here. While animal liberationists and deep ecologists agree that ethics must be extended beyond the human species, they differ in how far that extension can intelligibly go. If a tree is not sentient, then it makes no difference *to the tree* whether we chop it down or not. It may, of course, make a great difference to human beings, present or future, and to non-human animals who live in the tree, or in the forest of which it is a part. Animal liberationists would judge the wrongness of cutting down the tree in terms of the impact of the act on other sentient beings, whereas deep ecologists would see it as a wrong done to the tree, or perhaps to the forest or the larger ecosystem. I have difficulty in seeing how one can ground an ethic on wrongs done to beings who are unable to experience in any way the wrong done to them, or any consequences of those wrongs. So hereafter I will be concerned with a position based on consideration of the interests of individual sentient beings.

From the perspective of deep ecologists, the non-speciesist ethic I am advocating seems to stick too closely to traditional ethical viewpoints. It is, for example, compatible with classical utilitarianism, which judges acts as right or wrong by asking whether they will lead to a greater surplus of pleasure over pain than any other act open to the agent. As the great classical utilitarian writers – Jeremy Bentham, John Stuart Mill and Henry Sidgwick – all made clear, the boundaries of 'pleasure' and 'pain' do not stop at the boundary of our species (Bentham, 1948: ch. XVII, sec. 1, par. iv, p. 311; Mill, 1976: 131–32; Sidgwick, 1907: bk IV, ch. 1, p. 414). The pleasures and pains of animals must be taken into the calculation. This is not to say that a non-speciesist ethic concerned about individual animals must be a utilitarian ethic. Many different ethics are compatible with this approach, including an ethic based on rights, as Tom Regan (1983) has ably argued. Similarly, a feminist ethic based on the idea of extending our sympathy to others can reach a similar conclusion (e.g. Kheel, 1985).

The traditional view

While an ethic that includes all sentient beings as direct objects of our ethical concern is certainly not as radical a break from traditional ethics as some form of deep ecology that seeks to include all living things, in the context of the dominant Western ethical tradition it remains quite revolutionary. We all know the key passages in this tradition:

> And God said, Let us make man in our image, after our likeness: and let them have dominion over the fish of the sea, and over the fowl of the air, and over the earth, and over every creeping thing that creepeth upon the earth.
>
> So God created man in his own image, in the image of God created he him; male and female created he them.
>
> And God blessed them, and God said upon them, Be fruitful, and multiply, and replenish the earth, and subdue it; and have dominion over the fish of the sea and over the fowl of the air, and over every living thing that moveth upon the earth.
>
> (Genesis, i, 24–28)

Thus according to the dominant Western tradition, the natural world exists for the benefit of human beings. God gave human beings dominion over the natural world, and God does not care how we treat it. Human beings are the only morally important members of this world. Nature itself is of no intrinsic value, and the destruction of plants and animals cannot be sinful, unless by this destruction we harm human beings.

The traditional Judaeo-Christian view of the world is based on a creation myth that was decisively refuted more than a century ago. At least since Darwin, we have known that the forests and animals were not placed on earth for us to use. They have evolved alongside us. The assumptions that derive from that myth, however, are still with us. If we can succeed in clearing them away, the consequences for our way of living will be as far-reaching as any changes in human history have ever been.

Ethics across the species barrier

In any serious exploration of environmental values a central issue will be whether there is anything of intrinsic value beyond human beings. To explore this question we first need to understand the notion of 'intrinsic value'. Something is of intrinsic value if it is good or desirable *in itself*; the contrast is with 'instrumental value', value as a means to some other end or purpose. Our own happiness, for example, is of intrinsic value, at least to most of us, in that we desire it for its own sake. Money, on the other hand, is only of instrumental value to us. We want it because of the things we can buy with it, but if we were marooned on a desert island we would not want it. (Whereas happiness would be just as important to us on a desert island as anywhere else.)

Now consider any issue in which the interests of human beings clash with the interests of non-human animals. Since we are here concerned especially with environmental issues, I'll take as an example Australia's kangaroo industry, which is based on killing free-living kangaroos in order to profit from the sale of their meat or skins. As a community, Australians must decide whether to allow this industry to exist. Should the decision be made on the basis of human interests alone? For simplicity, I shall assume that none of the species of kangaroos shot is in danger of extinction. The issue therefore is one about whether, and to what extent, we consider the interests of individual non-human animals. So immediately we reach a fundamental moral disagreement – a disagreement about what kinds of beings ought to be considered in our moral deliberations. Many people think that once we reach a disagreement of this kind, argument must cease. I am more optimistic about the scope of rational argument in ethics. In ethics, even at a fundamental level, there are arguments that should convince any rational person.

Let us take a parallel example. This is not the first time in human history that members of one group have placed themselves inside a circle of beings who are entitled to moral consideration, while excluding another group of beings, like themselves in important respects, from this hallowed circle of protection. In ancient Greece, those they called 'barbarians' were thought of as 'living instruments' – that is, human beings who were not of intrinsic value, but existed in order to serve some higher end. That end was the welfare of their Greek captors or owners. To overcome this view required a shift in our ethics that has important similarities with the shift that would take us from our present speciesist view of animals to a non-speciesist view. Just as in the debate over equal consideration for non-human animals, so too in the debate over equal consideration for non-Greeks, one can imagine people saying that such fundamental differences of ethical outlook were not open to rational argument. Yet now, with the benefit of hindsight, we can see that in the case of the institution of slavery in ancient Greece, that would not have been correct.

Notoriously, one of the greatest of Greek philosophers justified the view that slaves are 'living instruments' by arguing that barbarians were less rational than Greeks. In the hierarchy of nature, Aristotle (1916 edn: 16) said, the purpose of the less rational is to serve the more rational. Hence it follows that non-Greeks exist in order to serve Greeks. No one now accepts Aristotle's defence of slavery. We reject it for a variety of reasons. We would reject his assumption that non-Greeks are less rational than Greeks, although, given the cultural achievements of the different groups at the time, that was by no means an absurd assumption to make. But more importantly, from the moral point of view we reject the idea that the less rational exist in order to serve the more rational. Instead we hold that all humans are equal. We regard racism, and slavery based on racism, as wrong because they fail to give equal consideration to the interests of all human beings. This would be true whatever the level of rationality or civilization of the slave, and therefore Aristotle's appeal to the higher rationality of the Greeks would not have justified the enslavement of non-Greeks even if it had been true. Members of the 'barbarian' tribes can feel pain, as Greeks can; they can be joyful or miserable, as Greeks can; they can suffer from separation from their families and friends, as Greeks can. To brush aside these needs so that Greeks could satisfy much more minor needs of their own was a great wrong and a blot on Greek civilization. This is something that we would expect all reasonable people to accept, as long as they can view the question from an impartial perspective, and are not improperly influenced by having a personal interest in the continued existence of slavery.

Now let us return to the question of the moral status of non-human animals. In keeping with the dominant Western tradition, many people still hold that all the non-human natural world has value only or predominantly in so far as it benefits human beings. A powerful objection to the dominant Western tradition turns against this tradition an extended version of the objection just made against Aristotle's justification of slavery. Non-human animals are also capable of feeling pain, as humans are; they can certainly be miserable, and perhaps in some cases their lives could also be described as joyful; and members of many mammalian species can suffer from separation from their family group. Is it not therefore a blot on human civilization that we brush aside these needs of non-human animals so as to satisfy minor needs of our own?

It might be said that the morally relevant differences between humans and other species are greater than the differences between different races of human beings. Here, by 'morally relevant differences' people will have in mind such things as the ability to reason, to be self-aware, to act autonomously, to plan for the future, and so on. It is no doubt true that, on average, there is a marked difference between our species and other species in regard to these capacities. But this does not hold in all cases. Dogs, horses, pigs and other mammals are better able to reason than newborn human infants, or humans with profound intellectual disabilities. Yet we bestow basic human rights on all human beings, and deny them to all non-human animals. In the case of human beings we can see that pain is pain, and the extent to which it is intrinsically bad depends on factors like its duration and intensity, not on the intellectual abilities of the being who experiences it. We should be able to see that the same is true if the being suffering the pain is not of our species. There is no justifiable basis for drawing the boundary of intrinsic value around our own species. If we are prepared to defend practices based on disregarding the interests of members of other species because they are not members of our own group, how are we to object to those who wish to disregard the interests of members of other races because they are also not members of our own group?

The argument I have just offered shows that while the dominant Western tradition is wrong on the substantive issue of how we ought to regard non-human animals, this same tradition has within it the tools – in its recognition of the role of reason and argument – for constructing an extended ethics that reaches beyond the species boundary and addresses the human–animal relationship. There is no objection of principle to this extension. The principle that must apply is that of equal consideration of interests. The remaining difficulties are about exactly how this principle is to be applied to beings with lives – both mental and physical – that are very different from our own.

Is a non-speciesist ethic also hostile to humanism?

I have argued that a non-speciesist ethic stands firmly within the broad framework of the Western ethical tradition, even though it extends the content of that tradition beyond all previous bounds. Some critics, among them the French philosopher Luc Ferry (1992), claim that both the ethic of animal liberation and the ethic of deep ecology are hostile to the best elements of Western ethics, and in particular to the humanist ideals of the Enlightenment. It is worth examining this criticism, both for its own sake and because it helps us to draw distinctions between humanism, animal liberation and deep ecology.

Ferry begins by setting out a familiar tripartite distinction in the ways in which people view nature. The first is human-centred, denying intrinsic value to anything outside our own species. The second, which he correctly identifies as my own position, regards all sentient beings as entitled to equal consideration of interests. The third, the stance of the deep ecologists, grants moral status to all of nature, including ecological systems as a whole.

In broad terms, this categorization is acceptable. But it misleads Ferry into thinking of the animal liberation movement and the deep ecology movement as being on the same continuum of thought. Of course, as we have already noted, both the animal liberation movement and deep ecologists challenge the idea that only human beings are of ethical significance. But for Ferry, the deep ecology movement represents much more than this. He finds in it links with the German Romantic movement and even with Nazi views of our relationship with nature. Ferry sees such viewpoints as indefensible, and also as dangerous, because they are a radical departure from the traditions of humanist civilization, which he sees as represented by the great declarations of the rights of human beings.

Whatever one may think of Ferry's critique of the deep ecology movement – and it is certainly open to serious objections – he is wrong to present the animal liberation movement as closer to deep ecology than it is to the Enlightenment tradition that he himself supports. As we have just seen, the animal liberation movement is in fact an extension and culmination of the very Enlightenment ideas of equality that Ferry so strongly champions. Consider what Ferry himself says about the Enlightenment tradition. He says that the French declaration of rights of 1789 marks a break with earlier conceptions of law, which were rooted either in the natural order, as in ancient Greece and Rome, or in a theological world view, as in medieval Europe (Ferry, 1992: 251). Ferry contrasts the abstract and universal humanism of 1789 with the counter-revolutionary Romantic tradition, which sees our moral obligations to others as dependent on their membership of a particular ethnic, national, cultural or religious community.

It is, however, precisely the abstract universalism of the Enlightenment, not the Romantic tradition, which is the basis of the animal liberation movement. Indeed, of all the historical texts quoted by modern animal liberationists, myself included, the one most often referred to is a celebrated footnote from the *Introduction to the Principles of Morals and Legislation* of the great Enlightenment thinker Jeremy Bentham, in which Bentham specifically refers to the fact that the French have discovered that 'the blackness of the skin is no reason why a human being should be abandoned without redress to the caprice of a tormentor' and goes on to look forward to the day when the number of the legs, or similar anatomical differences between humans and other animals, will be recognized as 'reasons equally insufficient for abandoning a sensitive being to the same fate' (Bentham, 1948: ch. 17). Here we have, in essence, the philosophy of animal liberation, and it is clearly a universalizing, Enlightenment idea.

The theories of twentieth-century ethicists often point in the same direction. R. M. Hare (1981) has argued that if we want our judgements to count as ethical judgements, they must be universalizable in form. By this he means that they must not contain proper names, personal pronouns or similar individual references. One test of whether we are prepared to universalize our judgements is to ask whether we would accept them if we had to live the lives of all of those affected by them – both those who lose and those who gain. This idea is a version of the Golden Rule – do unto others as you would have them do unto you – which has an honoured place in the Western tradition, as it does in several other traditions. Hare himself has accepted that his notion of universalizability applies to all sentient beings (Hare, 1993). The difficulties in putting yourself in the position of a non-human animal affected by your actions are scarcely greater than in putting yourself in the position of human infants, or other humans whose lives and ways of thinking and feeling are very different from our own. A non-speciesist ethic is therefore an extension of a humanist ethic, rather than something that has developed from a different direction altogether.

Rights and utility

In the last page or two I have referred more than once to declarations of rights for human beings, or to the fact that we bestow basic rights on human beings but not animals. Acceptance of the idea of 'human rights', combined with a rejection of speciesism, seems to lead directly to the idea of 'animal rights'. In what way, then, does my position differ from that of defenders of rights for animals, such as Tom Regan? (see his Chapter 10 in this volume or, for a fuller account of his position, Regan, 1983).

Utilitarians do not take rights as fundamental in ethics, but they can support declarations of rights, for human beings or for animals. They can regard such declarations or claims to rights as themselves based on utilitarian considerations. For example, it is plausible to believe that governments are often biased in their perception of what the public good requires, and come to see actions that are in their own interests as being in the public interest. Hence they may do terrible things to individuals, claiming that these are justified by the public good, when in fact they are only in the interests of the government or the people whose interests are served by it. For this reason, a society that recognizes certain basic rights as inviolable may better serve the interests of all its members than a society in which governments are able to do whatever *they* judge will best serve the interests of the whole society. Here we have a justification for basic rights, but one that is founded squarely on utilitarian considerations. A similar argument may lead to recognition of animal rights, for here human beings as a whole may be liable to deceive themselves into thinking that a course of action is in the best interests of all sentient beings when in fact it is in the interests of human beings alone. But precisely what rights this argument would justify bestowing on animals is a large question, too large to attempt to answer here.

The meanings of 'humanism'

Before we leave the topic of 'humanism', it needs to be said that it is a term with many meanings. While Ferry criticizes me for allegedly turning away from humanism, I have often been attacked by conservative Christian opponents of abortion and euthanasia as a 'secular humanist'. In the eyes of such people, secular humanism is a philosophy which holds that our ethical rules or principles come from human beings, and hence denies that they come from God. I am happy to be called a secular humanist in this sense. Since I do not believe in the existence of God, I cannot think of ethics as a system of divine commands. I do not think that all ethics is necessarily *human* in origin, since some non-human animals are capable of something akin to ethical behaviour. So ethics may be the product of social mammals, rather than humans specifically. Nevertheless, I readily grant that *our* systems of ethics are human conceptions or the products of human reasoning.

There are some who try to leap from the true statement that our systems of ethics are human conceptions to the conclusion that it is impossible, or meaningless, for such a morality to give rise to obligations on us in respect of non-human animals. But this is a blatant fallacy. One might as well argue that since our system of ethics is the product of human beings who are more than two years old, it is impossible, or meaningless, for such a morality to give rise to obligations on us in respect of infants. The fact that our systems of morality are the conceptions of human beings with post-infant capacities tells us nothing about who or what can be the *subject* of our morality.

One more form of 'humanism' needs to be mentioned. Historically, humanism can be traced to the Renaissance thinkers who rejected Christian teachings about the weakness and depravity of human beings. Instead they made man 'the measure of all things' and the centre of the 'Great Chain of Being' that stretched from the beasts below to the angels above. We might call this 'anthropocentric humanism'.

Anthropocentric humanism emphasizes the ways in which humans are distinct from animals. Hence although this line of thought was formed in opposition to medieval Christianity, in a more fundamental sense it is a continuation of the Christian view that human beings are special because we alone are made in the image of God and have an immortal soul. This convergence of anthropocentric humanism and Christianity is clearly apparent in the thought of Descartes, who argued that non-human animals are merely automata, incapable of suffering. Descartes's view was not widely accepted because it defies common sense to believe that animals are mere machines, and that their cries and howls when they appear to be suffering are really more like the ticking of a clock than like our own expressions of pain. Immanuel Kant's philosophical defence of anthropocentric humanism was less extreme, and for that reason in the long run more influential. Kant linked the concept of moral worth to the possession of reason and autonomy, which animals do not possess. Accordingly, Kant treats human beings as 'ends in themselves', but non-human animals as merely a means to human ends. As such, in his view we have no direct duties to them.

Anthropocentric humanism is unsound, both ethically and scientifically. It is time to move beyond anthropocentric humanism to an ethic based on a broad and compassionate concern for the suffering of others, and a rejection of all religious or ideological fanaticism. (Yes, I do acknowledge that there are fanatics among animal liberationists, as there are in any large movement, and they get a disproportionate share of media attention. They are, however, a tiny minority within the movement, and not an authentic representation of its philosophy.)

Does non-speciesism go far enough?

In many contexts, a non-speciesist ethic is so revolutionary that the question of whether it goes far enough is unlikely to arise. But at a conference on environmental justice, the question needs to be discussed.

I shall begin by attempting to forestall some misunderstandings of my position. Val Plumwood suggests, in her contribution to this volume (Chapter 12), that I admit to the ethical sphere 'the most human-like "higher animals"', which are claimed to be the only possessors among the non-humans of the supposedly defining human characteristic of awareness' (Plumwood, this volume, p. 190). There are two errors in this statement. First, by any normal usage of the term 'higher animal' – for example, using the term to mean mammals – I do not regard only higher animals as possessors of awareness. On the contrary, in *Animal Liberation* I wrote that 'The evidence that fish and other reptiles can suffer seems strong, if not quite as conclusive as it is with mammals' (Singer, 1975: 186). I also considered whether crustaceans may be able to feel pain, and said that they 'deserve the benefit of the doubt'. And I suggested that at least one mollusc, the octopus, is sentient (*ibid.*: 188). Hardly a case of higher animals only!

Second, it is not my view that awareness is the 'defining human characteristic', and even less is it my view that beings are only to be admitted to the moral sphere if they share a characteristic that defines humans, whatever that characteristic might be. I am not really very interested in finding 'defining human characteristics'. What would be the point? This is an exercise that matters only to those who think that there is some moral significance attached to being human. But that, as I hope the preceding pages have made clear, is exactly the view that I am opposing.

Plumwood herself seems to equivocate over whether the problem with a view like mine is that it includes in the ethical sphere only beings that have 'ethically relevant qualities like mind, consciousness and communication', or whether it is that the problem is one of 'failing to recognize that they can be expressed in many different, often incommensurable forms in an ethically and ecologically rich and diverse world' (Plumwood, this volume, p. 191). In this equivocation I see some evidence that Plumwood herself is not prepared to argue for equal moral status for beings that clearly lack the ethically relevant qualities she mentions. Instead, she wants to find it in places where I and other philosophers like me have not found it. To that I have no objection at all. It is simply a matter a waiting for the evidence to come in, and then assessing it. If good evidence is ever produced that trees or ecosystems possess consciousness, then *of course* trees and ecosystems will be included in the ethical sphere, along with those beings about whose awareness we already have good evidence. But it is not a good idea to mix such claims with the further claim that we should abandon the idea of ranking species altogether (Plumwood, this volume, p. 197).

A few pages later in her chapter, Plumwood adds further misunderstandings about my position when she says, in reference to it:

> Both Regan's rights approach and Singer's utilitarianism are strongly associated with a sentience-reduction position that would make a minimal extension of human-style ethics and liberal rights to a few species of

animals who are most like us, who qualify as 'persons'. Every being outside the 'person' category remains potential 'property', the dualistic class which is capable of being owned and treated instrumentally. Minimalism claims to be anti-speciesist but is not genuinely so in selecting for exclusive ethical attention those animals who closely resemble the human, any more than a culture which values women just in terms of their resemblance to men is genuinely non-androcentric.

(Plumwood, this volume, p. 199)

Any reader of *Practical Ethics* would know that I do not use the term 'person' in this way. For me 'persons' are beings capable of seeing themselves as existing over time, with a past and a future. They are, in a word, self-aware in a particular way. I think that some non-human animals are persons, but I emphatically and clearly state that while other animals may not be self-conscious, 'As long as sentient beings are conscious, they have an interest in experiencing as much pleasure and as little pain as possible. Sentience suffices to place a being within the sphere of equal consideration of interests' (Singer, 1993: 131). So the fact that a being is not a person certainly does not mean that it may be regarded just as property. As for the final sentence of this passage, it merely repeats the false claim that I select animals 'for exclusive ethical attention' on the basis of their resemblance to human beings. This is not the point of my ethical argument at all, and there is no comparison to the issue of women being valued just in terms of their resemblance to men.

It is regrettable that Plumwood does not take sufficient care to state correctly the positions of those she criticizes. But it would be much more helpful to her cause if she were to spend less time on criticism in general, and instead show exactly what her own position amounts to. Once we cease to rank some life-forms above others, as she would have us do, what follows about the choices we often must make between different life-forms? Do we cease to use antibiotics (even natural ones, like boiling water or garlic) because they kill bacteria? If not, on what basis do we decide that our lives, or those of other humans, or of an animal with an infection, are to be preferred to the lives of the bacteria? If we are supposed not to be indifferent to the destruction of plant lives (Plumwood, this volume, p. 200), should we give up eating alfalfa sprouts, and instead eat only fruits and seeds which do not require us to take the life of the plant? Until these questions receive well-grounded answers, radical opposition to ranking different forms of life is hard to take seriously.

Whereas I have defended the ethic of animal liberation by placing it within the broad framework of the Western tradition, and more specifically as derived from the humanism of the Enlightenment, Plumwood and others see that tradition as precisely the problem. They argue that it is Enlightenment humanism that is responsible for a civilization that has, for the first time in history, changed the climate of our planet, put a hole into the ozone layer and made species extinct at an unprecedented rate.

From a historical perspective, there is no denying the truth of these claims, but we need to look forward, not backwards. The real issue is what approach offers the best chance of getting us out of the mess we are in. Ironically, the environmental crisis is so grave that there is no problem in using quite a conventional ethic to argue for a radically different attitude to the environment. In many respects, even a traditional ethic limited to human beings would suffice. One could, entirely within the limits of the dominant Western tradition, oppose the mining of uranium on the grounds that nuclear fuel, whether in bombs or power stations, is so hazardous to human life that the uranium is better left in the ground. Similarly, many arguments against pollution, the use of gases harmful to the ozone layer, the burning of fossil fuels and the destruction of forests could be couched in terms of the harm to human health and welfare from the pollutants, or the changes to the climate that may occur as a result of the use of fossil fuels and the loss of forest. The fate of peasant farmers on low-lying lands in the delta regions of Bangladesh and Egypt may depend on whether citizens of the wealthy nations curb their greenhouse gas emissions. Even allowing for some uncertainty about the link between these gases and global warming, the imbalance in the interests at stake (on the one hand, the survival of 40 million people; on the other, such changes as restrictions on the use of private vehicles, or cutting our consumption of animal products produced by modern energy-intensive farming methods) is so great that there can be no doubt about the ethical course to take.

On other environmental issues, arguments about what is best for human beings can be supplemented by the introduction of consideration for the interests of non-human animals. The preservation of old-growth forests serves as an example here. Against the claim that cutting the forests down creates jobs and keeps small logging towns alive, one could argue in terms of the economic benefits of ecotourism, of the importance of forests for the preservation of our climate and the quality of our water supply, and of the loss to all future generations if they are unable to walk in an untouched forest. But this argument is strengthened by the recognition that forests are homes to millions of animals who will die from starvation and stress when the trees are felled. The suffering and death of these wild animals makes the clearing of the forests even worse than it would be if no sentient beings depended on them.

I see it as an advantage of these arguments that they are recognizable as being derived from the Western tradition. Like it or not, whether we come from the United States or India, from Russia or Japan, the Western way of thinking dominates our society, and older traditions are struggling just to preserve some by-ways of life in which they can still have an impact. Philosophically, too, alternative traditions are struggling, with no greater success, to make their positions coherent. By working within the Western tradition we can connect better with the way people think, and have a better chance of influencing the societies in which we live.

References

Aristotle (1916) *Politics*, London: J. M. Dent.

Bentham, J. (1948) *Introduction to the Principles of Morals and Legislation*, New York: Hafner.

Ferry, L. (1992) *Le Nouvel ordre écologique*, Paris: Grasset (translated as *The New Ecological Order*, tr. C. Volk, Chicago: University of Chicago Press, 1995.

Hare, R. M. (1981) *Moral Thinking*, Clarendon Press, Oxford.

—— (1993) 'Why I Am Only a Demi-vegetarian', in R. M. Hare, *Essays in Bioethics*, Oxford: Clarendon Press, pp. 219–35.

Kheel, M. (1985) 'The Liberation of Nature: A Circular Affair', *Environmental Ethics* 7 (2): 135–49.

Mill, J. S. (1976) 'Whewell on Moral Philosophy', reprinted in T. Regan and P. Singer (eds) *Animal Rights and Human Obligations*, Englewood Cliffs, NJ: Prentice-Hall.

Regan, T. (1983) *The Case for Animal Rights*, Berkeley: University of California Press.

Sidgwick, H. (1907) *The Methods of Ethics*, 7th edition, London: Macmillan.

Singer, P. (1975) *Animal Liberation*, New York: New York Review of Books/Random House.

—— (1990) *Animal Liberation*, 2nd edition, New York: New York Review of Books.

—— (1993) *Practical Ethics*, 2nd edition, Cambridge: Cambridge University Press, Cambridge.

10 Mapping human rights

Tom Regan

Introduction

Philosophers have written more about animal rights in the past twenty years than their predecessors wrote in the previous two thousand. Not surprisingly, disagreements abound. To begin with, among those who challenge the attribution of moral rights to animals are philosophers who operate within well-worn moral traditions in Western thought. Peter Singer (1975, and this volume, Chapter 9) and Carl Cohen (1986, 1996, 1997) are representative.[1] Singer follows in the tradition of the nineteenth-century English utilitarian Jeremy Bentham, who ridicules moral rights as 'nonsense upon stilts'. For both Bentham and Singer, not only non-human animals but humans too lack moral rights. Half true, maintains Cohen. Animals, he argues, most certainly do not have moral rights; but Bentham and Singer err, in Cohen's view, when they deny that humans have them. Nothing could be further from the truth: according to Cohen, not just some, *all* humans possess basic rights, including the rights to life and to bodily integrity.

As different as Singer and Cohen are in the conclusions they reach, they are importantly similar in how they approach the question of animal rights in particular and the more fundamental question of moral right and wrong in general. Both operate in what might be described as the Enlightenment tradition. Both assume that moral right and wrong are matters that in principle can be determined by the disciplined use of reason, just as both assume that the answers we seek must pay proper deference to the privileged moral position of certain individuals – individual human beings, in Cohen's case; individual sentient beings, in Singer's.

Despite their many differences, and easily lost in the storm of controversy, Singer and Cohen occupy common ground with philosophical advocates of animal rights (Pluhar, 1995; Regan, 1983; Rollin, 1981; Sapontzis, 1987). The latter also operate in the post-Enlightenment tradition; in other words, as is true of Singer and Cohen, philosophical advocates of animal rights also believe that moral right and wrong are matters that in principle can be determined by the disciplined use of reason *and* that the answers we seek must pay proper deference to the privileged position of certain individuals. Thus, Cohen's and Singer's quarrels with philosophical advocates of animal rights are of an 'intra-moral' nature. Even as they differ in the conclusions they reach, these philoso-phers all share a number of fundamental beliefs about the nature and conduct of moral philosophy.

Val Plumwood (1993, and this volume, Chapter 12) is representative of a different family of critics of animal rights. Like intramoral critics, Plumwood denies that animals have moral rights; but unlike these critics, Plumwood also denies certain assumptions these critics share with philosophical advocates of animal rights. In particular, whereas both these advocates and their critics believe that answers to moral questions must pay proper deference to the privileged moral position of certain individuals, Plumwood believes that accord-ing a privileged moral position to a select group of individuals – for example, all and only human beings (Cohen's view); or all and only sentient beings (Singer's view) – is part of the problem, not part of the solution. What is needed, in her view, is a radical, non-hierarchical reconceptualization of the place of humans and other beings in the larger community of life. Because, she argues, philosophical advocates of animal rights are committed to insisting upon the privileged moral position of individual animals, these philosophies are an impediment to philosophical and, indeed, moral progress.

Criteria of selection

Clearly, the arguments against extending a rights-based human ethic to non-human animals are rich and varied; just as clearly, nothing like a complete examination of these controversies can be attempted here. Let me explain my two reasons for examining those arguments I do.

First, as a matter of personal preference, I did not want to repeat any of my previously published responses to criticisms of animal rights – neither my responses to foundational critiques, including my replies to those philosophers, like Plumwood, who object to the Enlightenment prejudices allegedly embedded in the idea of 'the rights of the individual' (Regan, 1991), nor my responses to more specific criticisms of utilitarianism, including the form of utilitarianism favoured by Singer (Regan, 1980). Because I wanted to try to say something new (new for me, at least), I have nothing more to say on these matters on this occasion – and, thus, nothing further to say about the positions of Plumwood and Singer.

In addition to the personal interest I had in not repeating myself, my second interest was political in nature. I wanted to respond to those arguments that seem to be having the greatest influence among those who oppose the idea of animal rights in the larger world outside academic philosophy. Now, among the many possible candidates in this regard, Carl Cohen's criticisms of animal rights arguably satisfy this second basis of selection better than any other philosopher's. And since it happens to be true that I have not commented on his ideas before, and thus that I would not be repeating myself in responding to them, it is Cohen's arguments against the possibility of extending a rights-based ethic to non-human animals that I examine in what follows.

The idea of moral rights

All of us are familiar with the idea that human beings have certain basic moral rights, including such rights as life, liberty and bodily integrity. For some, like Cohen, this familiar idea is true; others, including Singer and Plumwood, think it false. And just as there is plenty of room for disagreement among those who think the former (for example, concerning how the idea of a moral right should be understood), so there is plenty of room for disagreement among those who deny that humans have them.

Notwithstanding these differences, all of us will agree that one way of understanding human ethics includes the attribution of basic moral rights to human beings; so if one were to attempt to show that no conception of human ethics can be 'mapped on to the consideration of the human–nature relationship',[2] one would be obliged to consider whether a rights-based human ethic, an ethic in which basic moral rights are attributed to human beings, can be 'mapped' on to some aspect of the human–nature relationship, the human–animal relationship in particular. Moreover, in this latter regard, one would be obliged to examine arguments that attempt to show *either* that a rights-based human ethic *cannot* be extended to non-human animals *or* that, while it might be possible to attempt such an extension, it *should not* be extended in this fashion.

On this occasion I explore only one of these options, limiting my remarks, as I do, to arguments that purport to show that a rights-based human ethic *cannot* be extended to non-human animals. Thus, even if, against all the odds, it should happen that everything I say is correct, every argument I formulate valid, everyone will recognize that I would not have shown that such an ethic can or should be extended in the manner under review. Merely to show the inadequacy of some arguments against the possibility of this extension does not prove that such an extension is either possible or desirable.

Cohen's view

Of all those philosophers who have criticized the idea of animal rights, none has enjoyed greater influence, especially among scientists who use non-human animals in their research, than Cohen. The genesis of this influence can be traced to a 'Special Article' that appeared in the *New England Journal of Medicine* (Cohen, 1986). For a philosopher's work to appear as the lead article in such a prestigious professional journal is no mean achievement; the pride of place accorded Cohen's essay reflects both the high regard in which he is held by members of the biomedical research community and the perceived importance of the issues he examines.

In this article Cohen goes beyond offering a justification of using non-human animals in biomedical research. Not only is it not wrong to use animals for this purpose, Cohen argues that it would be wrong not to use them. For example, he writes, 'If biomedical investigators abandon the effective pursuit of their professional objectives because they are convinced that they may not do to animals what the service of humans requires, they will fail, objectively, to do their duty' (Cohen, 1986: 868). And not only are researchers under a positive obligation to use them, Cohen argues that more rather than fewer animals should be used in certain circumstances. 'Should we not at least reduce the use of animals in biomedical research?' Cohen asks, then answers, 'No, we should increase it, to avoid when feasible the use of humans as experimental subjects' (*ibid.*).

Cohen's reasoning in support of continued widespread and possibly expanded reliance on non-human animals in biomedical research is of the utilitarian variety. 'The sum of the benefits [of using non-human animals in biomedical research] is utterly beyond quantification,' Cohen writes.

> Almost every new drug discovered, almost every disease eliminated, almost every vaccine developed, almost every method of pain relief devised, almost every surgical procedure invented, almost every prosthetic device implanted – indeed, almost every modern medical therapy is due, in part or in whole, to experimentation using animal subjects.
>
> (*ibid.*)

Along with these benefits to humans, of course, utilitarians must also consider the harms done to non-human animals. 'Let us do the weighing asked, by all means,' Cohen insists.

> The pain that is caused to humans (and to non-human animals) by diseases and disorders now curable, or one day very probably curable, through the use of labouratory animals, is so great as to be beyond calculation. What has already been accomplished is enough to establish that. What is now being accomplished, its benefits not yet in hand, would establish that truth with equal sureness even if only partially successful. And a fair weighing will put on the scales also those great medical achievements not yet even dreamed of but likely to be realized one day.
>
> (Cohen, 1996: 9, 39–40)

'[T]o refrain from using animals in biomedical research is, on utilitarian grounds, morally wrong' (Cohen, 1986: 868).

While allowing that a small handful of researchers may on occasion be guilty of varying degrees of misconduct, Cohen assures his readers that the work carried out by biomedical researchers who use non-human animals is otherwise beyond moral reproach. With adulation of their work so unqualified, and with a justification of animal model research at once so broad and deep, it is small wonder that many in the biomedical community see in Cohen their long-awaited saviour, sent to defend their good name and life-saving work against the abuse of such philosophical bullies as Peter Singer (1975), Bernard Rollin (1981), Steve Sapontzis (1987) and Evelyn Pluhar (1995).

Cohen has offered additional critiques of animal rights and presently has another in process (Cohen, 1996). There is little to choose between these works, in my opinion, and I shall treat them as together constituting what I call 'Cohen's view', a view which, while (as we have seen) it seeks to justify human use of other animals in biomedical research in particular, rests on a critique of animal rights in general, including animals living in the wild ('in nature') in particular. For it is just as logically impossible for non-human animals living in the natural world to have rights, given Cohen's position, as it is for non-human animals living in a laboratory to have them. That much granted, Cohen's view, if correct, would entail that a particular conception of human ethics – the one in which humans are represented as having basic moral rights – cannot be 'mapped on to the consideration of the human–nature relationship.'

Before turning to an examination of Cohen's critique, I want first to summarize the broad contours of his view, beginning with how he understands moral rights and why he thinks this is an important idea.

The importance of animal rights

'Rights trump interests' (Cohen, 1996: 4, 3). Cohen makes this point not once but several times. His meaning is clear. Moral rights are like invisible moral 'No Trespassing' signs. If I have a moral right to bodily integrity, for example, then you are not morally entitled to take one of my kidneys (to trespass on my right) simply on the ground that you will benefit from having it become one of your kidneys, or that someone else will benefit from having it become one of theirs. Similarly, if non-human animals have a right to bodily integrity, then humans are not morally entitled to take a kidney from a pig (to trespass on the pig's right) simply on the ground that they themselves or some other human being will benefit from having the pig's kidney become their kidney. Indeed, except in rare cases – for example, cases where it is necessary to hurt or kill an animal in self-defence – animals having rights would undermine the morality of our hurting or killing them in the course of advancing any of our interests, even important interests like health and life. As Cohen notes, 'if animals have any rights at all they have the right to be respected, the right not to be used like a tool to advance human interests . . . no matter how important those human interests are thought to be' (Cohen, 1996: 4, 1). If non-human animals have rights, in other words, research that utilizes them is wrong and should be stopped. It's not larger cages that would be called for; it's empty cages. That's what the existence and recognition of animal rights would require. Cohen even goes so far as to liken the use of non-human animals in the development of the polio and other vaccines to the use Nazi scientists made of Jewish children during the Second World War. '[I]f those animals we used and continue to use have rights as human children do, what we did and are doing to them is as profoundly wrong as what the Nazis did to those Jews not long ago' (Cohen, 1997: 92).

Cohen has good news for those scientists who use non-human animals in their research. These animals do not have rights. Indeed, they cannot have rights. Belief in animal rights he characterizes as a 'fanatical conviction' (Cohen, 1996: 4, 6), 'a profound and gigantic mistake' (*ibid.*), 'ill-founded and dangerous' (*ibid.*: 1, 8). Less dramatically, the belief that rats have rights, says he, is 'silly' (*ibid.*: 5, 4).

Of the many disparaging things Cohen says about the belief in animal rights, his characterization of it as 'dangerous' merits further comment. If belief in animal rights were ever to take hold, to the extent that the use of non-human animals in research ceased, Cohen believes that humans, not to mention other animals, would pay a terrible price in terms of loss of benefits – benefits which, in Cohen's view, as we have seen, are 'utterly beyond quantification'. Without the freedom to use non-human animals for experimental purposes, Cohen believes that none or, at most, vastly fewer of these benefits would be possible. That is what makes 'animal rights' the 'dangerous' idea it is; why steps must be taken to insure that 'research is not crippled by ignorant zealotry' (*ibid.*: 9, 40); and why Cohen clearly sees his critique of this 'fanatical conviction' as an important, honourable service to humanity.

Animal psychology, human uniqueness and rights

In some ways, Cohen does not make his task an easy one, as it would be if he followed today's neo-Cartesians, such as Peter Carruthers (1986), and denied a mental life to animals. To his credit, Cohen has a robust view of animal psychology – one that, in all essential respects, coincides both with the view I favour and, I think, with the settled convictions of all people of common sense. If for purposes of argument we simplify matters and limit the claims made about non-human animals to mammals and birds, then we can say that Cohen believes these animals not only are conscious, but have cognitive, communicative, affective and other noteworthy psychological capacities, including sentiency. These animals can reason. They have an emotional life. They can plan and choose. And some things they experience give them pleasure, while others cause them pain. In these and other respects, these animals are like us, and we, like them.

Still, there are important differences. In particular, Cohen believes that non-human animals are not morally responsible for their actions. Whatever they do, it is not and, indeed, it cannot be morally right; and neither is it, nor can it be, morally wrong. Moral responsibility, as is true of the ideas of moral right and wrong, apply to humans and – at least among terrestrial forms of life – to humans only.

While he has no doubt that humans have rights (humans 'certainly have rights,' he insists at one point; *ibid.*: 5, 1), Cohen never offers anything resembling an argument in defence of our having them. What he does instead is answer this (in his words) 'central moral question of all time' by summarizing some of 'the *kinds* of explanations of human rights that have been given by the greatest moral philosophers'. Cohen's list of the 'greatest' is eclectic, including philosophers as diverse as St Thomas Aquinas, F. H. Bradley, Thomas Dewey and Karl Marx, philosophers who, according to Cohen, despite their obvious differences, agree with him in thinking that there is something uniquely important about being human. 'Moral philosophers through the ages', Cohen writes,

> have not disagreed about *the essentially human locus of the concept of right.* Of the finest thinkers from antiquity to the present not one would deny – as the animal rights movement does seek to deny – that there is a fundamental difference between the moral stature of humans and that of animals.
>
> (*ibid.*: 5, 11)

When it comes to specifying what this 'fundamental difference' is, Cohen displays a certain fondness for Kant's views. He sides with Kant in extolling the

> unique capacity of humans to formulate moral *principles* for the direction of our conduct, to grasp the maxim of the principles we devise, and by applying these principles to ourselves as well as to others, to exhibit the moral *autonomy* of the human will.
>
> (*ibid.*: 4, 12)

Non-human animals, on both Kant's and Cohen's view, lack these capacities. On this point Kant and Cohen certainly seem to be correct. Even philosophers like Sapontzis, who discern the workings of a proto-morality in the loyal, courageous and empathetic behaviour of non-human animals, stop short of supposing that wolves and elephants trouble themselves over whether the maxims of their actions can be willed to be universal laws.

In general, then, Cohen believes – and with all this I agree – that humans and other animals resemble each other in many ways, and differ in others. Like us, other animals have a complicated, unified psychology, involving cognitive, affective, volitional and other capacities. But unlike humans, other animals lack the sophisticated abilities that make moral autonomy, agency and responsibility possible. In this respect humans are unique, at least among terrestrial creatures. So far, so good.

Duties to animals

Now, Cohen thinks we have duties to animals, not merely duties involving them. '[W]e humans have, and recognize that we have, many *obligations* to animals,' he writes (*ibid*.: 5, 1). In particular, 'the obligation to act humanely *we owe to them*' (*ibid*.: 5, 4). Mammals (at least) are 'morally considerable', and 'we humans surely ought cause no pain to them that cannot be justified. Nor ought we to kill them without reason' (*ibid*.: 5, 8). Again, 'we are obliged to apply to animals the moral principles that govern *us* regarding the gratuitous imposition of pain and suffering' (*ibid*.: 5, 4).

Fairly clearly, then, Cohen regards his position as what in the past I have called a direct, as distinct from an indirect, duty view (Regan, 1983: 150 ff.). Those who hold some version of the latter type (an indirect-duty view) believe that while we can have duties involving non-human animals, we have no duties to them. For example, if your neighbour mistreats your dog, you will be upset. And it is because it is wrong to upset you, not because of any wrong done to your dog, that an indirect-duty theorist might maintain that your neighbour has failed to do his duty with respect to your dog. On such a view, dogs and other non-human animals have the same moral standing, the same moral considerability, as sticks and stones. When assessed from the moral point of view, other-than-human animals, like sticks and stones, themselves count for nothing.

To the actual Carl Cohen's credit, he does not wish to align himself with such indirect-duty theorists as the early twentieth-century Jesuit, Joseph Rickaby, whose views are representative of the type. Writes Rickaby, 'We have . . . no duties . . . of any kind, to . . . animals, as neither to sticks and stones' (Rickaby, 1976). Not true, according to Cohen. If your neighbour mistreats your dog, the actual Carl Cohen would insist that your neighbour *has done something wrong to the dog*. And on this point, as on many others, Cohen certainly seems to be on the side of the angels. To support this judgement, consider the following two cases.

First, suppose your neighbour breaks your leg for no good reason. No one will deny that this hurts. A lot. Similarly beyond question is that your neighbour has harmed you. Directly. Of course, others might be upset, even outraged that you have been so badly mistreated. But even if no one is upset – even if everyone else in the neighbourhood is positively delighted about your misfortune – there is no question but that *you* have been harmed, that *you* have suffered a serious wrong.

Next, suppose your neighbour does the same thing, only this time it is your dog whose leg is broken for no good reason. Well, that hurts your dog. A lot. And your neighbour has harmed your dog. Directly. How all this can be true of both you and your dog, and it remain true that in breaking your leg your neighbour has done a direct wrong to you, but that in breaking your dog's leg your neighbour has not done a direct wrong to your dog – how this can possibly be true, I confess I do not understand. So far as I can see, the two cases are the same in all the morally relevant respects. Which is why, short of following the neo-Cartesians and denying that non-human animals feel pain, I do not see how it can possibly be true that your neighbour has a direct duty to you, not to cause you pain and harm for no good reason, but has no direct duty to your dog, to avoid doing the same thing. If I am right about this, then a credible Carl Cohen – a Cohen whose views about the type of duties humans owe to other animals – will have to agree that, when it comes to these animals, at least some of our duties are duties owed directly to them.

Now, this is something that, as we have seen, Cohen actually believes. I do not myself see how what he says about our moral ties to other animals permits any other interpretation. The actual Cohen subscribes to a direct-duty view. However, for reasons given below, a consistent Carl Cohen cannot believe what the actual Carl Cohen wants to believe. For (as I hope to show) if Cohen's arguments against animals having rights were successful, these same arguments would also show, *mutatis mutandis*, that we have no duties to them.

Before explaining the dilemma Cohen faces in this regard, something further needs to be said by way of a general characterization of his view, his understanding of the relationship between rights and duties in particular.

Duties and rights

From the fact (assuming it to be a fact) that we have duties directly to animals, Cohen argues that it does not follow that they have rights against us. This certainly seems to be correct. In the case of the moral ties that bind humans to one another, there are many cases where our duties do not have correlative rights. For example, I have a duty to render assistance to those in need, but no particular needy person has a right to demand that I discharge this duty by assisting her or him. Moreover, even when the duty is owed to a specific individual (for example, someone who has been especially kind or thoughtful), there need be no correlative right. 'A special act of kindness done to us', Cohen notes, 'may leave us with the obligation to acknowledge and return that kindness – but the benefactor to whom we are obligated has no claim of *right* against us' (*ibid.*: 5, 3).

It would therefore be a mistake, as Cohen is correct to see, to infer that animals have rights *simply on the ground* that we have direct duties to them, whether to animals in general or specific animals in particular. While it is true that all rights have correlative duties, it is false that all duties have correlative rights, let alone that all duties (in Cohen's words) 'arise from rights' (*ibid.*: 5, 2). On this point, Cohen will get no argument from me.

The problem is, he will get no argument from anyone. At least, not anyone I know of, including in particular those (less than the 'greatest') philosophers who maintain that animals have rights or that humans and other animals share an equal moral status. Even in the case of a position like mine, where duties are treated as the basis of rights, it is not claimed that rights are correlated with or arise from each and every duty. It is only when duties are (1) basic and (2) unacquired – only in the case of duties such as the duty of justice, for example, as distinct from the duty to keep a promise – that I maintain we may validly infer correlative rights. Possibly I am mistaken about this. Certainly I would be obliged carefully to consider an argument that purported to show that I am. But in the case of Cohen's argument, since it fails to address my views, there is nothing to consider. One does not show that no duties have correlative rights simply by showing that some duties do not have correlative rights.

The amorality–rights argument

How, then, does Cohen attempt to prove that animals do not and cannot have rights? Although he himself does not clearly distinguish between them, I think Cohen relies on at least three different arguments, each of which, although it is related to the other two, differs in important ways from them. The first argument grows out of the fact that animals are (in Cohen's judgement) *amoral*. This is an idea he introduces after asking us to imagine a lioness that kills a baby zebra. He writes:

> Do you believe the baby zebra has the *right* not to be slaughtered by that lioness? Does that lioness have the *right* to kill that baby zebra for her cubs? If you are inclined to say, confronted by such natural rapacity – duplicated a thousand, a million times each day on planet earth – that neither is right or wrong, and that neither has a *right* against the other, I am on your side. Rights are of the highest moral consequence, yes; but zebras and lions and rats are *amoral*; there is no morality for them; they do no wrong, ever. In their world there are no rights.
>
> (Cohen, 1997: 95)

Essentially, then, what we have is the following, the 'amorality–rights' argument:

1   Animals live in an amoral world (a world where nothing is right or wrong).
2   Those who live in an amoral world have no rights.
3   Therefore, animals have no rights.

Sapontzis's reservations about the moral agency of animals to the contrary notwithstanding, suppose we agree that animals are incapable of doing what is right and wrong. Concerning their interactions with one another, let us agree that it makes no sense to say that the baby zebra has a right not to be killed by the lioness or that the lioness has a duty not to kill the baby zebra. From this Cohen would have us infer that it makes no sense to say that the baby zebra, or any other animal for that matter, has a right against us – for example, the right not to be killed by us for no good reason. But not only does this not follow, a consistent, credible Cohen cannot believe that it does.

To make this clearer, consider a parallel argument, an argument that concerns duties, not rights, the 'amorality–duties' argument:

1   Animals live in an amoral world (a world where nothing is right or wrong).
2   Those who live in an amoral world have no duties.
3   Therefore, animals have no duties.

Cohen unquestionably accepts this argument's conclusion. Yet we know that a credible Cohen must believe that *we* have direct duties to animals. And we also know that the actual Carl Cohen does believe this. For Cohen, then, that *non-human animals have no duties to one another* is and must be perfectly consistent with *our* having duties to them.

Logically, the possibility of non-human animals having rights is no different. From the fact, assuming it to be a fact, that animals do not have rights against one another, it does not follow that they do not, let alone that they cannot, have rights against us. Possibly they do not. Possibly they cannot. But whether they do or do not, can or cannot, is not something that can be settled by insisting that, in their world, there is no right nor wrong, no duty, no respect for nor violation of the moral rights of others.

It bears noting that the logical point at issue here is independent of whether duties and rights are correlative. Suppose for the sake of argument that duties and rights never are correlative – that, in other words, one can never validly infer duties from rights, or vice versa. That would not affect the present criticism. The present criticism insists only that one cannot validly infer that (a) animals cannot have rights against us because (b) they do not or cannot have rights against one another, any more than one can validly infer that (c) we cannot have duties to them because (d) they do not or cannot have duties to one another. The criticism concerns, in other words, what follows from animals not having duties or rights, not what follows from someone's having a duty or someone's having a right, and thus quite independently of whether rights and duties are correlative.

The right-kind argument

Cohen has a second argument that stands or falls independently of the first. This one concerns 'the capacity for moral judgement'. If non-human animals are denied rights because they are unable to exercise this capacity, why is not the same thing true of 'many humans – the brain-damaged, the comatose, the senile?' To which Cohen replies as follows:

> This objection fails; it mistakenly treats an essential feature of humanity as though it were a screen for sorting humans. The capacity for moral judgment that distinguishes humans from animals is not a test to be administered to human beings one by one. Persons who are unable, because of some disability, to perform the full moral functions natural to human beings are certainly not for that reason ejected from the moral community. The issue is one of kind. Humans are of a kind that they may be the subject of experiments only with their voluntary consent. The

choices they make freely must be respected. Animals are of such a kind that it is impossible for them, in principle, to give or withhold voluntary consent or to make a moral choice. What humans retain when disabled, animals have never had.

(Cohen, 1986: 866)

Cohen's meaning is not admirably clear. Part of the lack of clarity concerns what it is that disabled humans 'retain'. Certainly it cannot be the capacities in question, since it is their lack of these very capacities that helps define their disabilities; and it makes no sense to say they retain what they lack.

What, then, do they retain? Well, presumably they retain their rights. And if we ask why, Cohen's answer turns on considerations about what kind of being human beings are. Possession of the capacities that make rights possible (for example, the capacity of moral judgment), in his view, defines the kind of being humans are. Not so in the case of non-human animals. To lack these capacities, while it is a defect for a human being, is nothing of the sort for other animals. Non-human animals simply are not the right kind of being to have rights; human beings – and all and only human beings – are. Thus do we have the 'right-kind' argument:

1 Individuals have rights if and only if they are the right kind of being.
2 All and only humans are the right kind of being (that is, have the capacities that make possession of rights possible).
3 Therefore, all and only humans have rights.

This argument is strongly counter-intuitive. If moral rights are rights *possessed by individuals*, by what manner or means can we fairly decide which individuals have them without resting our decision on the morally relevant capacities different individuals possess? Imagine that a university annually awards a scholarship in mathematics. Will anyone suggest that we may fairly decide who deserves it without bothering to ask about the mathematical abilities of the several candidates, considered individually? Why should our judgement concerning which individuals possess rights be any different?

To make my misgivings plainer, consider the following thought experiment. Imagine that a party of extraterrestrials (let us suppose it is E.T. and his clan) make contact with us humans. They give every indication of possessing those capacities Cohen believes are essential features of being human. But suppose the capacities they share are not essential, not typical of the species to which they belong. E.T. and his cohorts are genetic aberrations, cruising the universe in search of others of their kind.

Given Cohen's view, our extraterrestrials lack rights. Indeed, they cannot possibly have them. Why? Because those capacities they possess, as the individuals they are, are not essential to the kind of being they happen to be. 'Sorry, E.T.', we will have to say, 'but you and your friends don't have any rights.'

'But why?' E.T. asks; 'We have all the capacities you humans have.'

'Yes,' we reply, 'but we have them because they are essential to the kind of being we are, while your having them is some sort of genetic accident. So,' we say, 'we hope you won't mind if we do a little biomedical research on you while you're in the neighbourhood.'

On any credible account of rights, E.T. and his cohorts have them not because (or only if) they are of the 'right kind' but because of those rights-conferring capacities they have as the individuals they are. In this sense, contrary to Cohen, relevant criteria for possessing rights *do* serve as 'a screen for sorting', not of humans only but of all would-be candidates. And it is for this reason that, Cohen's 'right-kind' argument to the contrary notwithstanding, it is an open question whether any non-human animals qualify.

The community argument

Cohen has a third argument against the possibility of animals having rights. Unlike the amorality–rights argument, which builds on the allegedly amoral condition of non-human animals, and unlike the right-kind argument, which rests on allegedly essential features of kinds of beings, this third argument is grounded in how rights arise among those who have them. 'Rights', Cohen declares,

> are universally human; they arise in a *human moral world*, in a moral *sphere*. In the world of humankind moral judgments are pervasive, and it is the fact that all humans, including infants and the senile, are members of that community – not the fact that as individuals they have or do not have certain special capacities, or properties – that makes humans bearers of rights.
>
> (Cohen, 1996: 5, 14)

Other animals, alas, are not 'members of [this] community' and so lack rights, whatever their capacities might be. Bring forth whatever impressive list of capacities and achievements in the case of any non-human animal one might wish (communicative skills among non-human primates, the cleverness of cats, the sagacity of wolves) and compare these animals with a human bereft of all that rights arise only in a community of moral beings, and that therefore there are spheres in which rights do apply and spheres in which they do not.

(Cohen, 1997: 14–15)

I think any honest reading of this passage will have to conclude that what Cohen means is not as clear as one might wish. But let me try to explain what I think he means; then I will be able to explain why I believe he is mistaken.

As we have seen, Cohen states that rights 'arise' in a 'human moral world', 'in a moral sphere', in a 'community' (meaning, presumably, a 'moral community'). I take it that what Cohen means is that the very idea of a moral right assumes a social context without which this idea could not 'arise'. To explain: Cohen and I agree that rights place justified limits on how individuals may be treated; for example, we agree that our right to life entails that others are not at liberty to kill us except in very unusual circumstances (in self-defence, say). That being so, the idea of rights can 'arise' only if (a) individuals are living together and interacting with one another, and if (b) these individuals can understand what it means to have justified limitations on what they are free to do to one another. Human beings, Cohen thinks, satisfy both these conditions. Arguably, no other animals do.

True, non-human animals who live in groups in the wild satisfy condition (a); but it is implausible in the extreme to maintain that they also satisfy condition (b). Granted, *we* may be able to limit their liberty; but it is pure fancy, I think, to imagine that members of a pack of wolves, for example, can understand the idea of deliberately imposing justifiable limits on their behaviour. This is why zebras and gazelles, for example, cannot, I think, understand what a right is, and also why rights cannot 'arise' in a community of these animals.

The same is no less true of those animals who live in community with us. For example, Professor Cohen and his family share their life with a dog. And there can be no doubt that the members of the Cohen family lavish their love and affection on this lucky animal. But while it is true that the dog is a member of the Cohen community in one sense, this is not the sense in which the human members of the Cohen family belong to the human *moral* community. For the Cohen dog, like every other dog that has existed, currently is alive, or will one day roam the earth, never has, does not now, and never will understand, I think, that moral rights place justified limits on what individuals are free to do. This being so, the idea of rights can no more 'arise' among domesticated than it can among wild animals.

But what of those human beings (including, for example, infants, the insane and the seriously mentally disadvantaged) who, like both wild and domesticated animals, are unable to understand what moral rights are? Granted, these humans are members of the human community, not in some extended sense (as when we say, for example, that the Cohens' dog is an 'honorary' member of the Cohen family), but literally: they all *are* human beings. However, as is true of the 'honorary' canine member of the Cohen family, these humans do not now, and many of them never will, understand what a right is. Do these humans have rights?

As we have seen, Cohen says they do. He believes that while the inability of animals to understand what rights are is sufficient to exclude them from the class of right-holders, this same lack of understanding does not disqualify any human being. *Every* human being has rights because *all* are members of the community in which the idea of rights 'arises', whereas *every* non-human animal lacks rights because *none* belongs to this community. Clearly, then, the decisive criterion for possessing rights is not whether one understands what rights are; it is whether one is a member of the community in which rights 'arise'. And since the only community in which rights 'arise', at least in the terrestrial sphere, is the community composed of human beings, in that sphere being a human (belonging to the species *Homo sapiens*) is both a necessary and a sufficient condition of possessing rights.

I believe the foregoing captures both what Cohen believes and why he believes it. Let me summarize his argument as follows, the 'community argument':

1   All and only those individuals have rights who are members of communities in which the idea of rights arises.
2   Within the terrestrial sphere, the idea of rights arises only in the human community.
3   Therefore, within that sphere, all and only humans have rights.

Now, I think it should be fairly obvious why and how this argument goes wrong. Conceptually, there is a distinction between (a) the necessary and sufficient conditions of the origin or formation of an idea and (b) the scope of the idea. The former concerns how it is possible for an idea (to use Cohen's word) to 'arise', while the latter concerns the range of objects or individuals to which or to whom the idea may be intelligibly applied. The central point to recognize is that these two matters are logically separate in this sense: the scope of an idea is something that must be determined independently of considerations about the origin of the idea.

By way of example: as far as we know, ideas like that of central nervous system and genes 'arise' only among humans because only humans have the requisite cognitive capacities to form them. But the range of entities to which these ideas apply is not necessarily limited to all and only members of the community in which these ideas 'arise'. Indeed, not only is the scope of these ideas not necessarily limited to all and only humans; there are literally billions and billions of non-human animals to which the ideas are correctly applied – that have, that is, genes and a central nervous system.

Considered conceptually, discussions regarding rights are no different. We grant that, as far as we know, the idea of rights 'arises' only among humans because only humans have the requisite cognitive capacities to understand it. But the range of entities to which this idea applies is not necessarily limited to all and only members of the community in which the idea 'arises'. Logically, to make this inference would be to make the same mistake as inferring that wolves cannot have genes or that the Cohens' dog cannot have a central nervous system because these animals do not belong to a community in which these ideas arise.

Conclusion

Among those philosophers critical of animal rights, none has commanded a larger audience outside philosophy or exercised a greater influence than Carl Cohen. Although I do not question the sincerity of his beliefs concerning animal rights, it has been my object here to show how ill-supported those beliefs are. His central arguments (the amorality–rights argument, the right-kind argument and the community argument) are deeply, seriously and, in my opinion, irredeemably flawed.

Of course, recognizing how and why Cohen's arguments go wrong does not prove that animals *have* rights, any more than we can prove which animals have a central nervous system just by recognizing that the scope of this idea is not necessarily limited to all and only those individuals who belong to a community in which this idea 'arises'. But proving that animals *have* rights has not been the purpose of the present argument; if it had been, I would have been obliged to consider objections to this idea raised not only by Cohen but also by Peter Singer and Val Plumwood – to mention only two of the many other critics lined up against attributing rights to non-human animals. As I have been at pains to explain, my purpose on this occasion has been much more modest, limited as it has been to showing both that and why Carl Cohen's arguments fail to prove that other-than-human animals *cannot* have rights. That much granted, I conclude that Cohen has failed to show that a familiar conception of human ethics – the one where humans are represented as possessing basic moral rights – cannot be 'mapped on to the consideration of the human–nature relationship' in general, the human–non-human animal relationship in particular. My own positive account of one way in which this mapping can be done may be found elsewhere (Regan, 1983).

Notes

1 I gratefully acknowledge Carl Cohen's permission to quote from the manuscript of a work in his work-in-progress, *In Defense of the Use of Animals*.
2 In the words used to describe the plenary session to which Tom Regan, Peter Singer and Val Plumwood contributed at the conference on Environmental Justice at the University of Melbourne, 1–3 October 1997.

References

Carruthers, P. (1986) *Introducing Persons: Theories and Arguments in the Philosophy of Mind*, London: Croom Helm.

Cohen, C. (1986) 'The Case for the Use of Animals in Biomedical Research', *New England Journal of Medicine* 315 (14): 865–70.

—— (1996) 'In Defense of the Use of Animals', unpublished manuscript consisting of eleven short chapters, with each chapter beginning with its own page 1; quoted material cites the chapter first, then the page number(s).

—— (1997) 'Do Animals Have Rights?', *Ethics and Behavior* 7 (2): 91–102.

Pluhar, E. (1995) *Beyond Prejudice: The Moral Significance of Human and Non human Animals*, Durham, NC: Duke University Press.

Plumwood, V. (1993) *Feminism and the Mastery of Nature*, London: Routledge.

Regan, T. (1980) 'Utilitarianism, Vegetarianism, and Animal Rights', *Philosophy and Public Affairs* 9 (4): 305–24. Reprinted in T. Regan (1982) *All That Dwell Therein: Essays on Animal Rights and Environmental Ethics*, Berkeley, CA: University of California Press, 40–60.

—— (1983) *The Case for Animal Rights*, Berkeley, CA: University of California Press; London: Routledge.

—— (1991) 'Feminism and Vivisection', in *The Thee Generation: Reflections on the Coming Revolution*, Philadelphia: Temple University Press, pp. 83–103.

Rickaby, J. (1976) 'Of the So-Called Rights of Animals', in T. Regan and P. Singer (eds), *Animal Rights and Human Obligations*, Englewood Cliffs, NJ: Prentice-Hall, p. 179.

Rollin, B. (1981) *Animal Rights and Human Morality*, Buffalo: Prometheus Books.

Sapontzis, S. (1987) *Morals, Reason, and Animals*, Philadelphia: Temple University Press.

Singer, P. (1975) *Animal Liberation*, New York: Avon Books.

11 Indigenous ecologies and an ethic of connection

Deborah Bird Rose

Introduction

I aim to generate further dialogue that will enable Indigenous people's understandings of ecology to find conversational and practical ground in current world environmental justice debates. My work with dialogue is embedded both in Levinas's philosophy of ethical alterity and in my work with Indigenous people. Levinas teaches an ethic of human connectivity: 'consciousness and even subjectivity follow from, are legitimated by, the ethical summons which proceeds from the intersubjective encounter. Subjectivity arrives, so to speak, in the form of a responsibility towards an other' (Newton, 1995: 12).

Emil Fackenheim (1994: 129) articulates two main precepts for structuring the ground for ethical dialogue. The first is that dialogue begins where one is, and thus is always situated; the second is that dialogue is open, and thus that the outcome is not known in advance. Openness produces reflexivity, so that one's own ground becomes destabilized. In open dialogue one holds one's self available to be surprised, to be challenged, and to be changed. I intend this study to constitute an exploration both of some of the directions this dialogue may take and of the Fackenheim principle I advocate: situated availability. My work with Indigenous people leads me to understand dialogue in the broad sense of intersubjective mutuality, and thus to seek possibilities for mutual care in a system of connections and reciprocities that includes humans, non-human living things, and environments. I conclude with some thoughts on pragmatics for the restoration of connection and mutual care.

Sites

I have for many years been learning from Aboriginal people in Australia. The greater part of my research has taken place in the Victoria River District of the Northern Territory, where my learning focuses on intersubjectivity, ecology, and practices of colonization (Rose, 1991, 1992, 1996a, b). Here in the savannah regions of the monsoonal north, white settlers established broadacres cattle stations just over one hundred years ago. Overrunning the homes of the Indigenous hunter-gatherer peoples of the region, they first shot and hunted away the local peoples, and later pressed them into service on the cattle stations as an unfree and unpaid labour force. In the mid-1960s Aboriginal people in this region went on strike against the appalling system of oppression which ruled their lives, and as support for them was manifested throughout the nation, they began to articulate what to them had been an underlying purpose in all their forms of resistance: to regain control over at least some portions of their traditional homelands. In the 1990s, land rights have benefited some far more than others, but citizenship (granted in about 1969) has enabled people to participate in national and international struggles for equality. Indigenous people's cultural survival continues to be contested locally and nationally.

In this northern frontier, 'whitefella' culture (the culture of the conquerers) is represented most powerfully by extreme modernist and developmentalist conceptions of self, power, otherness and utility. Development discourse proposes that for our own good we channel our lives into practices which, we increasingly understand, generate the problems that make the continuity of life so precarious. In harnessing hope to violence they generate destruction and despair. My dialogue around environmental ethics takes shape as a movement running counter to these regimes of violence.

Monologue

Critical theory of recent decades has shown Western thought and action to be dominated by a matrix of hierarchical oppositions which provides powerful conceptual tools for the reproduction of oppression. In this matrix the world is formed around dualities: man/woman, culture/nature, mind/body, active/passive, civilization/savagery, and so on in the most familiar and oppressive fashion. In fact, however, these dualities are more properly described as a series of singularities because the pole labelled 'other' is effectively an absence. This point is articulated extensively by feminist theoreticians. Irigaray (1985), for example, shows that the defining feature of woman under phallocentric thought is that she is not man. Ecofeminists extend the analysis to include 'nature', and demonstrate that the same structure of domination controls women, nature, and all other living beings and systems that are held to be 'other' (Warren, 1990; Salleh 1992). Val Plumwood (1994: 74) speaks directly to the centrality of this structure: 'the story of the control of the chaotic and deficient realm of "nature" by mastering and ordering "reason" has been the master story of Western culture'. Stripped of much cultural elaboration, this structure of self–other articulates power such that 'self' is constituted as the pole of activity and presence, while 'other' is the pole of passivity and absence. Presence is a manifestation both of being and of power; absence may be a gap awaiting transfiguration by the active/present pole, or an enabling background – in either case, without power and presence of its own (Plumwood, 1997, and this volume, Chapter 12).

A critical feature of the system is that the 'other' never gets to talk back on its own terms. The communication is all one way, and the pole of power refuses to receive the feedback that would cause it to change itself, or to open itself to dialogue. Power lies in the ability not to hear what is being said, not to experience the consequences of one's actions, but rather to go one's own self-centric and insulated way. Plumwood (1993: 443) notes two key moves in sustaining dualism (as distinct from mutuality): dependency and denial. The pole of power depends on the subordinated other, and denies this dependence.

The image of bipolarity thus masks what is, in effect, only the pole of self. The self sets itself within a hall of mirrors; it mistakes its reflection for the world, sees its own reflections endlessly, talks endlessly to itself, and, not surprisingly, finds continual verification of itself and its world view. This is monologue masquerading as conversation, masturbation posing as productive interaction; it is a narcissism so profound that it purports to provide a universal knowledge when in fact its violent erasures are universalizing its own singular and powerful isolation. It promotes a nihilism that stifles the knowledge of connection, disabling dialogue, and maiming the possibilities whereby 'self' might be captured by 'other'.

At the margins, within the domain of the 'other', one knows that the world, life and people express themselves with rich and interactive presences that are invisible from the viewpoint of deformed power, except, perhaps, as disorder or blockage. The dismantling of this oppressive and damaging pole is a necessary step in moving towards dialogue. Dismantling will fail if it is confined to monologue; we must embrace noisy and unruly processes capable of finding dialogue with the peoples of the world and with the world itself. We must shake our capacity for connection loose from the bondage of the monological self.

Multiplicities

In the Victoria River District of the Northern Territory, Aboriginal people's cultural construction of subjectivity is organized through a multiplicity of presences. Unlike a system of singular presence and multiple absences, this is a system of multisited, multicentred cross-cutting and overlapping subjectivities. *Country* is the matrix for the relationships I will be discussing; the term gained its current connotations in Aboriginal English. Small enough to accommodate face-to-face groups of people, large enough to sustain their lives, politically autonomous in respect of other, structurally equivalent countries, and at the same time interdependent with other countries, each country is itself the focus and source of Indigenous law and life practice. To use the philosopher's term, one's country is a nourishing terrain, a place that gives and receives life (Levinas, 1989: 210).

Country is multidimensional: it consists of people, animals, plants, Dreamings; underground, earth, soils, minerals and waters, surface water, and air. There is sea country and land country; in some areas people talk about sky country. Country has origins and a future; it exists both in and through time. Humans were created for each country, and human groups hold the view that they are an extremely important part of the life of their country. It is not possible, however, to contend that a country, or indeed regional systems of countries, is human centred. To the extent that a country or region can be said to have a central focus, that focus is the system of interdependent responsibilities by which the continuity of life in the country and the region is ensured. A fundamental proposition in Indigenous law and society is that the living things of a country take care of their own. All living things are held to have an interest in the life of the country because their own life is dependent upon the life of their country. This interdependence leads to another fundamental proposition of Indigenous law: those who destroy their country destroy themselves.

This is the created world, brought into being as a world of form, difference, connection and responsibility by the great creating beings, called Dreamings. The origins of country – its living things, its internal organization and its relations to other equivalent countries – lie in Dreaming creation. In these terrains, consciousness and responsibility are manifested by all the participants in living systems. Subjectivity, in the form of consciousness, agency, morality and law, is part of all forms and sites of life: of non-human species of plants and animals, of powerful beings such as Rainbow Snakes, and of creation sites, including trees, hills and waterholes. Nourishing terrains are sentient.

Boundaries of mutual care are, on the one hand, geographical: people are 'born for country', and thus are born to responsibilities for that country. Equally, they are born into relationships in which their country is responsible for them. Hunting is central to the meaning and joy of life. It is a key form of nurturance (see Myers, 1996, on nurturance). The good hunters, women and men, are seen to have close, nurturing relationships with the world. Victoria River people say that their country gives them body and life; it takes care of them, they say. When people go hunting, fishing and getting plant foods, they call out to their deceased relatives who continue to live in the country, saying that the children are hungry and asking for food. Hunting, fishing and getting plant foods and medicines keep people in relationship with their country. Likewise, people take care of the country, seeking to ensure that other living things are also coming into growth and protection. They well recognize a self-interest in nurturing others so that others will be available to be hunted, fished or gathered. This interest promotes the continuity of country as a living system; it is understood to be moral. The continuities of life, locality, belonging and care are brought into being through the work and joy of nurturant provisioning.

Other relationships of mutual care cross-cut boundaries. Totemic relationships constitute a major system for linking living bodies into structured relationships of sameness and difference. In the Victoria River District there is a multiplicity of types of totems. Matrilineal totems, for example, link people to non-human species, and cut across boundaries of country. These people have or are the same flesh or 'meat' (*ngurlu*), and they share that sameness with each other and with the animal or plant with which they are associated: flying fox people share flesh with flying foxes, emu people with emus, and so on. Human and non-human kin are of the same flesh, and what happens to one has a bearing on what happens to the other. When an emu person dies nobody eats emus until the emu people tell them they can, and similarly when a flying fox person dies nobody eats flying foxes until the right people give permission. There are more variations than there is dogma, but there is a clear recognition that the lives of these beings are enmeshed in perduring relationships which bind people and certain animal or plant species together and thus differentiate them from others.

Country-based totems, inherited from one's father's father, mother's father and mother's mother's brother, link people with land, Dreaming tracks and sites, and the species of those Dreamings. The owlet nightjar, for example, is the country-based totem (Dreaming, *kuning*) for a small group of people who take this identity from their father's father.[1] They have responsibilities towards the sites of the owlet nightjar, and towards owlet nightjars generally, as well as owning the stories, songs, designs and sites. They are responsible for the flourishing of owlet nightjars in the world, and this means that they are responsible for their own flourishing as well. Nor are they alone in this responsibility. Owlet nightjars (the birds) also have responsibilities towards them (the people), but it is fair to say that as a human being one knows the most about human responsibilities.

A country-based totem is a singularity, and it is cross-cut. Countries are exogamous, meaning that countrymen must find their spouses in other countries.[2] The owlet nightjar people have other relationships and responsibilities from their mother's father, who was possum. These people are countrymen with owlet nightjar, and countrymen with possums. Countrymen take care of each other; to say that countrymen have responsibilities towards each other's interests is to include non-humans within the realm of law. The rule that countries are exogamous cross-cuts the singularity of country and country-based totems, generating kin relationships between countries and people.

In addition, differing types of totems cross-cut and overlap each other. Some nightjar people are emu people because their mother was emu; others are not. The members of a set of owlet nightjar people are thus both the same and different: the same by reason of being owlet nightjar, different by reason of their other, differing totemic relationships. As different categories cross-cut each other, the people and other species who are related in one context are unrelated, or differentiated, in another context.[3]

When Morgan died . . .

When Morgan died they sent him home from hospital saying there was nothing they could do to save him. Apparently he was another casualty of 'middle-aged death syndrome', a fatal condition that in Australia is almost exclusively confined to Aboriginal people. His family and friends, countrymen and in-laws, wanted to know why. Who was accountable, and was Morgan himself at fault? As he wasted away, we talked about the emu he had once shot, and wondered if that action was leading him towards death. He had a dream about snakes, and we talked about the Rainbow Snake he had shot. Then we wondered if it might have been a water snake after all, and we wondered if the water snake people had taken a set against him. His family negotiated with these others, and ascertained that these past actions were not the cause of death. Some time after the funeral, the family came to the conclusion that death was caused by the Australian Defence Force's use of his father country as a bombing range.

People who are countrymen share their being with their country, and when the country suffers, so do people. Likewise, when people die, their country suffers. People identify marks such as dead trees, scarred trees or scarred hills, for example, as having come into being because of the death of a person who was associated with that country. Similarly with Dreamings: when Dreaming sites are damaged, people die; when people die, their Dreamings are at risk.

This is also the case with respect to the ground itself. One instance occurred at a billabong which the white station owner had told his Aboriginal workers to enlarge. My friend and teacher Daly Pulkara told me that they had not wanted to cut into the earth at this place because they knew that it was particularly powerful. The earth bled when they started digging, Daly said, and he named some of the people who died from this action.

These examples go to show that it would be a mistake to regard the boundaries of the person as coterminous with the body, and it would equally be a mistake to believe that if other people share a person's body, that person is thereby violated. On the contrary, persons achieve their maturity and integrity through relationships with people, animals, country and Dreamings.

Implicit in this construction of the person is the idea that places, trees, waterholes, Dreaming sites and other animals are also subjects. Their being and becoming in the world exist in relation to other subjects, some of whom are human beings.

To be vulnerable

It seems that subjectivity is not confined by the boundaries of the skin, but rather is sited both inside, on the surface of, and beyond the body. Subjects, then, are constructed both within and without; subjectivity is located within the site of the body, within the bodies of other people and other species, and within the world in trees, rockholes, on rock walls, and so on. And of course location is by no means random; country is the matrix for the structured reproduction of subjectivities.

Much could be said about the tension, conflict and politicking involved in Aboriginal societies as people seek to determine in any given context the specific parameters of difference and sameness, exclusion and inclusion, autonomy and responsibility. I have sidestepped issues of political life in order to focus on ecological systems, but it ought not to be thought that the system I am describing creates what is loosely called 'harmony'. Rather, this multiplicity of social contexts provides innumerable opportunities to argue about social context, social responsibility and social action (see, for example, Sutton and

Rigby, 1982). Equally, however, this same multiplicity of contexts works to contain tension and conflict. The cross-cutting of categories and the multiple sites of subjectivity ensure that power is located throughout the system. Politics lies in the art of locating one's self in as many contexts as possible, rather than in accumulating contexts and collapsing them into a singularity.

Relationships of difference and sameness are crucial to this politics. Sameness makes inclusive relationships possible: the emus/people, the owlet nightjars/people, the kangaroos/people, and so on. Difference makes exclusion possible: emus are not kangaroos, and the people and species, earth and water, of one country are different from those of another country. Multiple sites of subjectivity cross-cut these boundaries, and while they do not and cannot extend indefinitely, they overlap with other sites of subjectivity which are cross-cut by others, which overlap with others.

The person who exists in others, and in whom others exist, is vulnerable to what happens outside their own skin, but that same person finds their power in the relationships which are situated beyond the skin. To share a subjectivity is to share a self-interest. Thus, duties of care are understood quite profoundly to be mutual and reciprocal. Totemic and country relationships distribute subjectivity across species and countries such that one's individual interests are folded within, and realized most fully in the nurturance of, the interests of those with whom one shares one's being. It follows that you cannot bring yourself into being; each living thing becomes itself in the world through the care of others in whose lives its own is enfolded. And while no individual is connected to all others, the overlap of connections sustains a web of interdependencies.

In sum, listening to Indigenous people in an ethic of situated availability throws open a world of connection and responsibility that is in no way confined either to particular categories of humans, or to human beings as a type of living things. In this system, living beings truly stand or fall together. The process of living powerfully in the world is thus based on nurturing the relationships in which one's subjectivity is enfolded. Care of one's country, one's people, one's Dreaming sites and one's non-human countrymen are just some of the actions through which people bring forth, and are themselves brought forth by, interacting subjectivities. Mutual care is neither infinitely obligatory, nor is it diffuse and undifferentiated. The process of mutual bringing forth is local and bounded, and embedded in responsibility and accountability.

Deep colonizing

Within the social and scholarly movements labelled deep ecology and ecofeminism, people are actively engaged in seeking to realize non-dominating, context-sensitive forms of relationship and responsibility, while at the same time generating critical insight into the forms of oppression that are destroying life's possibilities. There are many points of contact and convergence between the Aboriginal system I have described and feminist, ecofeminist and deep-ecology analyses. One of the ongoing issues concerns the relationship between abstract principles of justice and contextualized relationships of care. Warren's 'situated universals' (1997, and this volume, Chapter 8) and Benhabib's (1992: 3) 'interactive universalism', for example, both seek to link a tradition of universality with a tradition of contextualized care. Benhabib's argument is that both are necessary to an intersubjective ethic that can accommodate both structural equality and sensitivity to the contexts of embedded and embodied subjects. Ecofeminists take this further when they insist on a 'version of the world as active subject' (Haraway, 1988: 593). Mathews (1994: 146–47) articulates the philosophical ground of 'self' as process, and of 'becoming' as a process of connection. Her work is particularly inspiring because she shows the capacity of Western thought to experience its own understanding of 'ecocosm': 'the pattern that connects' selves within a system that itself is self-realizing.

These points of convergence are powerful, but the points of difference also matter. The system I have described structures difference within the context of face-to-face interactions, small-scale societies, continuing, non-random interactions with non-human species and environments, and communities of kin that include humans and non-humans. Such a system is not immediately translatable to societies of strangers, and to seek to borrow or reinvent such a system could well constitute exactly the kind of invasive appropriation that would undermine the moral ground of both dialogue and care. As I see it, the problem for Westerners is to acknowledge the brokenness of our intersubjectivities, and to recuperate connection without fetishizing or appropriating Indigenous people and their culture of connection.

Environmental ethics conceived in dialogue must be both situated and open; one begins where one is. In settler societies one is situated in a political economy of violence. The links between colonization and development have been subjected to excellent analysis by Shiva (1993), and my discussion depends on her insights. I use the term 'deep colonizing' to communicate some complex ideas about our contemporary period. While it is demonstrably the case that many formal relations between Indigenous people and the nation-states that encompass them have changed in recent decades, as have many of the institutions which regulate these relations, it is also the case that practices of colonization are very much with us. In Australia, as in other settler societies, many of these practices are embedded in the institutions that are meant to reverse processes of colonization. Colonizing practices embedded within decolonizing institutions must not be understood simply as negligible side effects of essentially benign endeavours. This embeddedness may conceal, naturalize, or marginalize continuing colonizing practices.

Furthermore, practices of colonization are so institutionalized in political and bureaucratic structures and policies that they are almost unnoticed. For example, in the vast pastoral leases of the Victoria River valley, cattle come first. Introduced pastures, thousands of miles of fences, the suppression of

Indigenous fire regimes, and many other factors aimed towards improving cattle productivity have massively degraded the rangelands. Aboriginal pastoralists are under considerable pressure to maintain the productivity of their businesses, and many Aboriginal pastoralists actively desire to run successful income-producing stations. The desire for economic success is consistent with current government policy. In the words of Senator Herron (1996), Minister for Aboriginal Affairs, government policy in 1996 was towards 'self-empowerment'. He explained, 'Self-empowerment means economic independence.' Pastoralists in the Victoria River district achieve economic independence through intensification of their use of the ecosystems upon which their enterprises depend. Thus, for Aboriginal people, the empowerment of one self (individual or group) is achieved at the expense of other selves: the non-human living beings and the environmental systems of the country with which people understand their own well-being to be connected.

In the late 1990s, government policy, the interests of multinational companies, the message of many Christian missionaries, the power politics within government-created Aboriginal institutions, and global economic forces converge in one arena: they promote a self-interest that rewards the isolated and dominating self, and they inflict pain and despair on those who understand their subjectivity to be located both within and without. Under the labels 'self-empowerment' and 'progress', deep colonizing goes hand in hand with what is called development, and frequently involves the destruction of places and species in which subjectivity is enfolded. The policy encourages people to act as isolated and private selves, and to make 'rational' decisions which require that they place self-interest in opposition to the interests of others (see Turner, 1989: 258).

We find here the diminishing of multiplicity to singularity, so that the person is increasingly disconnected from the world. We find multiple destructions, so that where a person would be connected (to their country, their Dreamings, their kin), there are increasingly absences – the trees are knocked down, the animals become extinct, the country blows away in the wind, and the people are dispersed. This is called development, and, most grievously perhaps, people are urged to engage in development as a matter of care for their own futures. Under these harmful regimes, what is being produced is an assimilated sameness: aggregates of disarticulated individuals whose hope for a life of dignity urges them to set themselves in opposition to the systems on which their livelihoods depend, and in which their well-being is enfolded.

This situation, arrived at in the name of progress, is a very clear-eyed and exact description of the condition under which most of us today struggle to sustain our lives. Under late capitalism we are these aggregates of individuals whose hope is channelled into violence. We know that self-interest entails shifting pain and damage elsewhere, and that even as we seek modestly to sustain our own lives, our destructive practices expand exponentially. It often seems that despair is our destiny as well as our daily temptation.

The paradoxical search for connection

We colonizers and settlers, the conquerors of new worlds, are paradoxically situated. Most of us are here because of hope. Our ancestors hoped to make better lives for themselves and their families. They hoped to build societies that would be more equitable and more generous than those they left behind. Their optimism was channelled into migration, settlement, hard work and a faith in progress, science and technology (and, for some, the guiding hand of God or the destiny of Empire). Their callous indifference to the dispossession, death and despair they generated for the Indigenous peoples and the ecosystems of their 'new' worlds, and for the many others who were caught up in their projects, rendered their whole enterprise hope-destroying right from the start.

Current conditions of damage and impending collapse are common knowledge. We live, as Bauman discusses in excellent detail, in a world of problems with no good solutions, a world of ambivalences and moral anguish in which every action bears the potential to spoil more than it can repair (Bauman, 1993: 225). Colonization produces these places torn and fractured by violence and exile. Colonizers and colonized, we all inhabit these death-scarred landscapes. We are here by hope, and we are here by violence.

Hope lets us direct our attention to the future, and invites us to work towards that which may be, rather than to devote our care to that which is. Situated dialogue, however, is embedded in attention to the histories that scar and divide, and therefore must be based in the kind of action that Fackenheim (1994: 310) characterizes as a 'turning toward'. I take this to mean action situated in the here and now, facing each other without knowing what we shall make of each other, our history, and the damaged places of our lives. 'Turning toward' is action founded in a passionate reaching out. We reach to each other because no singular self can mend the world, and because regaining an inter-subjectivity of mutually active care must be a step towards deconstructing the pitiless isolation and regimes of violence that have both dominated and seduced us.

I should imagine that hope can be uncoupled from violence, but I suggest that it is urgent for us to move beyond hope. By this I mean decentring tomorrow in favour of today, and decentring the 'I' in favour of 'you' and 'we'. We can thus begin one of the tasks of dialogue by making ourselves available to others. I believe that when the aspirations of settlers come into conflict with the aspirations of Indigenous people, there is a very clear principle to be applied within the dialogical ethic of situated availability. From the perspective of the settler, this principle is that the other must always come first. This position – the primacy of the other – does not obviate risk. In open dialogue the outcome is not known in advance by any party, so the primacy of the other determines the starting point but not the outcome (see also Rose, 1996b). This principle is, of course, the main precept in the work of Levinas (1989), and is based on the understanding that we only find our own personhood in the society of others.

I am advocating a practice of other-focused care that orients our attention into the here and now (Rose, 1996b), and focuses it on living beings and environments, on the sacred and damaged places of the world today. Our meeting ground today is the damaged world in which we now live. Rather than going into the so-called wilderness to find common ground, we may do better to find it in the midst of toxic waste, scald areas, salinity, damaged sacred sites, contested forests, feral invasions, and in parking lots, shopping malls, national parks and uranium mines.

And what of non-humans? I propose that in contestations between human designs and the interests (as best we can understand them) of other species, and of environmental systems such as rivers, seas and soils, we must again adopt the principle that the other comes first. We make ourselves vulnerable when we situate others first, whether those others are human or non-human, because we enter a world of mutuality in which there is no category of those who are exempt from suffering. To be deeply embedded in life is to stop free-riding on the pain of others.

There is a further paradox here. In outline, the problem is this: if I am able to understand that I am disconnected, and if I am pained by this existential loneliness and wish to find connection, then I may fall prey to the monological idea that I am responsible for reconnecting myself to the world. Indigenous ethics suggest that one's proper lifework is to care for others with whom one shares situatedness, and to care for one's self by being available.

Indigenous ethics speak to a world of sentient living beings whose passion for life is sustained in connection. A dialogical approach to connection impels one to work to realize the well-being of others. Remember that in a system of mutually embedded relationships of care, one can neither unfold nor enfold one's self. We depend on others to do this work. The path to connection, therefore, does not seek connection, but rather seeks to enable the flourishing of others. If one is to find one's self more deeply embedded in the inter-subjectivities of the living world, then paradoxically one can only act towards this state by taking care while keeping one's self available. In our reaching out in passionate concern for others, we might yet be surprised, challenged, claimed, and brought into connection.

Notes

Special thanks to Nicholas Low for organizing one of the most inspiring conferences I have ever attended, and to the other participants for their contributions. This chapter was developed in further discussions at the University of Adelaide (Women's Studies Program) and at the Pitzer College Annual Environmental Justice Symposium (1998). Freya Mathews, Philip Smith, Paul Faulstich and Jimmy Weiner were especially helpful in their comments. The final form took shape in conversations with Robert John Knapp. His spiritual gifts and the generosity with which he discussed my work with me led me to a much stronger analysis than I had previously articulated.

1 In the literature this type is usually referred to as a patrilineal totem, but as people in this region hold key responsibilities towards their mother's father's totems, country, songs, designs, and so on, and may also hold key responsibilities towards their mother's mother's totem, etc., the term 'patrilineal' is not quite accurate.
2 The term 'countrymen', in Aboriginal English, includes women and men.
3 I have not sought to include an exhaustive list of totemic relationships. Another category is that of 'skin' (subsection). This system differentiates children from their parents, so while skin identity depends on the identity of the mother (father, in some areas), one's identity is not the same as one's mother. Subsections are also linked to species (as well as to country, in some areas), generating sameness and responsibility. People who share their skin with a species take responsibility for that species, and they have the right to declare it taboo for hunting, meaning that no one in the region is allowed to hunt a particular species if the people who are linked to that species so decree.

References

Bauman, Z. (1993) *Postmodern Ethics*, Oxford: Blackwell.
Benhabib, S. (1992) *Situating the Self: Gender, Community and Postmodernism in Contemporary Ethics*, Cambridge: Polity Press.
Fackenheim, E. (1994 [1982]) *To Mend the World: Foundations of Post-Holocaust Jewish Thought*, Bloomington: Indiana University Press.
Haraway, D. (1988) 'Situated Knowledges: The Science Question in Feminism and the Privilege of Partial Perspective', *Feminist Studies* 14 (3): 575–99.
Herron, J. (1996) Address to students at St Ignatius College, 13 November.
Irigaray, L. (1985) *This Sex Which Is Not One*, Ithaca, NY: Cornell University Press.
Levinas, E. (1989) *The Levinas Reader*, ed. S. Hand, Oxford: Blackwell.
Mathews, F. (1994 [1991]) *The Ecological Self*, London: Routledge.
Myers, F. (1986) *Pintupi Country, Pintupi Self; Sentiment, Place, and Politics among Western Desert Aborigines*, Washington, DC: Smithsonian Institution Press.
Newton, A. (1995) *Narrative Ethics*, Cambridge, MA: Harvard University Press.
Plumwood, V. (1993) 'The Politics of Reason: Towards a Feminist Logic', *Australasian Journal of Philosophy* 71 (4): 436–62.
—— (1994) 'The Ecopolitics Debate and the Politics of Nature', in K. Warren (ed.) *Ecological Feminism*, London: Routledge, pp. 64–88.
—— (1997) 'Prospects for a Liberatory Political Conception of Nature', Paper presented at the Environmental Justice Conference, Melbourne, 1–3 October 1997.
Rose, D. (1991) *Hidden Histories: Black Stories from Victoria River Downs, Humbert River, and Wave Hill Stations*, Canberra: Aboriginal Studies Press.
—— (1992) *Dingo Makes Us Human*, Cambridge: Cambridge University Press.
—— (1996a) *Nourishing Terrains: Australian Aboriginal Views of Landscape and Wilderness*, Canberra: Australian Heritage Commission.
—— (1996b) 'Rupture and the Ethics of Care in Colonised Space', in T. Bonyhady and T. Griffith (eds) *Prehistory to Politics: John Mulvaney, the Humanities and the Public Intellectual*, Melbourne: Melbourne University Press.
Salleh, A. (1992) 'The Ecofeminism/Deep Ecology Debate: A Reply to Patriarchal Reason', *Environmental Ethics* 14: 195–216.
Shiva, V. (1993) 'Impoverishment of the Environment: Women and Children Last', in M. Mies and V. Shiva (eds) *Ecofeminism*, Melbourne: Spinifex.
Sutton, P. and Rigby, B. (1982) 'People with "Politicks": Management of Land and Personnel on Australia's Cape York Peninsula', in N. Williams and E. Hunn (eds) *Resource Managers: North American and Australian Hunter-Gatherers*, Canberra: Aboriginal Studies Press.

Turner, D. (1989) *Return to Eden: A Journey Through the Promised Landscape of Amagalyuagba*, New York: Peter Lang.

Warren, K. (1990) 'The Power and the Promise of Ecological Feminism', *Environmental Ethics* 12: 125–46.

—— (1997) 'Situated Universalism and Care-Sensitive Ethics', Paper presented at the Environmental Justice Conference, Melbourne, 1–3 October 1997.

12 Ecological ethics from rights to recognition

Multiple spheres of justice for humans, animals and nature

Val Plumwood

Introduction

We have been invited to address the question whether concepts of justice can and should be mapped on to non-humans. I think the answer to this question must be affirmative, but it is a complex affirmation that is required. My argument is a two-pronged one: I will argue, against those who insist we must confine ethical concepts to the human, that we can map a range of ethical stances that are components of justice on to non-human nature, and that there are important insights to be gained from doing so and to be lost from refusing to do so. But we face choices between different ways to make such mappings. I will also argue against closed mappings of justice that try to confine ethics to sentient or conscious beings, recognizing only those non-humans who are believed most closely to resemble humans. These positions may avoid the most extreme and blatant forms of species injustice, but they retain most of the problems of moral dualism and do little to help us change our perceptions or behaviour in ways relevant to the environmental crisis. For this diverse and complex terrain, we need to recognize multiple overlapping spheres of justice, rather than attempting to map on to the non-human sphere a single human-based concept of justice such as rights or utility (for accounts of multiple over-lapping spheres of justice see Midgley, 1983, and for the human case Walzer, 1983).

Non-human justice and moral dualism

We should not begin this inquiry with the assumption that we start from a condition of *tabula rasa*, that we have no conceptual mappings already or that they are neutral. On the contrary, we start out from a tradition that has consistently mapped non-humans on to human Others, and accorded both less than justice. Dominant traditions over at least twenty-five centuries have identified the human normatively with the rational, and both the non-human and the human Other with relative absence of reason and corresponding proximity to nature and the earth. Women have been consistently identified with lack of reason and with animals, and, by Hegel (1952: 263), with a plant-like form of existence. The humanistic revolution of the Enlightenment replaced the rational hierarchy built on a complex set of reason–nature dualisms with a simpler and starker mental and moral dualism between humans and non-humans. In the Cartesian mind–body dualism, for example, non-humans are hyper-separated from humans by their alleged lack of 'thought', and are subject to an extreme form of homogenization which consigns them uniformly to the same inconsiderable category as the least considerable and most instrumentalized among them, which for Descartes was the machine. Modern conceptions of nature, even those of supposedly liberatory versions of environmental ethics, have not fully broken with these traditions of human and rational supremacy, although they minimize our ability to render justice and our sensitivity to the Other, human and non-human.

Yet breaking from these traditions is what we must do in the interests of both justice and prudence. In the present context of ecological destruction, it would be wise for us to adopt philosophical strategies and methodologies that maximize our sensitivity to other members of our ecological communities and openness to them as ethically considerable beings, rather than ones that minimize ethical recognition or that adopt a dualistic stance of ethical closure that insists on sharp moral boundaries and denies the continuity of planetary life. The closed ethical strategy of moral dualism, I will show, appeals mostly for the wrong reasons, because it repeats a familiar but ambiguous political gesture, that of slightly expanding a privileged group while continuing or intensifying exclusion of a group of Others. The economic rationality of capital-ism, whose major defining features are the identification of rationality with egoism and competition and the related concern to maximize economic growth and property formation, is based on moral dualism and is a major support base for strategies that minimize ethical recognition of the other than human world. A closed ethical stance that minimizes the class of beings subject to ethical treatment at the same time maximizes the class of other beings that are available to be treated without ethical constraint as resources or com-modities. Since it is complicit in minimizing the beneficiary class, such a stance helps sanction injustice – the channelling of the world's ecological wealth for the benefit of fewer and fewer organisms. The prudential inappropriateness of these strategies and their destructive ecological effects are increasingly evident.

Questions of justice for non-human nature – including the question of ethical recognition and the critique of human-supremacist or anthropocentric values and ethical standards – have been intensely debated over the past three decades of environmental philosophy. I am among those environmental philosophers who say that Western culture is locked into an ecologically destructive form of rationality which is human-centred, or 'anthropocentric', treating non-human nature as a sphere of inferior and replaceable 'Others' (Plumwood, 1996). Human supremacism and anthropocentrism are incom-patible with justice to other species. Human supremacism in its strongest forms refuses ethical recognition to non-humans, treating nature as merely a resource we can make use of however we wish. It sees humans, and only humans, as ethically significant in the universe, and derives those limited ethical constraints it admits on the way we can use nature and animals entirely indirectly from harms to other humans. But just as other forms of supremacism and centrism, for example those based on race and gender, appear in various guises, so there are weaker and stronger, more obvious and more subtle forms of human supremacism and human-centredness.

The epistemic and moral dualisms associated with human-centredness are, I shall argue, harmful and limiting, even in their subtler and weaker forms. People under their influence, such as those from the Western cultural traditions in which anthropocentrism is deeply rooted, develop conceptions of themselves as belonging to a superior sphere apart, a sphere of 'human' ethics, technology and culture dissociated from nature and ecology. This self-enclosed outlook has helped us to lose touch with ourselves as beings who are not only cultural but also natural, embedded in the earth and just as dependent on a healthy bio-sphere as other forms of life. Thus we can come to take the functioning of the 'lower' sphere, the ecological systems which support us, entirely for granted, needing some grudging support and attention only when they fail to perform as expected. The moral dualist assumption that ethics and value are exclusively concerned with and derived from the human sphere is an ethical expression of this same project of human supremacy and self-enclosure (see Abram, 1996). Although these ethical exclusions have in the modern age helped Western culture achieve its position of dominance, by maximizing the class of other beings that are available as 'resources' for exploitation without constraint, they are now, in the age of ecological limits, a danger to the survival of human and planetary life.

Despite our contemporary context of accelerating human destruction of the non-human world, some philosophers and traditionalists have been reluctant to censure even strong forms of human supremacism. Thus according to what is perhaps the strongest recent statement of this kind, that of William Grey (1993), there is no call for an ethic which extends moral consideration beyond the human sphere, even to animals. (For other defenders of anthropocentrism see Passmore, 1974; Mannison, 1980; Norton, 1991; Thompson, 1990.) Others are critical of these strong forms, but nevertheless cling to subtler forms which remain anthropocentric and are overly restrictive in their ethical recognition of non-humans.

The most human-like 'higher animals', which are claimed to be the only possessors among the non-humans of the supposedly defining human charac-teristic of awareness (Singer, 1997), may be admitted to the ethical sphere, but the door is firmly closed against all others. This strategy is aptly termed 'Minimalism'. It aims to enlarge the human sphere of justice rather than ethically to integrate human and non-human spheres, a strategy which, as I argue below, must result in minimal admissions to the privileged class. It minimally challenges anthropocentric ranking regimes that base the worth of a being on their degree of conformity to human norms or resemblance to an idealized 'rational' or 'conscious' humanity; and it often aims explicitly at minimal deviations from the prevailing political assumptions and dominant human-centred ethic into which they are tied.[1] It tends to minimize recognition of diversity, focusing on ethically relevant qualities like mind, consciousness and communication only in forms resembling the human and failing to recognize that they can be expressed in many different, often incommensurable forms in an ethically and ecologically rich and diverse world (see Rogers, 1997). I contrast this Minimalist ethical stance of closure with a more generous ecojustice stance of openness and recognition towards non-humans which acknowledges ethical diversity and critiques anthropocentric moral dualism as the 'Othering' of the non-human world, a form of injustice that parallels racial and gender injustice in making the Other radically less than they are or can become.

Moral dualism makes an emphatic division of the world into two starkly contrasting orders, consisting of those privileged beings considered subject to full-blown ethical concern as 'humans' or 'persons', and the remainder, considered beneath ethical concern and as belonging to an instrumental realm of resources (or, in the prevailing political context, of 'property') available to the first group. Both the positions I recommend rejecting in this chapter are moral dualisms; that is, the traditionalist human-supremacist position that refuses any extension of ethics beyond the class of humans, and also the minimalist animal rights variation on this that refuses any extension of ethics beyond the class it considers conscious. Typically, moral dualism organizes moral concepts so that they apply in an all-or-nothing way: for example, a being either has a full-blown 'right' to equal treatment with humans, or it is not subject to any form of ethical consideration at all. As I will show below, there are good reasons to reject moral dualism. We have many opportunities to organize the ethical field differently; some ethical concepts and practices of recognition and justice, for example, can be applied to humans and also to non-human animals and nature more generally. A relatively uncontroversial example is the ethical practice of care or guardianship, which should not be interpreted as a purely instrumental practice for either humans or non-humans (see Plumwood, 1999). And ethically relevant qualities such as mind, communication, consciousness and sensitivity to others are organized in multiple and diverse ways across life forms that do not correspond to the all-or-nothing scenarios assumed by moral dualism.

In both the human and the non-human case, a politics of conflict can be played out around these moral dualisms, in which the moral exclusion of the class defined as 'resource' is represented as a benefit to or even a moral duty of less fortunate members of the 'person' class, and the rejection of moral dualism is represented as depriving persons of resources that are rightfully theirs.[2] Much humanist rhetoric has involved policing exaggerated boundaries of moral considerability and forming a pan-human identity akin to racist and macho (male-bonding) identities, building solidarity within the human group through creating an inferiorized non-human out-group of 'Others' that the pan-human identity is defined against. The exclamation 'What are we – animals? – to be treated like this!' both implicitly appeals to such an identity and implies that ill-treatment is appropriate for animals. Moral dualism helps to construct concern for non-human nature in this conflictual way, as a deficit of attention or concern for some less privileged group, although the remorseless conflict scenario this assumes can usually be reconceived in complementary rather than competitive ways.[3] As in the case of conflicts within the sphere of human justice, we have, I believe, an overriding, higher-order obligation to try to circumvent and reduce or eliminate such justice conflicts where possible, and to avoid multiplying and reinforcing them. This translates into an obligation to favour, where they are available, complementary over competitive constructions of justice spheres, other things being equal.

We need then to attend to the ways in which the human and non-human spheres of justice, although not free of some limited and sometimes manufactured conflicts of this kind, can be constructed not as competitive but as complementary approaches which need and strengthen each other. Thus we should note that moral dualism is also a moral *boomerang* which too often returns to strike down humanity itself when allegedly 'lower' orders of humans are assimilated to nature and to animals, as they have been systematically throughout Western history. Conversely, many forms of ethical practice and sensitivity to others are not only not especially sensitive to whether these others are human or non-human, but can actually be strengthened and deepened generally when we refuse the arbitrary exclusion of non-human others and the self-impoverishment and blunting of sensibilities exclusion involves. That is one reason why opening to and caring for non-human others can be a general ethical learning, healing and development practice for both children and adults (especially wounded individuals). When we act to reject moral dualism, we can open up ways to reflect critically and sympathetically on the ethical status of the act of exclusion itself, on our own identities, and on ethical practices of boundary-breaking; these meta-lessons are among the most important for human and non-human spheres of justice alike.

One reason for rejecting moral dualism is that its stance of closure unnecessarily blunts our sensitivity to the excluded class and those assimilated to them, and this can involve prudential hazards as well as injustices. It is in our interests as well as the interests of the Other to adopt a less impoverished ethical stance and view of the Other. Thus, by refusing recognition to nature we lose not only an ethically but also a prudentially crucial set of connections which link human and non-human movements for liberation and justice. By blunting our sensitivity to nature and animals we lose a prudentially important set of insights which can help us to reflect on our limitations as human actors and observers and correct crucial blindspots in our relationships with the more-than-human world. Further, the attempt to articulate various forms of recognition for nature and to counter anthropocentrism is important for practical activism in a number of ways, and also affects the way political alliances between groups can be formed. Such a recognition is crucial for the birth of the new communicative and care paradigm for the human–nature relationship which must now, in an age of ecological limits, take the place of the mechanistic paradigm associated with the past centuries of human expansion and conquest.

Mapping the Othering of nature

Human supremacism has relied on a range of conceptual strategies of injustice which have been employed also within the human sphere to support supremacism of nation, gender and race – especially the conception of nature and animals as inferior 'Others' to the human. Chief among these are Otherization strategies that construct a major boundary or gulf which cannot be bridged or crossed, for example that between an elite, morally considerable group and an out-group defined as 'mere resources' for the first group that need not or cannot be considered in similar ethical terms. In the West especially, this gulf is usually established by constructing non-humans as lacking in the department Western rationalist culture has valued above all else and identified with the human, that of mind, rationality or spirit, or what is often seen as their outward expression: language and communication. The excluded group is conceived instead in the reductive terms established by mind–body or reason–nature dualism, as mere bodies, and thus as servants, slaves, tools or instruments for human needs and projects. This construction remains common today, especially among scientists.

Similarly, centric and reductionistic modes of conception as Other continue to thrive. The forms I outline below are the precursors of many forms of injustice in our relations with non-humans, preventing the conception of non-human Others in ethical terms, distorting our distributive relationships with them, and legitimating insensitive commodity and instrumental approaches. Distributive injustices to non-humans include the use of so much of the Earth for human purposes that they cannot survive or reproduce their kind. Not only do these modes of conception herald other forms of injustice, there is injustice in each of these denials and reductive modes of conception themselves, an injustice which varies in character according to the kind of being. There is injustice for any communicative and ethical being in being treated systematically in ways that refuse recognition of their capacity for communication, both within and between species. There is injustice for any striving and intentional being in being constructed in reductive terms as an ethical nullity in the fashion of Singer, or as 'mere' body or mechanism, because such construction opens its referent to treatment as radically less than it is.[4] The framework of moral dualism blocks recognition of these injustices from within. Ecofeminists have articulated a framework for recognizing these anthropocentric injustices as parallel to those we recognize for humans in the cases of human Othering that appear in contexts of colonization, in androcentrism and also in ethno/ Eurocentrism.

Radical exclusion

We meet here first *hyper-separation,* an emphatic form of separation that involves much more than just recognizing difference. Hyper-separation means defining the dominant identity emphatically against or in opposition to the subordinated identity, by exclusion of their real or supposed qualities. The function of hyper-separation is to mark out the Other for separate and inferior treatment. Thus 'macho' identities emphatically deny continuity with women and try to minimize qualities thought of as appropriate for or shared with women. Colonizers exaggerate differences (for example through emphasizing exaggerated cleanliness, 'civilized' or 'refined' manners, body covering or alleged physiological differences) between what are defined as separate races (see Gould, 1981). They may ignore or deny relationship, conceiving the colonized as less than human. The colonized are described as 'Stone-Age', 'primitive', as 'beasts of the forest', and contrasted with the civilization and reason attributed to the colonizer.

Similarly, the human 'colonizer' treats nature as radically Other, and humans as emphatically separated from nature and from animals. From an anthropocentric standpoint, nature is a hyper-separate lower order lacking any real continuity with the human. This approach tends to stress heavily those features which make humans different from nature and animals, rather than those they share with them, as constitutive of a truly human identity. Anthropocentric culture often endorses a view of the human as outside of and apart from a plastic, passive and 'dead' nature which lacks agency and meaning. A strong ethical discontinuity is felt at the human species boundary, and an anthropocentric culture will tend to adopt concepts of what makes a good human being which reinforce this discontinuity by devaluing those qualities of human selves and human cultures it associates with nature and animality. Thus it associates with nature inferiorized social groups and their characteristic activities; women are historically linked to 'nature' as reproductive bodies, and through their supposedly greater emotionality. Indigenous people are seen as a primitive, 'earlier stage' of humanity. At the same time, dominant groups associate themselves with the overcoming or mastery of nature, both internal and external. For all those classed as nature, as Other, identification and sympathy are blocked by these structures of Othering.

Homogenization/stereotyping

The Other is not an individual but a member of a class stereotyped as interchangeable, replaceable, all alike, homogeneous. Thus essential female and 'racial' nature is uniform and unalterable (see Stepan, 1993). The colonized are stereotyped as 'all the same' in their deficiency, and their social, cultural, religious and personal diversity is discounted (Memmi, 1965). Their nature is essentially simple and knowable – unless they are devious and deceptive, not outrunning the homogenizing stereotype (Said, 1979). Homogenization is a major feature of pejorative slang, for example in talk of 'slits', 'gooks' and 'boongs' in the racist case, and in similar terms for women.

The famous presidential remark (allegedly made by Ronald Reagan), 'You've seen one redwood, you've seen them all', invokes a parallel homogenization of nature. An anthropocentric culture rarely sees nature and animals as individual centres of striving or needs, doing their best in their conditions of life. Instead nature is conceived in terms of interchangeable and replaceable units, as 'resources', rather than as infinitely diverse and always in excess of knowledge and classification. Anthropocentric culture conceives nature and animals as all alike in their lack of consciousness, which is assumed to be exclusive to the human. Once they are viewed as machines or automata, minds are closed to the range and diversity of their mindlike qualities. Human-supremacist models promote insensitivity to the marvellous diversity of nature, since they attend to differences in nature only if they are likely to contribute in some obvious way to human interests, conceived as separate from nature. Homogenization leads to a serious underestimation of the complexity and irreplaceability of nature. These two features of human–nature dualism, radical exclusion and homogenization, work together to produce in anthropocentric culture, a polarized understanding in which the human and non-human spheres correspond to two quite different substances or orders of being in the world.

Backgrounding and denial

Women's traditional tasks in house labour and childraising are treated as inessential, as the background services that make 'real' work and achievement possible, rather than as achievement or as work themselves. The colonized have been denied as the unconsidered background to 'civilization', the Other whose prior ownership of and agency and trace in the land is denied, represented as inessential as their land and their labour embodied in it is taken over as 'nature'.

Nature is represented as inessential and massively denied as the unconsidered background to technological society. Since anthropocentric culture sees non-human nature as a basically inessential constituent of the universe, nature's needs are systematically omitted from account and consideration in decision-making. Dependency on nature is denied, systematically, so that nature's order, resistance and survival requirements are not perceived as imposing a limit on human goals or enterprises. For example, crucial biospheric and other services provided by nature and the limits they might impose on human projects are not considered in accounting or decision-making. We only pay attention to them after disaster occurs, and then only to fix things up. Where we cannot quite forget how dependent on nature we really are, dependency appears as a source of anxiety and threat, or as a further technological problem to be overcome.

Incorporation

In androcentric culture, the woman is defined in relation to the man as central or normal, often conceived as a 'lack' in relation to him. The colonized too is judged not as an independent being or culture, but as an 'illegitimate and refractory foil' to self (Parry, 1995), as negativity or *lack* in relation to the colonizer (Memmi, 1965), devalued as an absence of the colonizer's chief qualities ('backward', 'lack of civilization'), usually represented in the West as Reason. Differences are judged as deficiencies, grounds of inferiority. The speech, voice and culture of the colonized is recognized or valued only to the extent that it is assimilated to that of the colonizer. Similarly, rather than according nature the dignity of an independent Other or presence, anthropocentric culture treats nature as Other as merely a refractory foil to the human. Defined in relation to the human or as an absence of the human, nature has a conceptual status that leaves it entirely dependent for its meaning on the 'primary' human term.

Thus nature and animals are judged as 'lack' in relation to the human/colonizer, as negativity, devalued as an absence of qualities said to be essential for the human, such as rationality. We consider non-human animals inferior because they lack, we think, human capacities for abstract thought, but we do not consider those positive capacities many animals have that we lack – remarkable navigational capacities, for example. Differences are judged as grounds of inferiority, not as welcome and intriguing signs of diversity. The intricate order of nature is perceived as disorder, as unreason, to be replaced where possible by human order in development, an assimilating project of colonization. Where the preservation of any order there might be in nature is not perceived as representing a limit, nature is available for use without restriction.

Instrumentalism

In anthropocentric culture, nature's agency and independence of ends are denied, subsumed in or remade to coincide with human interests, which are thought to be the source of all value in the world. Mechanistic world views especially deny nature any form of agency of its own. Since the non-human sphere is thought to have no agency of its own and to be empty of purpose, it is thought appropriate that the human/colonizer impose his own purposes. Human-centred ethics views nature as possessing meaning and value only when it is is made to serve the human/colonizer as a means to his or her ends. Thus we get the split characteristic of modernity in which ethical considerations apply to the human sphere but not to the non-human sphere. Since nature itself is thought to be outside the ethical sphere and to impose no moral limits on human action, we can deal with nature as an instrumental sphere, provided we do not injure other humans in doing so. Instrumental outlooks distort our sensitivity to and knowledge of nature, blocking humility, wonder and openness in approaching the more-than-human, and producing narrow types of understanding and classification that reduce nature to raw materials for human projects.

Injustice does not take place in a conceptual vacuum, but is closely linked to these desensitizing and Othering frameworks for identifying self and other. The centric structure imposes a form of rationality, a framework for beliefs, which naturalizes and justifies a certain sort of self-centredness, self-imposition and dispossession, which is what Eurocentric and ethnocentric colonization frameworks as well as androcentric frameworks license. The centric structure accomplishes this by promoting insensitivity to the Other's needs, agency and prior claims as well as a belief in the colonizer's apartness, superiority and right to conquer or master the Other. This promotion of insensitivity is in a sense its function. Thus it provides a wildly distorted framework for perception of the Other, and the project of mastery it gives rise to involves dangerous forms of denial, perception and belief which can put the centric perceiver *out of touch* with reality about the Other. Think, for example, of what the Eurocentric framework led Australian colonizers to believe about Aboriginal people: that they had no religion, had a single culture and language, were ecologically passive 'nomads' with no deep relationship to any specific areas of land, and so on. The framework of centrism does not provide a basis for sensitive, sympathetic or reliable understanding and observation of either the Other or of the self. Centrism is (it would be nice to say 'was') a framework of moral and cultural blindness.

To counter the Othering definition of nature I have outlined, we need a depolarizing reconception of non-human nature which recognizes the denied space of our hybridity, continuity and kinship, and is also able to recognize, in suitable contexts, the difference of the non-human in a non-hierarchical way. Such a nature would be no mere resource or periphery to our centre, but another and prior centre of power and need, whose satisfaction can and must impose limits on our own conception of ourselves, and on our own actions and needs. The nature we would recognize in a non-reductive model is no mere human absence or conceptually dependent 'Other', no mere pre-condition for our own star-stuff of achievement, but is an active collaborative presence capable of agency and other mindlike qualities.[5] Such a biospheric other is not a background part of our field of action or subjectivity, not a mere pre-condition for human action, not a refractory foil to self. Rather, biospheric others can be other subjects, potentially ethical subjects, and other actors in the world – ones to which we owe a debt of gratitude, generosity and recognition as prior and enabling presences.

The reconception of nature in the agentic terms that deliver it from construction as background is perhaps the most important aspect of moving to an alternative ethical framework, for backgrounding is perhaps the most hazardous and distorting effect of Othering from a human prudential point of view. When the other's agency is treated as background or denied, we find also various failures of justice: we give the other less credit than is due to them, we can come to take for granted what they provide for us, to pay attention only when something goes wrong, and to starve them of resources. This is a problem for prudence as well as for justice, for where we are in fact dependent on this other, we can gain an illusory sense of our own ontological and ecological independence, and it is just such a sense that seems to pervade the dominant culture's contemporary disastrous misperceptions of its economic and eco-logical relationships. So one of several severe defects of the minimalist frame-work I now discuss is that it retains backgrounding for the great majority of the individuals, species and systems of nature it does not count as conscious, treating them as mere pre-conditions for consciousness, an insentient Other defined entirely relatively to the sentient One as the 'ground and home of con-scious life'. Feminists especially have a good idea of what it means for one organism to be regarded as a background condition for the comfort and pro-vision of another life, rather than as having some claim to these good things oneself on behalf of one's own life.

The Otherization of nature bears on a key question of justice: the concern with obstacles to justice, especially forms of partiality and self-imposition that prevent us giving others their due (on the question of partiality as an obstacle to justice see Marilyn Friedman, 1993). The approach to justice suggested by the analysis of anthropocentrism is one of studying up rather than studying down, shifting the onus of proof from inclusion to exclusion and moving the ethical focus from the evaluated item and the dubious question of their 'qualifications' for ethical inclusion and attention (studying down) to the differ-ent and largely neglected question of the ethical stance of the human evaluator (studying up) and its own moral status. What requires critical philosophical engagement in the context of anthropocentric culture is self rather than other, the limits imposed by the human rather than the nature side of the ethical rela-tionship, the ethical stance of closure rather than the ethical stance of openness.

Minimalist methodologies of closure

A major project for a non-anthropocentric form of culture would be that of developing ethical and epistemic frameworks that can give non-humans a non-derivative, non-secondary or non-instrumental place. I do not think that this ethical project can be carried out in terms of the Minimalist programme out-lined by Peter Singer (this volume, Chapter 9, and Singer, 1975) or is best expressed in terms of the concept of rights developed by Tom Regan (this volume, Chapter 10, and Regan, 1983). I share with Singer and Regan an active opposition to the dominant humanistic assumption that ethics is effec-tively confined to the human sphere; the historic exclusion of non-humans from ethics needs to be strongly contested, and I appreciate the vigorous argu-ments they have put forward over a long period to this effect. I think that a case can be made for extending the kinds of ethical treatment appropriate to persons to our nearest relatives in the primate family, and that it is imperative to extend many elements of human-style ethics to other sentient beings (on the extension to primates see Goodin *et al.*, 1997[6]). But I do not see that the possibility of such an extension provides any good reason why we should maintain the stance of closure towards other non-humans. An effective challenge to moral dualism entails recognizing the continuity of all life forms and contesting the full framework of human–nature dualism, involving the ethical exclusion not only of animals but of nature itself.

Singer (this volume, Chapter 9) explicitly advocates Minimalism: a minimal extension of recognition to a few animals most like we humans, and a Minimalist methodology urging minimal departures from the status quo of humanism. Both Regan's rights approach and Singer's utilitarianism are strongly associated with a sentience-reduction position that would make a minimal extension of human-style ethics and liberal rights to a few species of animals that are most like us, that qualify as 'persons'. Every being outside the 'person' category remains potential 'property', the dualistic contrast class which is capable of being owned and treated instrumentally. Minimalism claims to be anti-speciesist but is not genuinely so in selecting for exclusive ethical attention those animals that closely resemble the human, any more than a culture which values women just in terms of their resemblance to men is genuinely non-androcentric. Minimalism continues to see consciousness in singular and cut-off terms, and discounts the great variety of forms of sentience and mind – hence Singer's conviction that trees have no form of sentient or aware life, which runs counter both to what is disclosed by any reasonably attentive observation and to scientific evidence. Minimalism is not able to recognize consciousness as just one among many relevant differences among species, differences which are largely incommensurable as to value rather than hierarchically ordered along the lines of resemblance to the human.[7] Rather, Minimalism makes consciousness the basis for an absolute ethical positioning of all species within a hierarchy based on human norms. Minimalism does not really dispel speciesism, it just extends and disguises it.

Philosophically, Minimalism attempts a minimal deviation from rationalism and Cartesianism – where rationalism has as its central tenet the doctrine that the rational (usually identified with elite humans) is the only thing that fundamentally counts in the universe. The Cartesian division of the world was based on the assumption that the criterion of consciousness (which Descartes equated, through various equivocations on the concept of 'thought', with calculative reason) picked out just the class of the human. Animals were automata entirely lacking consciousness.[8] The sentience-reduction position minimally corrects Descartes' error of identifying consciousness with the human, but it retains both the exclusion of the original rationalist doctrine that reason is all that matters in the universe, and retains the same logic of exclusion. It shifts the boundary to a new point but still leaves far too much outside, and has the same intense emphasis on the need for a boundary between what counts and what does not.[9] Thus the ethical closure characteristic of Minimalism accepts the Cartesian overemphasis on consciousness (which derives from the Cartesian interpretation of rationality as the self-transparency of the rational subject), and declares with a similar exclusiveness of focus that 'consciousness in all its forms is what fundamentally counts in the universe' (Taylor, 1996: 260). This claim is objectionable on many counts, one of which is its identification of a *single* end or source for all value (a defect it shares with utilitarianism). A moral dualist approach, it gives some higher animals access to moral consideration on a par with humans, but leaves the great majority still outside in the cold, subject to continued instrumentalization as the 'ground and home of conscious life'.

Although, as Singer notes (this volume, Chapter 9), rationalism has traditionally been used to exclude slaves – and of course many other subordinated humans such as women, along with animals – from the category of the fully human, and to interpret this 'human' category in elitist ways, Singer is content with a very minimal dislocation of these rationalist assumptions which leaves most of non-human nature just where it was before, in a sphere of ethical exclusion.[10] Singer defends his exclusion of allegedly non-conscious vegetable life with the claim that 'if a tree is not sentient, it makes no difference to the tree whether we chop it down or not' (Singer, 1997, this volume, p. 146). But if a tree is a striving (teleological) and adaptive being, it must select some states to strive for (for example, life, best development) and others to avoid (death). There is an important sense therefore in which death does 'matter' or 'make a difference' to a tree, and to any living being, since it is not the case that all states are indifferent to it. Singer's claim incorrectly and arbitrarily ties the concept of 'making a difference' to consciousness rather than to teleology and intentionality, thus producing an unnecessarily dualized account of the field. Elsewhere he states of plants, 'Such a life is a complete blank; I would not in the least regret shortening this subjectively barren form of existence' (Singer, 1980). Singer claims to acknowledge continuity, but as these statements make clear, Singer's methodology organizes the diverse and continuous field of mind and ethics into a polarized, simple 'on–off' form – that of consciousness or nothing (a 'complete blank', 'no conscious experience at all'), in just the same way as Descartes.

But we do not have to choose such a dualistic framework. We can adopt towards plants, for example, the more graduated and diversity-sensitive Intentional Stance, recognizing them as organized, intentional and goal-directed beings which value their own lives and strive to preserve them in a variety of challenging circumstances, and opening to the many possibilities for intentional construction and explanation they can present to us. Even if we accept their non-consciousness (in the terms of Singer's narrow and singular concept of awareness), this does not exclude an ethical approach to plant life. We can and should, I think, have great respect and reverence for plants and be grateful to them for the many ways in which they support our lives. This means, among other things, that we must never count their lives for nothing, or treat their deaths or destruction lightly or casually as of no consequence or significance.[11] We can honour them both individually, and in species and ecological community terms as great time-travellers and teachers, and be open to and grateful for the wisdom they have to give us. We do them an injustice when we treat them as less than they are, destroy them without compunction, see them as nothing more than potential lumber, woodchips or fuel for our needs (a form of incorporation), fail to attend adequately to them, radically dissociate from them and deny their organization as intentional (and perhaps communicative) beings, thus adopting the stance of ethical closure or dismissal.

Singer's Minimalism is also a political position urging minimal departure from prevailing liberal, humanistic and Enlightenment assumptions and from the present system of economic rationality (this volume, Chapter 9, p. 156).[12] But surely an ecological society will require more than minimal departures from these systems, none of which have been innocent bystanders in the development of the rational machinery which is bringing the stripping of the planet for the benefit of a small elite of humans to a high point of rational refinement. Indeed, Singer concedes that dominant Western traditions are the source of the problem, even as he delegitimates any exploration of conceptual alternatives to them. From the standpoint of spokesperson for animal liberation, he endorses the current attempt to identify 'shallow' forms of ecology with humanism and a progressive affirmation of modernism, and to blacken any deeper form of ecological thought which challenges the dualism of humans and nature in a more thoroughgoing way through an alleged association of deep ecology with Romanticism, anti-modernism, anti-humanism and even Nazism. But the crudeness of such an analysis is apparent when we consider that all these positions have multiple faces which must render such simple equations suspect. The analysis ignores the oppressive aspects of humanism and modernity, the diversity and liberatory aspects of some forms of Romanticism, and the well-documented complicity of the worst aspects of Nazism with modernism and rationalism (see Baumann, 1989; Proctor, 1993).

Politically, the gesture of Minimalism reproduces the polarizing Cartesian double gesture characteristic of moral dualism. As we have seen, Descartes replaced the graduated (although blatantly reason-centred and hierarchical) framework inherited from Aristotelian rationalism with a more humanistic and apparently 'democratic' one that reorganized the field in terms of greater equality among the privileged class of humans, extending it to include all humans equally and freeing it from the classical rationalist hierarchy that positioned women as less rational than men and placed slaves alongside domestic animals as minimally rational and considerable beings. However, the Cartesian double gesture at the same time offset this first revolutionary gesture of progressive humanism by a second gesture that intensified the moral dualism between humans and their homogenized contrast category of nature. Singer retains this double gesture of equality and exclusion which opens the door to some, at the price of renewed exclusion and homogenization of others; his moral dualism extends the privileged class slightly to include some animals as persons, but otherwise leaves the Cartesian framework minimally disturbed.

The Minimalist double gesture is a complex and familiar political gesture combining inclusion and exclusion comparable in many ways to those forms of feminism which aimed at admitting white women to the class of political

rights-holders on the basis of their resemblance to white men and their supposed dissimilarity and superiority to excluded categories such as black people
(see Plumwood, 1995a, b). Such forms of polarizing recognition via assimila-
tion to the elite group are of necessity highly limited in the class to which they
can be extended. To the extent that they do not challenge or loosen up the
dualistic criteria for inclusion in the privileged group, but only squeeze more
items into it, their ultimate effect must be an enlargement of the elite, and a
retention of conceptual strategies of erasure and denial for excluded groups,
instead of the recognition of the kinship of all living things in ethical consideration and biological exchange we need for a truly ecological ethic. As many
feminists have urged, we need to learn some new political gestures, including
some non-polarizing and unambiguously inclusionary ones that affirm non-
hierarchical difference and acknowledge solidarity through both diversity and
continuity (refer to the essays in Bok and James, 1992).

Regan, rights and vegetarian justice

To the person in the street, the concept of animal rights is synonymous with a
challenge to the idea that animals do not count ethically, but to philosophers
the concept of rights carries a lot of additional and more complex baggage.
Even environmental philosophers who have been clear about wanting to challenge human supremacism have been sharply divided about rights: although
some, especially Tom Regan, have mounted their challenge in terms of the
rights framework, many others have found the extension of the rights framework to the non-human world problematic. I am inclined to join them in
doubting that the concept of rights offers the best way to challenge the framework I have been discussing and to include non-humans in the picture. The
rights framework has rhetorical appeal but has many limitations as an activist
concept, even for the case of the treatment of animals, that is capable of focusing and theorizing the widespread popular opposition to injustice to animals.
The fact that the rights concept is difficult to apply beyond the animal sphere
means that it tends to support moral dualism.

Serious problems which afflict animal rights theory include its tendencies to
excessive individualism, cultural universalism and moral dualism. These faults
are not only philosophical but political weaknesses. They limit its application
for an ecological ethic, limit its effectiveness for animal justice activism, and
also for evolving theory that is able to integrate animal justice concerns with
those of environmental and human justice movements. These tendencies often
surface in rights-based arguments for an absolutist, uncontextualized form of
vegetarianism as the principal movement strategy, one in which a focus on the
carnivorousness of the individual consumer (sometimes it seems as a form of
original sin) takes the place of a more multiple, contextually and politically
sensitive set of strategies and a stronger focus on the responsibility of cultural,
political and economic systems. Rights theory as normally interpreted supports
a closed rather than an open ethical stance, and maintains person–property

dualism; on the rights and 'subject of a life' criteria, most non-human animals – and certainly all of nature – are just as strongly excluded from ethical consideration and just as open to exploitation as they were under the Cartesian dispensation.

The individualistic tendencies of rights theory emerge from the way it takes a complex set of social and political conditions and congeals them into a singular and apparently simple and stereotypical quality of an individual whose rights are violated by another individual who is responsible for their violation. As Benton (1993: 92) notes, rights theory proceeds as if 'the moral status of animals were a function of the kinds of beings they are, independently of the diverse relationships in which they stand to human moral agents and their social practices'. It is not the focus on the integrity and autonomy of the individual right-holder which is itself problematic here so much as the disincentive this framework provides for any more complex and structural social thinking. This is one reason why the rights approach has been able to turn attention away from the structural origins of the atrocities daily committed against animals in the factory farm and commodity framework to the question of the virtue of individual consumers who make a choice of eating meat, no matter how obtained. It is also one reason why the rights focus, as Benton also notes, tends to be practically not very effective in the many contexts which cannot be reduced to this formula, for example cases of moral conflict. In the context of the ecological community with its mixed relations of carnivores and others, conflicts between rights to eat and rights not to be eaten are in fact the standard case. The politics of individual virtue which issues from the rights approach also promotes an excessively polarizing focus on absolute abstention from rather than reduction of meat consumption, even though, as Mary Midgley (1983) notes, the former is likely to be less effective than the latter in actually reducing animal suffering.

The cultural absolutism fostered by the rights approach and openly advocated by leading rights theorists also supports the vegetarian cultural imperialism and ethnocentrism which assimilates all planetary meat-eating practices to those of North American grain feeding and its alternatives, and is insensitive to the culturally variable practices and meanings involved in the use of other animals as food. Although the points about eating lower down the food chain and the focus on land-use systems are important, absolutist vegetarians quote as decisive and universally applicable statistics drawn from the North American context, comparing the ecological costs of meat and grain eating, a comparison which is supposed to dispose of the problem of conflict between animal rights and ecological ethics. Such universalist comparisons assume that grain production for human use is always virtually free of ecological costs or costs to animal life (whereas it is in many arid-land contexts highly damaging to the land and to biodiversity), and ignore the fact that for much of the world the statistics they cite are largely irrelevant because food animals are not grain-fed (as an example of such an unqualified and ethnocentric approach, see David Waller 1998). Although contextual vegetarianisms that are more sensitive to cultural difference are possible (see Plumwood, 1997), they tend to be ignored because the rights approach fosters an ethical and cultural absolutism which insists that the only alternative to its North American-focused absolutism is an indiscriminate cultural relativism (for an argument that the alternatives of absolutism and relativism present a false choice see Mathews, 1994). The outcome is too often a hegemonic universalism, as the vegetarian imperative is applied everywhere, even to those indigenous peoples for whom hunting is an ecological necessity.

Because they base moral consideration on stringent standards for ethical inclusion that demand qualifications equivalent to those of the human 'subject-of-a-life', standards which cannot easily be extended beyond some higher animals, most rights approaches support a moral dualism similar to Singer's which divides the field in terms of rights or nothing. This division corresponds to a person–property dualism, or a dualism of conscious and non-conscious life in which only the former deserves moral consideration. In principle it seems possible for a position allowing rights to animals to supplement its animal rights morality by some other account of the basis of moral considerability for non-conscious forms of life, and thus to avoid the worst effects of moral dualism. In practice this almost never happens. Especially where rights theorists make a close connection to vegetarianism, they are tied to an exclusionary imperative.

The exclusionary imperative derives from the reliance of rights-based forms of vegetarianism on the (unstated) assumption that only those beings not admitted to the class of rights-holders can ever ethically become food. Given this assumption, arguments for vegetarianism can be based on a simple demonstration that certain animals should be treated as rights-holders, and the vegetarian conclusion follows (with the help of *modus ponens*). But any position which has thus equated availability as food with moral exclusion is thereby committed to an exclusionary imperative, since it is forced to insist on a substantial outclass of living beings that are morally excluded in order to locate any viable form of eating which allows an ethical basis for human survival. Mostly this exclusionary imperative takes the form of insisting on the complete moral exclusion of plants and other allegedly non-conscious forms of life; that is, a position similar to Singer's. The other options that are opened up by rejecting the equation of availability as food with moral exclusion, namely some form of ethical eating of beings within the morally considerable class, remain unexplored. They are not usually consistently available to the rights-based vegetarian because they tend to undermine the apparent simplicity of the rights strategy. Once the equation of food with moral exclusion is abandoned, arguments for vegetarianism become more complex and uncertain. It is no accident, then, that rights-based forms of vegetarianism are so often involved in strong forms of moral dualism that deny the ethical continuity of planetary life.

To sum up: rights positions are essentially also a form of Minimalism, tending to remain trapped in the Western strategies of denial and radical exclusion which create further out-groups, redrawing the boundary of moral dualism in a different place, at the border of animality rather than humanity, rather than dissolving that dualism and taking moral consideration 'all the way down'. Rights accounts have assisted recent animal rights theory in making powerful critiques of the reduction of and injustice to animals involved in human supremacism, and in effectively contesting the dominant human-centred assumption that ethics, mind and communicative ability are confined to the human sphere. But leading forms based on the rights approach have been less successful than one might hope at critiquing their own privileged biases of gender, class and ethnicity, or at evolving theory that is able to integrate its concerns adequately with those of environmental and human justice movements. Rights theory retains a modified form of person–property dualism that shifts the boundary of ethical consideration slightly to enlarge the class of persons, but which leaves the basic logic of human exclusion and hyper-separation unchallenged beyond the new boundary and reproduces the same dynamic of dualistic exclusion with a slightly enlarged base class, the almost-human.

I have argued that such moves to limit and withhold recognition encounter the same kinds of problems as comparable strategies in other liberation movements that aim to have some privileged group from among the class of the excluded join the master group. If some animals are admitted to rights because they are included in the category of persons in recognition of their newly emphasized resemblance to the human and discontinuity from other animals, we can recognize in this the same double gesture some elite forms of liberal feminism have made in arguing that women should be admitted to the privileged class of political rightholders in virtue of their discontinuity with allegedly 'lower groups' such as black slaves, and their similarity to the master group, elite white men. The strategy of extending the category of persons without recasting the person–property dualism in terms of which the person category is constructed is bound to fail as an attempt to elevate animals, for exactly the same reasons that similar liberal feminist strategies were/are bound to fail. The door opens to admit a few, but closes to keep the rest outside where they were. One boundary of moral dualism is momentarily penetrated, but the rest remain in place or new ones are constructed.

Care and the rational economy – linking our strategies

Minimalism and moral dualism are bad politics and bad methodology. They are bad politics because their half-hearted and ahistorical analysis stops short of enabling us to grasp the important connections between forms of oppression that could provide the basis for a common framework of opposition and coalition. They are bad methodology because at the level of intellectual analysis at least, we should determine our stance here not by a mind-shackling principle of minimal departure from the status quo, but in terms of the potential for multiplication of alliances and for articulating a common struggle. This must be a struggle especially against the machinery of rational egoism and self-maximization which, in the form of a maximizing economic rationality of global reach, is now disrupting so catastrophically the fabric of human and non-human life.

The gulf Minimalism retains between those privileged biospheric groups admitted to the ethical sphere, as rights-holders or as 'conscious' beings, and the many who remain instrumentalized as Other, as mere resources for the first group, corresponds in the dominant economic framework to the gulf discerned between the few who can own and those many who can ethically be owned, traded and commodified. The ethical dualism Minimalism insists we must retain is, then, a key element in supporting the rationalist economic regime which is commodifying the world, which Minimalism would also vary only minimally. The regimes of factory farming are the product of this self-maximizing calculus, of the rationalist economy stripped of all 'ridiculous' and 'corrupting' human emotion and compassion. These regimes aim always to extract the most from the other who is the resource, relegated in this ethical dualism to the role of the pre-ethical condition for the privileged ethical life. The rationalist economic calculus which divorces 'rational' and 'efficient' political and economic life from care, compassion, social and ecological responsibility is the ultimate modern expression of the West's ancient rationalist opposition between reason and emotion, male and female, culture and nature, in which it has now ensnared the entire globe and all its species.

In this regime, spheres of injustice are interlinked and strengthened as economic power increasingly translates into political and communicative power, and ecological costs are transferred on to those who already suffer economic, communicative and social deprivation (see Plumwood, 1998a). The linking rationality of competition and maximization penetrates even into contexts where they are irrational because the predominance of relations based on dependency means that appropriate rationalities must be based on mutual thriving, trust, care and communication. These spheres include especially those of ecology and those of political, social and personal life. As ethical choice is individualized and confined to the shrinking and increasingly instrumentalized area of the 'domestic', ecological and social forms of care are disempowered as the responsibility of the individual (usually female) consumer, who is rarely able to exercise care in any meaningful or systematic way in the absence of time, real information, real choice and a real ability to influence the underlying political relationships responsible for growth, production, design and planning.

As Marilyn Waring (1988) observed, the natural forest which climbed the hill she overlooked 'counted for nothing' unless it was destroyed to make a plantation or was somehow potentially commodified into marketable, replaceable units. The rationality which counts nature and ecological relationships as background, as 'complete blanks', and which also 'counts for nothing' disposable human categories of 'non-contributors', is programmed into the rationalist market and its background institutions. The Minimalist and rights strategies are complicit with these developments.

Will environmental ethics continue to support the arrogant arrangements that treat nature in terms of a commodity model, demurring only for a minimal extension of personhood for a few higher animals, hopefully protected from commodification? Or will it realize that a framework in which the non-human other only gains relief from ruthless reduction to 'replaceable resource' status if it is placed in some exceptional category of protection is the wrong framework? If it is the framework which has given us collapsing fisheries, forest destruction and growing human misery all around the world, should we aim only at a *minimal* dislocation of and departure from it?

Many feminists have emphasized as alternatives care perspectives that stress emotional and dispositional forms of care for nature, as a more-than-instrumental basis of concern. As feminist theorists of care have pointed out, in the service of the opposition rationalism presupposes between emotion and reason, between ethics and economics, women have been denigrated as only half-ethical beings, and our civilization has evolved conceptions of both reason and ethics which exclude what women have stood for (see Tronto, 1993; Held, 1993).

In a rationalist economy which defines its hardness in opposition to the symbolic woman, as Other, and which increasingly invades every corner of our lives, we should not be surprised that care and compassion for others are increasingly inexpressible in the public 'rational context' that is defined against the domestic sphere. As Joan Tronto (1993) has pointed out, the associated hyper-separation between politics and ethics makes politics a cynical, manipulative numbers game, and leaves ethics individualized, etherealized and disempowered.

In this context, we can contrast an integrative ecofeminist ecosocialist position with the Minimalist approaches in the case of animal justice: in the Minimalist position complicit with the liberal framework, the questions of animal and environmental justice are discussed in a confined, depoliticized way as a matter of individual rectitude and self-limitation about food choices. The responsibility belongs simply with each individual consumer to make the choice to be a vegetarian. In the alternative integrative feminist approach I am outlining, we bring considerations of ethics and politics together to understand the current typical plight of domestic animals in the spheres of advanced neo-liberalism as framed by a pet–meat dualism, which is constructed in terms of the public–private, person–property and reason–emotion dualisms that help to define liberalism and whose intensification (in the form of intensified exclusion of the social and of care or responsibility for others) helps define neo-liberalism.

The specific form of pet–meat dualism characteristic of factory farming is the collaborative product of the public–private dualism of liberalism (with its confinement of care to the zone of personal relationships), the neo-liberal rationality of self-maximizing egoism expressed in maximizing economic relationships based on dualized conceptions of reason in opposition to emotion and ethics, and the ideology of human supremacism expressed in the dualism of person–property. Untangling these conceptual and social knots demands careful analysis which deconstructs polarized binaries and reconstructs social spheres on the basis of a different form of rationality, not an attempt to play on the politics of individual rectitude and to expand the privileged class of personhood while leaving everything else just as it is.[13]

We need to understand that we are not just dealing here with an absence of ethics in the animal sphere, but also with a disempowerment of ethics and communicative experience generally in neo-liberal societies, and hence with something that affects both humans and non-humans. As the rationality of the economic sphere spreads to other spheres, the rationalist machinery is coming for us too. It has already constructed our work and much of our own lives in the same oppressive terms that demands that we leave emotional expression, self-direction and creativity, along with love and communication, in the carefully limited zone marked 'personal', and discard care for others as inefficient in the relentless drive for economic competition. Experiences of care can be shared with household members, including animals, but the public and economic spheres are increasingly occupied by a narrow egoism and by work structures less and less unlike those of factory-farmed animals. Surely the fate of these animals gives us a glimpse into some human futures too, and into aspects of the human present and past.

An integrative feminism that refuses the solution of moral dualism allows us to see ways to draw together our struggles around that crucial – but not singular – site for contest that is the struggle against the systemic excision of care and justice from our economic and political lives. This is a struggle to defeat the rationalist machinery which is making nearly all of us into Others to ourselves, into less than we could be. The growing exclusion of justice, care and responsibility from the economic sphere in the interests of global competition affects all of us in different ways, but these different ways can still bring us together in the struggle for a fair and just form of economic life as it affects both human and non-human spheres.

Notes

1 According to Brian Norton (1991) an instrumentalist stance towards non-humans will solve the problem of disunity between different parts of the green movement by minimizing departures from the status quo. Not only is Norton's 'unity' attained by effectively eliminating from consideration all the more imaginative visions, including those of animal justice, deep ecology and ecosocialism, it is subject to the objections raised in this chapter to Minimalism.

2 See Tom Regan's critique of Cohen in this volume (Chapter 10). Just as poor whites were seen to be further deprived by the liberation of slaves, and working-class men by the liberation of women, so our duty to underprivileged humanity may be said to require the continued treatment of animals as mere resources, and of trees as mere fodder for timber mills.

3 Expressions of concern for misused animals are often greeted with the comment, 'How can you be so concerned about these (mere) animals when there are all these oppressed humans who need your concern?' Thus Murray Bookchin (1992: 362) comments that to be concerned with animal suffering and death in nature 'cheapens the meaning of real [human] suffering and death'. To the extent that such remarks point up some (real rather than imaginary) excessive exclusiveness of concern with animals, they can have a point, but this exclusiveness is not demonstrated by the

mere fact of concern with animals. Such arguments for excluding non-humans from a justice focus are of course circular, and the further assumption that moral concern is a limited good such that concern for one group must mean less for another group is, as Mary Midgley (1983) points out, quite unwarranted.

4 The term 'construction' is appropriate here because a reductive conception of a being will often result in their becoming reduced, by denying them the opportunity to develop potentials they would otherwise have. Reductive construction correspondingly stunts potentials and sensitivities in those who so reduce them (see Weston, 1996). The construction locution is also useful in circumventing the false dichotomy often presented between materialism and idealism. Thus reductive views of animals help construct the practices of the factory farm, while the reductive practices of factory farming help to select and maintain reductive views of animals. Conflicting as well as supportive relations are also possible.

5 Gary Snyder (1990: 109) presents an alternative account of the theory of evolution in active terms: 'It would appear that the common conception of evolution is that of competing species running a sort of race through time on planet earth, all on the same running field, some dropping out, some flagging, some victoriously in front. If the background and foreground are reversed, and we look at it from the side of the "conditions" and their creative possibilities we can see these multitude of interactions through hundreds of other eyes. We could say food brings a form into existence. Huckleberries and salmon call for bears, the clouds of plankton of the North Pacific call for salmon, and salmon call for seals and thus orcas. The Sperm Whale is sucked into existence by the pulsing, fluctuating pastures of squid, and the open niches of the Galapagos Islands sucked a diversity of bird forms and functions out of one line of finch.'

6 The moral extension proposed is problematic to the extent that it is based exclusively on the overall similarity of simians to humans, on their being 'our nearest relatives'. The assumption seems to be that simians are clearly superior to all other species in ethically relevant capacities, including cognitive capacities. However, the experimental data cited by Rogers (1997) on the cognitive abilities of some bird species suggest that their abilities in some areas may exceed that of simians (including humans), leading to problems about just who does count as a 'relative' and why we should be prepared to recognize only relatives. This points up the difficulties in employing a purely familial ethical metaphor here.

7 One measure of genuine ethical progress in this century has involved declining the nineteenth-century colonial obsession with imposing *a priori* ethical rankings on diverse human groups – especially races, genders and civilizations. In my view it is time we began to apply the same principles of non-ranking to other species (see Luke, 1995; Plumwood, 1998a, b).

8 This idea is now largely discredited among philosophers, although it clings on remarkably among scientists, especially those making experimental use of animals. Scientists often claim that the attribution of consciousness and any other mental states to animals is 'anthropomorphic'. This claim is of course entirely circular as an argument for keeping the mental and mentalistic identified with the human, as Mary Midgley (1983) shows.

9 It may be that some limits are imposed by the viable logic of concepts of respect and other ethical concepts themselves, but this kind of closure does not require that we draw a special boundary of moral exclusion in minimizing ways (see Birch, 1993; Plumwood, 1993, 1998a, b; Luke, 1995).

10 Some other rationalist elements of Singer's position such as the strongly calculative rationality of his utilitarianism and his strong exclusion of feeling as ethically relevant have been critiqued by feminist thinkers (e.g. McKenna, 1995). One might also include for critique the adoption of a Cartesian concept of consciousness, as

the self-transparency of the rational subject, as the criterion of moral considerability. For further discussion of Singer and Regan's moral extensionism see Cuomo (1998).

11 Personally, to my sensibilities, Singer's proclaimed indifference to plant lives is deeply shocking. I have concern for plants and old trees I know well as individuals and can feel regret and loss when they die or are harmed. I cannot see why my experience is any less rational, ethically valid or relevant than Singer's, and I certainly cannot see any rational case, in the present dire ecological context, for making a virtue out of callousness and insensitivity to other species' lives.

12 Humanism, like these other positions, is multifaceted and includes potentially positive elements such as human solidarity. But the position has long been open to distortion and subversion in several respects: first through the tendency to build concepts of human equality and solidarity on an exclusionary form of bonding which defines the human in dualistic opposition to its Other, the dualistic contrast class of the non-human; second, by the legacy of an older elite-based rationalism which continues to whisper its interpretations of leading concepts into the receptive ear of Enlightenment, converting its disarming declarations of equality into programmes for the benefit of a rational meritocracy. Included here is the idea of impartiality, universality and objectivity as the exclusion of care, compassion and emotionality. Third, the doctrines of equality and justice these positions have enunciated have been subverted by the insistence on a *boundary* to their inclusiveness. All these human-supremacist features rebound against the project of human solidarity, and have been mobilized against those human groups associated with the excluded nonhuman class. Thus the third element has long done battle with the first element of equality and solidarity. None of these problematic elements of humanism can be adequately challenged in the Minimalist programme, as we have seen.

13 As I argue in Plumwood (1997), reason–emotion, public–private and person–property dualisms construct the contrast between the private pets – with whom so many of us share intense subjective lives – and the reductively conceived and atrociously treated 'factory-farmed' animal now subject to the modernist polarity that construes 'rational' economic relationships in alienated, masculinist and narrowly instrumental terms as hyper-separated from moral and affective familiar relationships, and affective relationships as occurring in a highly circumscribed 'private' sphere of altruism supposedly untainted by economic considerations. The hyper-separation between the 'pet' animal and the fully commodified animal is intensified in neo-liberalism as the latter becomes subject to the rationally instrumentalized mass-production regime of the factory farm or laboratory. This boundary-shifting strategy is the expression of the complementary polarity of the subjectivized and underemployed 'pet' animal is both legally and in popular culture a *de facto* person subject to care and compassion, and the outrageously overexploited animal commodified in the market. These tend now to monopolize the roles these dualisms have left open, leaving other possible roles invisible and all animals more marginal to human lives.

References

Abram, D. (1996) *The Spell of the Sensuous*, New York: Pantheon.
Baumann, Z. (1989) *Modernity and the Holocaust*, London: Verso.
Benton, T. (1993) *Natural Relations*, London: Verso.
Birch, T. (1993) 'Moral Considerability and Universal Consideration', *Environmental Ethics* 15: 313–32.

Bok, G. and James, S. (eds) (1992) *Beyond Equality and Difference: Citizenship, Feminist Politics and Female Subjectivity*, London: Routledge.

Bookchin, M. (1982) *The Ecology of Freedom*, Palo Alto: Cheshire Books.

Cuomo, C. J. (1998) *Feminism and Ecological Communities*, London: Routledge.

Friedman, M. (1993) *What Are Friends For?*, Ithaca: Cornell University Press.

Goodin, R., Pateman, C. and Pateman, R. (1997) 'Simian Sovereignty', *Political Theory* 25 (6): 821–49.

Gould, S. J. (1981) *The Mismeasure of Man*, New York: W. W. Norton.

Grey, W. (1993) 'Anthropocentrism and Deep Ecology', *Australasian Journal of Philosophy* 71 (4): 463–75.

Hegel, G. W. F. (1952) *Philosophy of Right*, tr. T. Knox, Oxford: Clarendon Press.

Held, V. (1993) *Feminist Morality*, Chicago: University of Chicago Press.

Luke, B. (1995) 'Solidarity across Diversity: A Pluralistic Rapprochement of Environmentalism and Animal Liberation', *Social Theory and Practice* 21 (2): 177–206; also in R. S. Gottlieb (ed.) (1997) *The Ecological Community*, London: Routledge, pp. 333–58.

McKenna, E. (1995) 'Feminism and Vegetarianism: A Critique of Peter Singer', *Philosophy in the Contemporary World* 1 (3).

Mannison, D. (1980) 'What's Wrong with the Concrete Jungle?', in D. Mannison, M. A. McRobbie and R. Routley (eds) *Environmental Philosophy*, Canberra: Australian National University.

Mathews, F. (1994) 'Cultural Relativism and Environmental Ethics', *EWG Circular Letter*, August.

Memmi, A. (1965) *The Coloniser and the Colonised*, New York: Orion Press.

Midgley, M. (1983) *Animals and Why They Matter*, Athens, GA: University of Georgia Press.

Norton, B. (1991) *Towards Unity among Environmentalists*, New York: Oxford University Press.

Parry, B. (1995) 'Problems in Current Theories of Colonial Discourse', in B. Ashcroft, G. Griffiths and H. Tiffin (eds) *The Post-colonial Studies Reader*, London: Routledge.

Passmore, J. (1974) *Man's Responsibility for Nature*, London: Duckworth (2nd edn 1980).

Plumwood, V. (1993) *Feminism and the Mastery of Nature*, London: Routledge.

—— (1995a) 'Feminism, Privacy, and Radical Democracy', *Anarchist Studies* 3: 97–120.

—— (1995b) 'Has Democracy Failed Ecology? An Ecofeminist Perspective', *Environmental Politics*, 4 (4): 134–68 (Special Issue 'Ecology and Democracy').

—— (1996) 'Anthropocentrism and Androcentrism: Parallels and Politics', *Ethics and the Environment* 1 (2): 119–52.

—— (1997) 'Babe: The Tale of the Speaking Meat', *Animal Issues* 1 (1).

—— (1998a) 'Ecojustice, Inequality and Ecological Rationality', in J. Dryzek and D. Schlosberg (eds) *Debating the Earth: The Environmental Politics Reader*, Oxford: Oxford University Press.

—— (1998b) 'Intentional Recognition and Reductive Rationality: A Response to John Andrews', *Environmental Values* 7 (4): 397–422.

—— (1999) 'Knowledge in an Ethical Framework of Care', *Australian Journal of Environmental Management* (supplement to vol 5), pp. 27–38.

Proctor, R. (1993) 'Nazi Medicine and the Politics of Knowledge', in S. Harding (ed.) *The Racial Economy of Science*, Bloomington: Indiana University Press, pp. 344–58.

Regan, T. (1983) *The Case for Animal Rights*, Berkeley: University of California Press.

Rogers, L. (1997) *Minds of Their Own: Thinking and Awareness in Animals*, Sydney: Allen & Unwin.

Said, E. (1979) *Orientalism*, New York: Vintage.

Singer, P. (1975) *Animal Liberation*, New York: New York Review of Books/Random House.

—— (1980) 'Animals and the Value of Life', in T. Regan (ed.) *Matters of Life and Death*, New York: Random House.

—— (1997) 'Ethics across the Species Boundary', Paper given at the Conference 'Environmental Justice: Global Ethics for the 21st Century', University of Melbourne, 1–3 October.

Snyder, G. (1990) *The Practice of the Wild*, New York: North Point Press.

Stepan, N. L. (1993) 'Race and Gender: The Role of Analogy in Science', in S. Harding (ed.) *The Racial Economy of Science*, Indianapolis: Indiana University Press, pp. 359–76.

Taylor, A. (1996) 'Animal Rights and Human Needs', *Environmental Ethics* 18: 249–64.

Thompson, J. (1990) 'A Refutation of Environmental Ethics', *Environmental Ethics* 12: 147–60.

Tronto, J. C. (1993) *Moral Boundaries: A Political Argument for an Ethic of Care*, London: Routledge.

Waller, D. (1998) 'A Vegetarian Critique of Deep and Social Ecology', *Ethics and the Environment* 2 (2): 187–98.

Walzer, M. (1983) *Spheres of Justice*, New York: Basic Books.

Waring, M. (1988) *Counting for Nothing*, Auckland: Allen & Unwin.

Weston, A. (1996) 'Self-Validating Reduction: Toward a Theory of Environmental Devaluation', *Environmental Ethics* 18: 115–32.

Part III
Global political justice

13 Indigenous peoples, the conservation of traditional ecological knowledge, and global governance

Henrietta Fourmile

Introduction

Indigenous knowledge and biodiversity are complementary phenomena essential to human development (Warren, 1992: 1). The world's indigenous peoples are the custodians of much of the planet's biodiversity, which in many cases they have nurtured and developed over many millennia. However, the role of indigenous peoples and their knowledge in the conservation and sustainable use of this biodiversity has gone unacknowledged. Yet this knowledge, far more sophisticated than previously assumed, offers new models for development that are both ecologically and socially sound (Posey, 1985: 139–40). While indigenous knowledge continues to be exploited for commercial gain by alien governmental institutions and transnational corporations alike under a Western system of intellectual property law which does not protect the communal rights of indigenous peoples in their traditional knowledge, innovations and practices, signs of change are emerging in a body of new international environmental treaties which promise not only to generate new standards of protection for indigenous knowledge and benefit sharing, but also to encourage greater participation and collaboration by indigenous peoples in conservation partnerships for the maintenance of Earth's biological heritage.

Indigenous peoples

Indigenous peoples live in many different geopolitical contexts. It is very difficult to define who they are as a class of peoples. However, a number of characteristics have been identified which are common to most indigenous peoples worldwide, namely that they:

- have a strong and abiding identification with the lands and seas which they have traditionally occupied and used, irrespective of any alienation of those lands which may have taken place over time;
- have a historical continuity with pre-invasion and pre-colonial societies that developed on their territories; in other words, they are the direct descendants of what are sometimes referred to as the First Nations peoples;

- possess distinctive cultural characteristics which distinguish them from the prevailing society in which they live. These cultural characteristics extend to distinctive religious, political, social, legal, economic, educational and health structures;
- constitute a non-dominant part of the population of the countries in which they live. This non-dominance is also frequently characterized by higher rates of unemployment, incarceration, poverty, drug dependence and ill health, as well as shorter life expectancies and lower educational achievement; and
- maintain a powerful sense of self-identification as indigenous (Daes, 1995).

On the basis of these characteristics, it is estimated that there are currently at least 300 million people worldwide, or about 4 percent of the entire world population, who identify as indigenous (Colchester, 1994: 6, citing data compiled by the International Work Group on Indigenous Affairs, Copenhagen). Indigenous peoples inhabit areas of North and South America, Northern Europe, Asia, Africa, Australia and the Pacific. They occupy a wide geographical range from the polar regions to the deserts, savannahs and forests of the tropical zone. The geographic distribution of indigenous peoples is estimated to be as follows:

- 150 million in Asia – including more than 67 million in China, 50 million in India and 6.5 million in the Philippines;
- 30 to 80 million in Latin America (the smaller number represents government figures, the larger those of indigenous leaders);
- 3 to 13 million in North America (depending on whether Chicanos and Métis are included);
- several million in Africa;
- over 8 million in the Oceanic region, including 6 million in New Guinea, 300,000 in both Australia and New Zealand, with the balance inhabiting the islands of Polynesia, Micronesia and Melanesia; and
- in the polar region, 125,000 circumpolar Inuit and 60,000 Scandinavian Saami peoples.

Indigenous peoples constitute much of the planet's cultural diversity. If language is taken as an indicator (even if an imperfect one) of cultural diversity, it is estimated that about three-quarters of the world's six thousand languages are spoken by indigenous peoples. They include groups as disparate as the Quechua descendants of the Inca civilization in Bolivia, Ecuador and Peru, who collectively number more than 10 million, and the Gurumalum band of Papua New Guinea, who number fewer than ten. In New Guinea alone, over six hundred languages are spoken among a population of only 6 million, and in India there are more than 1,600 languages (Durning, 1992).

Many of the areas of highest biological diversity on the planet are inhabited by indigenous peoples, providing an inextricable link between biological diversity and cultural diversity. In the Declaration of Belém, for example, it is asserted that 'native peoples have been stewards of 99 percent of the world's genetic resources' (International Society of Ethnobiology, 1988, in Posey and Dutfield, 1996: 2). Of the nine countries which together account for 60 per cent of human languages, six of these centres of cultural diversity are also biologically megadiverse countries with exceptional numbers of unique plant and animal species and high levels of endemism. In what has been dubbed the 'Biological – 17', the nations that are home to more than two-thirds of the Earth's biological resources are also the traditional territories of most of the world's indigenous peoples (Mittermeier *et al.*, 1997).

International recognition of the rights of indigenous peoples

Indigenous peoples are also, for the most part, the stateless peoples of the world. We constitute what is sometimes called the Fourth World, comprising peoples who are 'spoken for' by the governments of what were once referred to as the First, Second and Third Worlds who have jurisdiction over us and whose sovereignty is recognized in the fora of the United Nations. As a class of people, we remain probably the most discriminated against in the world, despite the foundations for the recognition of our rights being laid in the mid-1960s by Article 1.1 of the International Covenant on Economic, Social and Cultural Rights (Centre for Human Rights, 1994: 8–19), Articles 1.1 and 27 of the International Covenant on Civil and Political Rights (*ibid.*: 20–40) and Article 1 of the International Convention on the Elimination of All Forms of Racial Discrimination, ICEAFRD, (*ibid.*: 66–79). Article 1.1 of the two human rights covenants states that:

> All peoples have the right of self-determination. By virtue of that right they freely determine their political status and freely pursue their economic, social and cultural development. Article 27 of the International Covenant on Civil and Political Rights, requires that: in those States in which ethnic, religious or linguistic minorities exist, persons belonging to such minorities shall not be denied the right, in community with the other members of their group, to enjoy their own culture, to profess and practice their own religion, or to use their own language.

The ICEAFRD, in Article 1, states that:

> In this Convention, the term 'racial discrimination' shall mean any distinction, exclusion, restriction or preference based on race, colour, descent, or national or ethnic origin which has the purpose or effect of nullifying or impairing the recognition, enjoyment or exercise, on an equal footing, of human rights and fundamental freedoms in the political, economic, social, cultural or any other field of public life.

Despite the clear and unambiguous references to the recognition and rights of 'peoples' above in Article 1.1 of the two human rights covenants, this has not as yet been extended to include the recognition of indigenous peoples as 'peoples' and therefore entitled to the rights expressed in these and other instruments. Consequently, indigenous peoples are still not recognized under international law as legal entities or active subjects (Simpson, 1997: 53). It also means that we have no standing in international law as peoples.

It is only recently that our human rights have begun to be effectively recognized. The International Labour Organization's controversial Convention No. 169 notwithstanding, it is in the area of international environmental law that our rights have been achieving greatest recognition.[1] This recognition comes most notably in such instruments as the Rio Declaration on Environment and Development (Principle 22) and its programme of implementation, Agenda 21 (Chapter 26), the Convention on Biological Diversity, the Authoritative Declaration with Non-Legally Binding Force of Principles for a Global Consensus on Management, Conservation and Sustainable Development of All Types of Forests (generally referred to as the Statement of Forest Principles), and such regional treaties as the Nuuk Declaration on Environment and Development in the Arctic.[2] Of these, only the Convention on Biological Diversity is legally binding. The other instruments, however, have undeniable moral force and provide practical guidance to states in their conduct.

The value of such instruments rests on their recognition and acceptance by a large number of states, and, even without binding legal effect, they may be seen as declaratory of broadly accepted principles within the international community (Centre for Human Rights, 1994: xii). Our rights are also recognized in a number of framework declarations which are in the process of being developed, such as the draft United Nations Declaration on the Rights of Indigenous Peoples,[3] the draft Declaration of Principles on Human Rights and the Environment (Article 14) and the draft International Covenant on Environment and Development (Articles 11.4, 12.6, 43.2 and 44.1) (United Nations Meeting of Experts on Human Rights and the Environment, 1994; Commission on Environmental Law of IUCN and International Council of Environmental Law, 1995).

Indigenous peoples worldwide have also articulated their concerns in many 'counter'-declarations, statements and charters of their own, such as the Charter of the Indigenous Tribal Peoples of the Tropical Forests (Indigenous-Tribal Peoples of the Tropical Forests, 1992, in Secretariat for the Convention on Biological Diversity, 1996a: 82–85), the Mataatua Declaration on Cultural and Intellectual Property (First International Conference on the Cultural and Intellectual Property Rights of Indigenous Peoples, 1993, in Secretariat for the Convention on Biological Diversity, 1996a: 86–88), the Treaty for a Lifeforms Patent-free Pacific and Related Protocols (Peteru, 1995) and The 'Heart of the Peoples' Declaration (North American Indigenous Peoples Summit on Biological Diversity and Biological Ethics, 1997).

Indigenous world views

Generally speaking, indigenous peoples worldwide have a view of the world which places them within the natural order in which all living things are inter-related and interdependent. Essential to this natural order is the notion of reciprocity: if you care for the land and the life it sustains, it will care for you. The Eurocentric division between 'man' and 'nature' as expressed in the original categories of the World Heritage Convention, 'natural' and 'cultural', is notably absent. Thus indigenous peoples' world views could be characterized as being ecocentric, rather than technocentric. The latter view expresses the notion that nature exists to serve humanity and can be acted upon, owned, objectified, exploited, manipulated and understood through primarily scientific and technological means. Similarly, where ecological or environmental problems arise these can be addressed through the application of the appropri-ate technology – the 'technological fix', as it were.

Integral to this indigenous respect is the sense that the Earth is the Mother, the Creator of all things. This is expressed in paragraphs 31 and 32 of the Kari-Oca Declaration and the Indigenous Peoples' Earth Charter:

> Indigenous peoples were placed upon our Mother, the earth by the Creator. We belong to the land. We cannot be separated from our lands and territories. Our territories are living totalities in permanent vital rela-tion between human beings and nature. Their possession produces the development of our cultures.
>
> (World Conference of Indigenous Peoples on Territory, Environment and Development, 1992, in Secretariat for the Convention on Biological Diversity 1996a: 78)

Or Article 3 of the Charter of the Indigenous-Tribal Peoples of the Tropical Forests states:

> Our territories and forests are to us more than an economic resource. For us, they are life itself and have an integral and spiritual value for our com-munities. They are fundamental to our social, cultural, spiritual, economic and political survival as distinct peoples.
>
> (Indigenous-Tribal Peoples of the Tropical Forests, 1992, in Secretariat for the Convention on Biological Diversity [SCBD], 1996a: 82)

Thus, in the holistic world view of indigenous societies, the division between 'man' and 'nature' does not exist. Instead, indigenous peoples have evolved complex relationships based on systems of eco-kinship with the elements of the world that surround them, often expressed through totemic relationships with various species, and religious ceremonies that involve the celebration of human–nature relationships.

Traditional ecological knowledge

Knowledge is the information held in human memories that is accessible, by recall and the practice of learned skills, in a useful way in day-to-day life. In the context of traditional knowledge it is often used to mean wisdom, which implies a blend of knowledge and experience integrated with a coherent world view and value system. *Traditional* means handed down from one generation to another and, in the case of traditional knowledge, usually means knowledge that has been accumulated by societies in the course of long experience in a particular place, landscape or ecosystem. It can be contrasted with *cosmopolitan* knowledge, which is drawn from global experience and combines 'Western' scientific discoveries, economic preferences and philosophies with those of other widespread cultures (SCBD, 1996c: 8–9).

Working definitions of traditional knowledge stress the links between traditionality, cultural distinctiveness and the local environment to which each culture has adapted. To clarify this linkage, we can imagine a semi-isolated indigenous community developing an increasingly distinctive set of cultural features over time. This process is driven partly by the group responding to its environment, and partly by the insights and creativity of its members. Embedded in the distinctive culture will be much information concerning the physical and biological processes of the landscape, its seasons, soils, plants and animals, and their relationships, behaviours and forms of usefulness (SCBD, 1996c: 9).

This body of environment-related knowledge is generally referred to as traditional ecological knowledge (TEK), a term used to describe those aspects of traditional knowledge that are directly related to management and conservation of the environment. It has been defined by the Dene Cultural Institute as:

> a body of knowledge built by a group of people through generations living in close contact with nature. It includes a system of classification, a set of empirical observations about the local environment, and a system of self-management that governs resource use.
>
> (Johnson, 1992: 4)

While various other definitions exist (see, for example, Gadgil *et al.*, 1993: 151), TEK has been given sufficient airing in the literature and government policy documents to be considered an established concept (see, for example, Johannes, 1989; Inglis, 1993; Williams and Baines, 1993; Lalonde, 1993; Berkes *et al.*, 1997; and New Zealand Conservation Authority, 1997: 89–90).

Thus, TEK is far more than a simple compilation of facts. It is the basis for local decision-making in areas of contemporary life, including natural resource management, nutrition, food preparation, health, education, and community and social organization. TEK is holistic, inherently dynamic, constantly evolving through experimentation and innovation, fresh insight, and external stimuli. Notions of traditional societies which see them as unchanging and static should be resisted. As the Four Directions Council points out:

What is 'traditional' about traditional knowledge is not its antiquity, but the way it is acquired and used. In other words, the social process of learning and sharing knowledge, which is unique to each indigenous culture, lies at the very heart of its traditionality. Much of this knowledge is actually quite new, but it has a social meaning, and legal character, entirely unlike the knowledge indigenous peoples acquire from settlers and industrialised societies.

(Four Directions Council, 1996, quoted in SCBD, 1996b: 17–18)

Others have stressed that tradition is a 'filter through which innovation occurs', and that many local indigenous farming communities maintain a tradition of invention in which it is the 'traditional methods of research and application, not always particular pieces of knowledge', that persist (Pereira and Gupta, 1993).

We might take forest ecosystems around the world as an example. For indigenous peoples, forests are more than just a source of timber. Most traditional peoples who inhabit forests or areas close to forests rely extensively upon hunted, collected or gathered foods and resources, a significant portion of which are influenced by humans to meet their needs. These species, sometimes called 'semi-domesticates' or 'human-modified species', form the basis for a vast treasury of useful species that have systematically been undervalued and overlooked by science, yet provide food and medicinal security for local communities around the world. Such useful species provide most of the foods, medicines, oils, essences, dyes, colours, repellents, insecticides, building materials and clothes needed by each local community. Indigenous peoples plant forest gardens and manage the regeneration of bush fallows in ways which take advantage of natural processes and mimic the biodiversity of natural forests. In many countries there are ancient forest groves which are sacred places dedicated for rituals, which may also be used for burial sites, and as sources of medicinal plants. Such sites have been found to have conservation importance for the communities and to provide other environmental benefits.

It must also be recognized that much of the world's crop diversity is also in the custody of indigenous farmers who follow age-old farming and land-use practices that conserve biodiversity and provide other local benefits. Polyculture is the norm in farming systems in Africa and other parts of the world, 'a traditional strategy to promote diet diversity, income generation, production stability, minimization of risk, reduced insect and disease incidence, efficient use of labor, intensification of production with limited resources, and maximization of returns under low levels of technology' (Altieri, 1987: 73). These ecologically complex agricultural systems associated with centres of crop genetic diversity include traditional cultivars or 'landraces' that constitute an essential part of our crop genetic heritage, and non-domesticated plant and animal species that serve humanity as biological resources (Oldfield and Alcorn, 1991: 37). However, it is not only in terrestrial biomes that indigenous peoples conserve biodiversity. They carry out a range of practices that are effective in the conservation of coastal and marine biological diversity, including such biodiversity-rich biomes as mangrove systems and coral reefs. Elements of such practices may include application of TEK, customary authority over resources and environments, and social regulation of exploitation (Johannes and Ruddle, 1993).

Thus there are numerous categories of traditional knowledge among indigenous peoples, which clearly have great potential for application in a wide range of sustainability strategies. Indigenous peoples conserve biological diversity, and in many cases provide other environmental benefits; for example, soil and water conservation, soil fertility enhancement, and management of game and fisheries (see, for example, Hecht and Posey, 1989). In summarizing the nature of TEK, it can be seen that it is made up of the following linked features:

- information about the various physical, biological and social components of a particular landscape;
- rules for using them without damaging them irreparably;
- relationships among their users;
- technologies for using them to meet the subsistence, health, trade and ritual needs of local people; and
- a view of the world that incorporates and makes sense of all the above in the context of a long-term and holistic perspective in indigenous and local-community decision-making (SCBD, 1996c: 10).

In reference to the last feature, the long-term perspective of such communities is often measured in accordance with a span of seven generations. For example, for North American indigenous peoples, each generation has a responsibility to ensure the survival of its lifeways for the seventh generation (for further discussion of this principle see Clarkson *et al.*, 1992).

The paramount quality of TEK is its profound knowledge of relationships. If the defining character of an ecosystem is not only the array of species which comprise it but their relationships one to another, then TEK is true ecological knowledge because it is the relationships between species, and between species and humans, that are the most salient character of TEK. It is this quality of knowledge which is usually lacking in scientific knowledge about ecosystems, particularly those of the more remote and inaccessible regions of the planet commonly inhabited by indigenous peoples. Scientists often come to know a lot about a few species, particularly those which might be of commercial value, but rarely comprehend the full complexity of complete ecosystems. This kind of knowledge can only be gained by generations of people who have lived in a particular environment for perhaps many thousands of years accumulating knowledge and passing it on from generation to generation.

An example of the depth of knowledge about a particular ecosystem is provided by the Inuit. This example concerns their knowledge of sea ice. Inuit have always spiritually respected the sea ice as a living form with influence on the daily lives of both humans and animals. Their strong reliance on sea ice for travelling and hunting is reflected in their knowledge of its processes, character- istics and annual cycles. The formation of sea ice varies with weather and currents. Different ice conditions determine which sea mammals are present in winter, which areas can support hunting and travel, and the type of spring break-up. If, for example, the floe-edge ice does not break off during the spring tides, the landfast ice will continue to expand and thicken, altering the habitat for ringed seals, bearded seals and eider ducks in particular.

Inuit use distinct terms to describe each different ice condition through five stages of development. Seven ice conditions are linked directly to early ice for- mation; twenty-five are related to development of landfast ice; and thirty-four to ice developing from the floe edge, including those associated with marine currents and 'ice joints'. One is relevant only to spring, and four others occur only after spring break-up (in McDonald *et al.*, 1997: 15).

The traditional practice of seasonal burning as an environmental manage- ment tool in Australia, as practised by Aboriginal communities, provides an excellent example of ecosystem management for the benefit of both humans and species. Also known as 'firestick farming', this practice was used for many thousands of years to sustainably manage virtually all the different ecosystems found on the continent. While evidence suggests that when first used the prac- tice radically altered the nature and species composition of ecosystems and probably contributed to the extinction of a number of species over time, never- theless a balance was reached such that ecosystems and humans became mutually interdependent. Essentially the practice consisted of a system of cool patch burnings carried out on a seasonal rotational basis. These patch burnings were generally carried out in the tropics after the wet season in the cool months of the early dry season when there was enough fuel to burn. Each patch was generally defined by the area which would burn within one day. The fire would be lit about mid-morning and burn through until put out by the evening dew. Thus areas would be burned without threatening most of the faunal species, which could take refuge in neighbouring unburned areas. The fires would burn up useless dry grasses, destroy seedlings while not harming mature trees, and enable the woodlands to be kept open so that the next generation of grasses could grow when germinated by the next rains (for a detailed account of traditional burning practices in central Australia see Latz, 1995).

The benefits to the Aboriginal peoples were that regular cool fire patch burn- ings (a) helped cleanse the country of harmful insects and parasites (ticks, spiders, scorpions, etc.) to make it habitable; and (b) by keeping the woodlands open, they made available a regular supply of succulent grasses for the favoured food species (the various species of kangaroo and wallaby, emu, scrub turkey, and so on).

In terms of the benefits for each ecosystem, the controlled burning prevented the build-up of undergrowth which, if ignited by lightning, could result in uncontrollable wildfires capable of burning out huge areas of otherwise fire-tolerant species and destroying fauna unable to escape such large-scale fires, and it maintained ecosystem biodiversity.

Having experienced wildfires which have caused extensive destruction of property, resources and, on occasion, lives, conservation authorities throughout Australia are now once again seriously considering implementing traditional fire regimes for the management of nature and timber reserves and water catchment areas. For example, the Wet Tropics Management Authority, the agency which manages the Wet Tropics of Queensland World Heritage Area, is looking to collaborate with Aboriginal traditional owners in the area to carry out such burnings. In tropical Australia, where some national parks have been returned to the traditional owners, but retained as national parks, such fire regimes are a regular part of park management.

The simple truth is that the world's indigenous peoples have had to manage sustainably the environments in which they live, or else perish.

The value of traditional ecological knowledge

The value of indigenous knowledge of traditionally used species has long been known; for example, to locate plants from which medically useful compounds can be extracted or synthesized. A number of essential drugs used in modern-day medical practice are plant-derived and owe their origins to indigenous usage of such plants, for example, atropine (anticholinergic), codeine (antitussic/ analgesic), colchicine (antigout), digitoxin/digoxin (cardiotonic), vincristine (antitumour), morphine (analgesic), quinine/artemisinin (antimalarial), reserpine (antihypertensive) and physostigmine (cholinergic) (Akerele, 1995: 78). In the United States nine out of ten prescription drugs used are based on natural compounds from plants, fungi, animals and micro-organisms; that is, based on the products of biodiversity (President's Committee of Advisers on Science and Technology, Panel on Biodiversity and Ecosystems [PCAST], 1998: 5).

Scientists are becoming increasingly aware of the sophistication of TEK among many indigenous and local communities. For example, the Shuar people of Ecuador's Amazonian lowlands use 800 species of plants for medicine, food, animal fodder, fuel, construction, fishing and hunting supplies. Traditional healers in South-east Asia rely on as many as 6,500 medicinal plants, and shifting cultivators throughout the tropics frequently sow more than 100 crops in their forest farms (Durning, 1992). East African farmers

> recognise in maize, as in potato cultivars, important differences in taste, texture, storability, marketability, disease and pest resistance, and response to moisture stress. At least nine possible end uses, many of them simulta-

neously relevant on a single farm, help determine the maize genotypes east African farmers prefer.

(Haugerud and Collinson, 1991: 6)

In India, local farmers have evolved over 100,000 varieties of rice (Shiva, 1998b: 1). Recently, the first World Food Prize laureate, Dr M. S. Swaminathan, commenting on criticisms of the Green Revolution as being too restrictive in crop variability, noted that 'while the Incas had sustained themselves on as many as a thousand crops, in the Green Revolution era, only four or five crops were available for humankind' (in Worthington, 1996: 10). Over 30,000 species of medicinal plants are reported to be used globally and across cultures, providing health security in the context of primary health care to an estimated 80 per cent of the world's 5 billion inhabitants (Akerele, 1995: 78, citing World Health Organization statistics). In the context of health care it is relevant to observe that in developing countries, Western-trained medical doctors are scarce compared to traditional medical practitioners. In Ghana, for example, statistics indicate that there is an average of one medical doctor for every 20,625 people, but one traditional medical practitioner for every 224 people. Thus traditional medical practitioners and midwives, using traditional medicines and remedies, play an important role in primary health-care programmes (Cunningham, 1993: 27).

Thus indigenous and local communities carry out a major duty to the Earth through their innovative practices of not only conserving but extending biodiversity. While biotechnology is projected as increasing food production fourfold, small locally based agro-ecosytems, by stressing total ecosystem productivity rather than that of the single-species monocultures so heavily promoted by proponents of Western techno-scientific agriculture, achieve productivity hundreds of times higher than large or conventional farms (Shiva, 1998b: 2, 1998a; GRAIN, undated). As Berkes *et al.* (1997: 11) conclude, many traditional management systems contribute to the conservation of biodiversity by the use of a greater number of varieties, species and landscape patches than do modern agricultural and food production systems (citing Sporrong, 1997), and by monitoring and responding to ecosystem change (Gadgil *et al.*, 1993; Berkes *et al.*, 1995).

The global role of indigenous peoples in the management of the world's biodiversity

It is widely considered axiomatic that biological diversity cannot be conserved without cultural diversity (Anil Gupta in Diversity, 1996: 17; Declaration of Belém in Posey and Dutfield, 1996: 2; Reid *et al.*, 1992: 23), that indeed, long-term future world food and medicinal security will depend on maintaining this inextricable link between cultural diversity and biodiversity, that one cannot be maintained without the other. There is also a growing realization that cultural diversity is as important for civilization's evolution as biodiversity is for biological evolution (see, for example, COBASE, 1998): the promotion of monocultures in both cases poses a serious threat to human survival. At a workshop on 'Drug Development, Biological Diversity and Economic Growth' convened by the National Cancer Institute at Bethesda, Maryland, in 1991, one of the conclusions reached was that 'Traditional knowledge is as threatened and is as valuable as biological diversity. Both resources deserve respect and must be conserved' (in Cunningham, 1993: 19).

Yet human cultural diversity is threatened on an unprecedented scale. It has been estimated that half the world's languages – the storehouses of peoples' intellectual heritages and the framework for their unique understandings of life – will disappear within a century. It is the fear of many elders that their knowledge will not survive them (Society for Research and Initiatives for Sustainable Technologies and Institutions, undated: 5). As Eben Hopson, founder of the Inuit Circumpolar Conference, once stated, 'our language contains the memory of four thousand years of human survival through conservation and good management of our Arctic wealth' (Inuit Circumpolar Conference, undated: 8).

According to UNESCO (1993), nearly 2,500 languages are in immediate danger of extinction. If language extinction is some indicator of the disappearance of cultural diversity, we may note that in Australia, for example, it is predicted that by the year 2000, only 20 of the original estimated 250–300 indigenous languages will remain viable, representing a language loss of the order of 90 per cent over the two hundred years since European colonization (Schmidt, 1990: 1–3). In a mega-biodiverse continent, this represents a staggering loss of invaluable cultural knowledge patiently accumulated over a period spanning some 60,000 years. This should also be seen in the context of a rate of genocide of the same order, whereby the 1788 pre-European contact population of around 1 million Aborigines was reduced to about 60,000 by 1927 (over 90 per cent of the Aboriginal population were destroyed). The world, therefore, has already lost a vast store of TEK from this one continent alone.

It is also worth mentioning that, apart from the macadamia nut, whose commercial introduction to the world took place elsewhere (O'Neill, 1995: 12–13), Australia has made no contribution from its terrestrial biodiversity to the staple foods which now largely constitute the diet of the world's population. In fact, it has taken nearly two centuries for the new nation of European colonizers to seriously evaluate the commercial potential of what are termed Australia's native bushfoods – what Aboriginal peoples refer to as 'bush-tucker'. In a press release in 1995, the Australian Native Bushfood Industry Committee (ANBIC) set a growth target for the industry from an estimated $15 million in 1995 to $100 million by 1998, with an ultimate goal 'to take bushfoods to local and export markets underpinning them as mainstream food ingredients of our Australian food culture for the 21st century'. The industry has spawned a number of associations, networks and periodicals, such as the *Australian Rain-*

forest Bushfood Industry Association Newsletter. Despite its clear potential for Aboriginal participation, it remains to be seen whether this potential is realized.

Fortunately, the world has formally recognized in its environmental treaties, such as the Convention on Biological Diversity (CBD), the critical role that indigenous peoples have to play in sustainably managing crucial areas of bio-diversity. Initially there was considerable disillusionment, particularly among indigenous peoples, with the CBD. Primarily this was caused by the omission of the word 'peoples' after the term 'indigenous', the CBD's affirmation of *state* sovereignty over natural resources and the processes involved in determining access to them, the subjection of Article 8(j) to national legislation – thereby striking at the very heart of, particularly, indigenous peoples' struggle for the recognition of their rights to self-determination (and for land and natural resource rights), and concern over how the term 'embodying traditional lifestyles' might be interpreted by states (see, for example, International Alliance of Indigenous-Tribal Peoples of the Tropical Forests and International Work Group for Indigenous Affairs, undated: 113–29). There was also un-certainty as to how the contracting parties might accommodate indigenous and local community interests in implementing the CBD. However, while these very real grievances remain, initial fears about the role of indigenous and local communities in the implementation of the CBD have been largely dispelled by subsequent actions and decisions of the Conference of the Parties to the Convention (see, for example, Decisions III/14, III/17 arising from the third meeting of the Conference of the Parties, United Nations Environment Programme, 1997: 47–51, 55–58).

Indigenous peoples' representatives co-chaired both working groups with Conference of the Parties Bureau members and were able to 'sit at the table' with contracting party delegates and negotiate directly with them to produce recommendations for the establishment of a programme of work and the establishment of an *ad hoc* open-ended intersessional working group on the implementation of Article 8(j) and related provisions at the Madrid Workshop on Traditional Knowledge and Biological Diversity in November 1997 (Secretariat for the Convention on Biological Diversity, 1997a, b). The working group was subsequently established by decision IV/9 of the Conference of the Parties at its fourth meeting, in Bratislava in May 1998, and indigenous and local community representatives will be able to participate in the working group as observers. Contracting parties, however, are also encouraged to include indigenous and local community representatives in their delegations, in which case they would be able to participate (admittedly, to the limits determined by their countries) in the decision-making processes of the parties themselves at both the meetings of the working group and the Conference of the Parties.

Consequently, the CBD is now increasingly being looked upon by indigen-ous and local communities worldwide as the most important international instrument through which to gain protection for their natural resources, knowledge, traditions and lifestyles. Many indigenous groups have focused on the CBD because the language of Article 8(j), while extremely limited compared to indigenous aspirations, is one of the most powerful references to our interests in international law (Bragdon *et al.*, 1998: 9). While never intended by its drafters as a human rights instrument, the CBD nevertheless has, for indigenous and local communities, many of those attributes. Couched within parameters which require relevance to the conservation and sustainable use of biological diversity, are provisions which, *inter alia*, require that:

- the traditional knowledge, innovations and practices relevant to the conservation and sustainable use of biodiversity of indigenous and local communities be respected, preserved and maintained [Article 8(j)];
- such knowledge, innovations and practices should only be used with the approval of their holders [Article 8(j)];
- where they are used, the holders of such knowledge, innovations and practices should equitably share in the benefits arising from their use [Article 8(j)];
- customary uses of biological resources in accordance with traditional practices should be protected and encouraged [Article 10(c)]; and
- traditional knowledge and technologies be accorded the same treatment as other forms of knowledge and technologies that can contribute to the conservation and sustainable use of biodiversity [Articles 17.2 and 18.4].

All this is within the framework of a legally binding international instrument. However, the requirements to 'respect, preserve and maintain' traditional knowledge, and to secure the approval of its holders, can take place only within the context of acknowledging and protecting the intellectual property rights (IPRs) of indigenous and local communities in their knowledge, innovations and practices, either by using and adapting existing IPR regimes, or through the creation of new mechanisms and alternatives which will guarantee such acknowledgement and protection.

In order to achieve the objectives of the Convention, the linkages between Article 8(j) and other provisions dealing with such thematic areas as identification and monitoring (Article 7), incentive measures (Article 11), conservation *in situ* (Article 8), access to genetic resources (Article 15), and access to and transfer of technology (Article 16) must be established so that they can be effectively implemented (Secretariat for the Convention on Biological Diversity, 1997a). It is also expected that contracting parties to the Convention will in addition take account of these articles and their linkages in their sectoral policies concerning, for example, agro-biodiversity, forestry, fisheries and aquaculture, and rangeland management.

The CBD does not recognize the rights of indigenous peoples as such; any possibility of such recognition was deliberately eliminated by the use of the term 'indigenous and local communities' in the wording of the CBD. With regard to the CBD, this also means that indigenous peoples have no legal standing to pursue remedies under the CBD either as individuals ('persons') or as 'peoples' for any alleged breach of the Convention, regardless of whether they are directly affected (Simpson, 1997: 53). However, a number of countries and regional organizations are giving recognition to various indigenous and local community rights in direct response to the CBD. For, example, in recognition of the communal ownership of TEK, the Philippines requires that the prior informed consent of indigenous and local communities be given in accordance with local customary laws before any bioprospecting can be carried out on their territories (Barber and La Vina, 1997: 128). Similarly, the fifty-three member-state Organization of African Unity's Scientific, Technical and Research Commission has prepared draft legislation on community rights and access to biological resources which provides that 'local communities shall at all times and in perpetuity be the lawful and sole custodians of the relevant knowledge, innovations and practices' (Article 5.2). The draft also provides for a two-tiered regime governing prior informed consent with regard to access to genetic resources and associated knowledge, namely that of the state as well as the concerned local community, and subjects the giving of such consent to a detailed set of conditions and requirements to be met by the bioprospector (Articles 4.2 and 5.3) (Organization of African Unity Scientific, Technical and Research Commission, 1998b). For a more detailed analysis of the Draft Legislation see Fourmile (1998).

The recognition of indigenous interests necessary to biodiversity conservation

It is one thing to have formal international recognition of our role in biodiversity conservation and management; it is another to translate this recognition into national reality. For example, with regard to the CBD, such recognition as is contained in Article 8(j) is subject to national legislation and has been condemned, in reference to such words as 'promote' and 'encourage', for its 'soft' language (see, for example, Simpson, 1997: 53). This suggests, in a worst-case scenario, that a national government with jurisdiction over indigenous peoples can ignore the Convention with respect to its obligations to those peoples. However, Downes argues that while such clauses give countries flexibility, they do not give them complete discretion (Downes, 1995). Chandler (1993: 141) further argues that the 'caveat', 'subject to . . . national legislation' in relation to Article 8(j) makes clear that parties can maintain the general legal concepts and structures that they use to govern indigenous affairs, such as legal definitions of indigenous tribes, but it does not absolve them from obligations to implement the general provisions of the article in accordance with the CBD.[4] These uncertainties do not, of course, prevent states from taking measures in this area to implement these obligations in good faith (Mann, 1997: 29).

If indigenous peoples are going to contribute in any meaningful way to biodiversity conservation, the following minimal conditions must be met. We must:

- have security of tenure over our traditional land and marine estates;
- exercise our right of self-determination;
- have our sovereignty over natural wealth and resources – including our right to our own means of subsistence – recognized;
- be able to practise our own cultures and languages;
- have our customary laws respected; and
- have our systems of TEK accorded the same status and levels of protection as Western categories of knowledge.

With regard to the last point, preambular paragraph 9 to decision III/14 regarding the implementation of Article 8(j) of the Conference of the Parties to the CBD recognizes that 'traditional knowledge should be given the same respect as any other form of knowledge in the implementation of the Convention' (UNEP, 1997: 48).

The protection of indigenous knowledge

Particularly since the Rio Earth Summit in 1992, worldwide concern has been expressed over the exploitation of the knowledge of indigenous communities by commercial interests, which have thus far been seeking free access to what they consider to be public-domain knowledge of plant resources and their uses, and modifying this public property superficially and transferring it into the private domain of intellectual property rights (IPRs). This is particularly the case regarding the patenting of life forms and the recognition of plant breeders' rights. Indigenous peoples widely regard life in all its forms as sacred, and life, therefore, cannot be owned. Such opposition is expressed, for example, in the Phoenix Declaration of Indigenous Peoples of the Western Hemisphere, Regarding the Human Genome Diversity Project (Indigenous Peoples of the Western Hemisphere, 1995, in Cultural Survival Canada, undated: 4), which states that 'We oppose the patenting of all natural genetic materials. We hold that all life cannot be bought, owned, sold, discovered or patented, even in its smallest form' (see also Peteru, 1995). This so-called public-domain knowledge, defined according to criteria laid out in standard intellectual property law, is in fact the communally owned knowledge of indigenous peoples, governed and regulated by their customary laws with regard to its access, use, dissemination, and so on. One of the main consequences of the incompatibility of 'Western' systems of IPR laws with local customary systems can be seen in the way in which corporations gain access to, use, benefit from and ultimately control components of traditional knowledge. Dutfield highlights the processes involved:

> One of the main difficulties with protecting TEK under IPR laws is that much TEK cannot be attributed to a single community, much less to an individual. In fact TEK is frequently shared openly, in which case it is the

public domain and therefore unprotectable under patent law. . . . As a result, scientists visiting indigenous communities can collect TEK from the local peoples and acquire copyright protection for their compilations of this knowledge, whether or not they acknowledge the intellectual contributions of their local informants. Company researchers may then read this work, investigate a certain piece of information in it, add knowledge of their own, and patent the result.

(Dutfield, 1997: 7)

The recognition of our intellectual property rights to our communal knowledge is long overdue.

Intellectual property rights systems were designed to meet two fundamental and interrelated objectives: first, to act as an incentive for investment; and second, to facilitate technology transfer and access. Of the basic forms of intellectual property, patents and plant breeders' rights are of the most relevance to the conservation and sustainable use of biodiversity and of concern in the protection of indigenous intellectual property in our TEK. Another form, farmers' rights, has been defined but effective mechanisms for its recognition and protection are still in the process of being developed under the umbrella of the Food and Agriculture Organization's International Undertaking on Plant Genetic Resources (Commission on Genetic Resources for Food and Agriculture, 1997). Existing international instruments such as the Paris Convention for the Protection of Industrial Property (governing patents) and the International Convention for the Protection of New Varieties of Plants, and agreements such as the Trade-Related Aspects of Intellectual Property Rights (TRIPs), do not enable any explicit recognition of the contribution of TEK involved in the development by indigenous peoples of the wide variety of plants developed for foods, medicines, adhesives, cosmetics, dyes, poisons and other purposes (Fourmile, 1998; Williams, 1997; Dutfield, 1997; Simpson, 1997).

It can be argued that the CBD does, however, recognize communal rights to knowledge, innovations and practices by virtue of its reference to indigenous and local communities in Article 8(j) even if it does not explicitly recognize our property rights in such knowledge. It is also clear under the provisions of Article 8(j) that indigenous peoples should be adequately compensated for the use of their knowledge when benefits arise out of its use. The Conference of the Parties to the Convention has been instrumental in inviting other UN agencies, such as the World Intellectual Property Organization (WIPO) and the World Trade Organization (WTO), to examine the issue of the protection of indigenous knowledge (see, for example, Decision III/17 of the third Conference of the Parties: UNEP, 1997: 55–58).

With respect to decisions arising from its most recent meeting, Decision IV/9, in addition to the establishment of the *ad hoc* open-ended intersessional working group on the implementation of Article 8(j), deals in large part with matters concerning the protection of traditional knowledge and establishing a working relationship with WIPO to address the issue. WIPO has initiated fact-finding missions to identify and explore the intellectual property needs, rights and expectations of holders of traditional knowledge in order to promote the contribution of the intellectual property system to their social, cultural and economic development. These missions use as their reference point the 'traditional knowledge, innovations and practices of indigenous and local communities embodying traditional lifestyles' of Article 8(j) of the CBD (WIPO, 1998a: 1; 1998b).

Among the options for protection of indigenous knowledge now being considered are:

- Contracting parties to the CBD to exercise their rights under Article 15.5 to also require that the prior informed consent of indigenous and local communities in relevant circumstances must be obtained before bioprospecting contracts can be issued. Examples include the Andean Pact Common System on Access (Rosell, 1977); and the Philippines Presidential Executive Order No. 247 of 1995, 'Prescribing Guidelines and Establishing a Regulatory Framework for the Prospecting of Biological and Genetic Resources, Their By-products and Derivatives, for Scientific and Commercial Purposes; and for Other Purposes' (Barber and La Vina, 1997).
- Modification or adaptation of existing IPR regimes to accommodate protection for indigenous communally based knowledge systems; for example, by requiring disclosure of country/indigenous community of origin of knowledge and biological samples used in research leading to the application for a patent as part of the normal patent application process; acceptance by patent offices of evidence of traditional knowledge as prior art – which could stop the issue of a patent, as in the case of turmeric (Secretariat for the Convention on Biological Diversity, 1996b: 21; Dutfield, 1997: 11; Tobin, 1997);
- *Sui generis* legislation; that is, national governments implement parallel laws which deal with indigenous IPRs in a way which is satisfactory to the indigenous peoples concerned. Proposed models for such laws include the Principles and Guidelines for the Protection of the Heritage of Indigenous People, elaborated by Mrs Erica-Irene Daes, Special Rapporteur of the Sub-Commission on Prevention of Discrimination and Protection of Minorities (Daes, 1996); the Third World Network's Conceptual Framework and Essential Elements of a Rights Regime for the Protection of Indigenous Rights and Biodiversity (Nijar, 1996a); and the Organization of African Unity Scientific, Technical and Research Commission's (1998b) *Draft Legislation on Community Rights and Access to Biological Resources.*
- The recognition of indigenous traditional resource rights through the exercise and recognition of a 'bundle of rights', as expressed through current and emerging international and domestic jurisprudence, in national and subnational legal systems (Posey and Dutfield, 1996).

- Institutional self-regulation through industry-based codes/guidelines for ethical conduct, such as the Manila Declaration (Seventh Asian Symposium on Medicinal Plants, Spices and Other Natural Products, 1992); the Guidelines for Equitable Partnerships in New Natural Products Development, of the People and Plants Initiative of the World Wide Fund for Nature and UNESCO, 1993, and the US National Institutes of Health and the National Cancer Institute, Conclusions of the Workshop on Drug Development, Biological Diversity and Economic Growth (National Institutes of Health, 1992).
- Biodiversity agreements (between biodiversity prospectors, governments and indigenous communities); these may take the form of legally binding contracts, material and information transfer agreements, licensing agreements, memoranda of understanding, etc., and would be based on the prior informed consent of the relevant community, be subject to mutually agreed terms and include benefit-sharing arrangements (see Posey and Dutfield, 1996: 67–74; Grifo and Downes, 1996).
- Indigenous community-administered access regimes as available to the Inuit in the Canadian Northwest Territories under the Scientists Act 1974 (Mann, 1997: 20–25).
- Use of existing statutory mechanisms (for example, access and permit regimes, capacity to make by-laws, biodiversity management plans, regional agreements) which may be available to indigenous communities under a raft of national and subnational laws dealing with cultural heritage, land tenure, nature conservation, environmental protection, economic development and community self-governance.
- Accommodation of indigenous customary law systems within national legal frameworks, which could be used to govern access to TEK and natural resources (for a review of these options see Fourmile, 1998).

Indigenous communities may also be able to use the common law as a source of remedy for protection against abuse of their TEK in those countries which have such a tradition. While there are a number of common-law principles which could be applied, such as those governing unconscionable behaviour and unjust enrichment, actions for breach of confidentiality, 'passing off' and unfair competition have brought relief. And, as a last resort, indigenous communities may be able to use procedures available to them under various international treaties to enforce protection for their TEK. One such measure is the use of the optional complaint procedures available under Article 14(1) of the ICEAFRD in those countries which have made a declaration recognizing the competence of the Committee on the Elimination of Racial Discrimination (CERD) to receive and consider communications from individuals or groups within their jurisdiction who claim to have had their rights as set out in the Convention violated. Under Article 5, the Convention guarantees a range of civil, political, economic, social and cultural rights (Fourmile, 1998).

Conservation partnerships with indigenous peoples

As Downes (1992: 3) points out, development activities that work with and through indigenous knowledge and organizational structures have several important advantages over projects that operate outside them. Indigenous knowledge provides the basis for grassroots decision-making, much of which takes place at the community level through indigenous organizations and associations where problems are identified and solutions to them are determined. Solution-seeking behaviour is based on indigenous creativity leading to experimentation and innovations as well as the appraisal of knowledge and technologies introduced from other societies.

Conventional Western scientific resource management has come under criticism because it is equilibrium-based or has an underlying assumption of ecological stability (Berkes *et al.*, 1997: 17, citing Holling, 1986; Gunderson *et al.*, 1995; Holling *et al.*, 1997). Resource management from a stability point of view may be characterized in terms of rules and regulations made by technical experts, often from a central bureaucracy, and enforced by agents who are not themselves resource users; emphasis on steady states and the maintenance of predictable yields, such as maximum sustainable yield; focus on controlling the resource to increase the predictability of yields; and the use of primarily quantitative techniques, such as stock assessment.

Several analyses have illustrated that such management seems to cause a gradual loss of resilience, of variability and opportunity, and moves the ecosystem towards thresholds and surprises (Berkes *et al.*, 1997: 17, citing as examples Regier and Baskerville, 1986; Ludwig *et al.*, 1993). Loss of resilience is often masked by the development of fossil fuel-dependent technologies to maintain yields, such as bigger fishing vessels or artificial fertilizers. It can also be masked through support from socio-economic infrastructures that make it possible to maintain a business-as-usual strategy when faced with ecological disturbance. Examples include capital markets that provide loans and financial insurance to fisherman and farmers in periods of resource crisis, thereby removing incentives for building an ecological knowledge base (Berkes *et al.*, 1997: 17–18).

TEK and practice, in contrast to scientific ecology, may be characterized as 'resource management from a resilience point of view' (Berkes *et al.*, 1997: 18). It is carried out using rules which are locally crafted and socially enforced by the users themselves. Resource use tends to be flexible, with, for example, resource rotations, species switching, multiple-species management regimes, protection of specific habitats (often a function of sacred areas), and protection of species at various stages of vulnerability in their life cycles. The inhabitants of such ecosystems have accumulated an ecological knowledge base that helps respond to environmental feedbacks, such as the catch per unit of effort, that helps monitor the status of the resource, and a reliance on a diversity of resources for livelihood security, thus keeping options open and minimizing risk. It is carried out using qualitative management wherein feedbacks of resource and ecosystem change indicate the direction in which management should move (for example, by adjusting levels of harvesting of a species) rather than towards a quantitative yield target. Such systems have evolved because of the necessities of survival over long periods of time (the seven generation principle) and are a consequence of historical experience with disturbance and ecological surprise, as well as not having access to modern technology and socio-economic infrastructures with which disturbance can be exported in space and time (Berkes *et al.*, 1997: 18, citing Holling *et al.*, 1997). Thus TEK could be viewed as a 'library of information' on how to cope with dynamic change in complex systems. It connects the present to the past and re-establishes resilience (Berkes *et al.*, 1997: 18, citing Gunderson *et al.*, 1997). Building ecological knowledge to understand qualitative changes in complex systems has thus been a means for improving a group's chances of survival.

Scientific ecology and traditional ecological knowledge are, however, potentially complementary (Berkes *et al.*, 1997: 17). This is recognized by the contracting parties to the CBD, where case studies have been called for under a number of decisions of the Conference of the Parties which examine the integration of scientific and traditional knowledge and management regimes in the management and sustainable use of biodiversity (for example, Decision III/ 14 of the third Conference of the Parties and Decision IV/9 of the fourth Conference of the Parties). There is now a plethora of studies which attest to successful collaborations between science and TEK (see, for example, Secretariat for the Convention on Biological Diversity, 1997c). As the US Minerals Management Service concluded in its case study on traditional knowledge, 'The blending of TEK and Western Science will change forever the way scientific research, inquiry, and assessment is done in the Arctic' (US Minerals Management Service, in Secretariat for the Convention on Biological Diversity, 1997c).[5]

Threats to indigenous rights: biotechnology, biopiracy and the new world trade order

The capacity of indigenous peoples to carry out their duty to the Earth and the life it sustains is now being placed in further jeopardy by advances in biotechnology, the increasing incidence of biopiracy – which takes advantage of culturally based intellectual property regimes – and the emerging new world trade order, which seeks to entrench both.

Advances in genetic engineering and biotechnology are now seen to have real and potentially great economic value in programmes of national development based on what may be termed the 'life industry' involving agrochemicals, plant breeding, food and beverages, pharmaceuticals, cosmetics and veterinary products. The life industry has arisen through mergers and cooperative agreements among corporations to profit from the manipulation and ownership of living organisms. With the development of biotechnology and the increased use of Western-derived intellectual property systems, the previously discrete agrochemical, seed, pharmaceutical and food industries increasingly depend upon a similar set of technologies and laws which allow the monopoly control of living organisms (Rural Advancement Foundation International, undated: 75, 68–70). The giant US-based corporation Monsanto is emerging as a classic example of this phenomenon. One of the largest life industry corporations, it was an agrochemical and chemical giant before it started genetically engineering crops, such as 'Round-Up Ready Soya and Cotton', which are resistant to its own brand of herbicide, 'Round-Up', the world's top-selling herbicide.

The most widespread application of genetic engineering in agriculture is herbicide resistance – the breeding of crops to be resistant to herbicides. When introduced to indigenous and local community farming systems, this will lead to increased use of agrichemicals, thus increasing environmental problems. It will destroy the biodiversity that is the sustainable and livelihood base of rural populations. What are weeds for corporations like Monsanto are the food, fodder and medicine for such communities (Shiva, 1998b: 2). Monsanto has recently acquired Cargill Seeds, which buys 80 per cent of all US produce, and has moved to wholly own the Delta and Pine Land Company, the company that, in conjunction with the US Department of Agriculture, developed and patented the 'terminator technology' which suppresses plant seed germination (Ho, 1998: 2).

The purpose of the terminator gene is to protect corporate patents on seeds, to prevent farmers from saving seeds for replanting, and thus it provides a biological means to enforce patent ownership. The implications of such technology are enormous, and, if pursued, could threaten the food security of an estimated 1.4 billion local community farmers in Latin America, Africa and Asia, who depend upon saved seed and their own breeding skills in adapting other varieties for use on their (often marginal) lands (RAFI, 1998b). Critics have already pointed out that the terminator genes are bound to spread from the genetically engineered plants and 'truly terminate our food supply' (Ho, 1998: 1). As Ho emphasizes, the technology also involves genes from the most infectious genetic parasites that are capable of scrambling genomes, and hence it is potentially extremely hazardous to human and animal health. This brings home to us that what is at stake is not just whether we should accept genetically engineered foods. It is life, our life-support and our value system as human beings that are being placed under unaccountable corporate monopoly (*ibid.*).

The following figures provide some example of the scale of the economic value of the life industries. With regard to pharmaceuticals, for example, in the United States their commercial value exceeds $36 billion annually. Globally, about 80 per cent of the human population relies on traditional medical systems, and about 85 per cent of traditional medicine involves the use of plant extracts. Over-the-counter plant-based drugs have an estimated market value of $20 billion per year in the United States and $84 billion worldwide (PCAST, 1998: 5). Extractions from wild species in biodiversity's 'genetic library' account for approximately 50 per cent of annual increases in crop productivity accomplished by biotechnological and agricultural research and development.

At present, just over 100 plant species directly or indirectly contribute 90 per cent of the global human food supply, with only three – rice, corn and wheat – supplying 60 per cent, but thousands of plant species are cultivated or consumed from the wild somewhere on Earth. Some of these may be more nutritious or better suited to certain widespread growing conditions than are species currently widely cultivated (*ibid.*). Scientists estimate that fewer than 30 per cent of species that occur in developed countries like the United States have been discovered and described, while worldwide the total is estimated to be fewer than 15 percent. Many described species have useful properties; it is therefore reasonable to predict that some of those which have yet to be discovered also have beneficial attributes (genetic resistance to disease, food value, compounds that could become pharmaceuticals, etc.) that could be employed in sustainable agricultural and industrial development (*ibid.*: xvi).

In addition to the general concern regarding the hazards of genetic engineering biotechnology on human health and biodiversity (see, for example, Ho, 1998b; Ho *et al.*, 1998) is the influence of corporate dollars on the integrity of the research carried out to trial and test many of these new genetic creations. There is substantial evidence that tensions between research ethics and the corporate imperative for profit are especially pronounced when life industry companies underwrite the costs of clinical trials for their new products, as they frequently do in the United States and Canada. Researchers who are funded by companies to carry out trials on their behalf, for example, are also frequently bound by non-disclosure clauses in their contracts so that they are unable to publish research results which reflect negatively on products. Alternatively, if their research reveals 'problems' with a certain product due for market release, they (or their university) may be denied lucrative research contracts from the company in the future (Foss and Taylor, 1998).

It has now been realized that indigenous peoples and their knowledge provide short-cuts to many of the most potentially useful species and that we are, after all, the best situated to manage that biodiversity (see, for example, Warren *et al.*, 1995; Clarkson *et al.*, 1992). The use of traditional knowledge reportedly increases the efficiency of screening plants for medicinal properties by more than 400 percent, significantly cutting research costs and the time involved (Balick, 1990). And Brush argues that to acquire a comprehensive base of knowledge for genetic resource conservation, 'the genetic establishment must accept a mandate to be concerned not only with germplasm but also the knowledge systems that produce it' (Brush, 1989: 22). In spite of the recognition of the importance of our knowledge to the life industry, international intellectual property systems based on Western notions which recognize intellectual property as private property still refuse to acknowledge or accommodate our communally based intellectual property rights in our traditional knowledge (Secretariat for the Convention on Biological Diversity, 1996b: 20). Consequently, the phenomenon of biopiracy, whereby Western corporations are stealing centuries of collective knowledge and innovation carried out by indigenous and local communities worldwide, is reaching epidemic proportions (Shiva,

1998b: 1; RAFI, 1996). Some of the more noteworthy examples of biopiracy include cases concerning turmeric (Agarwal and Narain, 1996), endod – also known as the African soapberry plant (RAFI, 1993a), Bolivian quinoa (*Chenopodium quinoa*) (RAFI, 1996: 1–2), natural-coloured cottons of the Americas (RAFI, 1993b) and chickpeas (RAFI, 1998a).

Likewise, we are left out in the cold in most of the world's trading agreements, such as the World Trade Organization's Agreement on Trade-Related Aspects of Intellectual Property Rights (TRIPs) (see, for example, Nijar, 1996b; Williams, 1997; Dutfield, 1997; Simpson, 1997) and the OECD-inspired draft Multilateral Agreement on Investment (MAI). The latter is seen by some as the blueprint for the future global economy and is intended to lock in the benefits of the substantial investment regime liberalization that has been achieved in recent years and to roll back measures that still discriminate against foreign investors. OECD member governments are determined to negotiate an agreement based on 'high standards', which refers primarily to the 'quality' of the investment environment, namely rules that will provide the highest degree of market access and legal security for investors and their investments (Witherell, 1996: 2). Successful conclusion of the MAI will mean that much of the world's investment flows will be covered by a comprehensive framework of international rules of the game (*ibid.*).

It is proposed that the MAI should be legally binding and contain provisions regarding its enforcement; that it should apply to all levels of government. Private enforcement against state and local government laws is a major goal of the MAI. In the WTO setting, if a provincial law is found to violate the agreement, it provides only a 'diplomatic' remedy against the offending national government, such as the reimposition of tariffs by the complaining government. In the MAI, negotiators propose giving foreign investors access to domestic courts to enforce remedies against state laws (Singer *et al.*, 1997: 11).

It is also proposed to give foreign investors the right to seek monetary damages for both past and future losses, and to seek rulings on whether domestic laws violate investor protections under the MAI. Foreign investors may also have a choice of using international arbitration panels or domestic courts to seek damages. While international arbitration panels might be friendlier fora for foreign investors, access to domestic courts gives them the ability to seek injunctions to stop enforcement of offending laws, a power that international arbitrators do not have (Singer *et al.*, 1997: 23). The implementing legislation for both NAFTA and the WTO states that if a provision of the agreements *conflicts* with federal law (which includes the Constitution), then that provision has no effect. In contrast, the MAI proposal takes the approach that domestic law applies to the extent that it is *consistent* with the MAI (*ibid.*: 24). This constitutes a direct assault on national sovereignty.

The MAI, as currently proposed, will have several impacts on indigenous communities, especially those which lie within the jurisdiction of OECD member countries. In those self-governing indigenous communities which have local government status under national or provincial laws, the MAI will affect their abilities to create and enforce local laws governing such matters as access to land and natural resources, negotiate and enforce labour contracts and employment conditions, and take special measures to protect their environment. Transnational corporations (e.g. bioprospecting and logging companies) will be able to take local indigenous councils to court and seek monetary damages against them if they feel that such local laws unduly hinder their activities.

The MAI as proposed could impact on any *sui generis* laws instituted to specifically protect indigenous rights or interests (such as cultural heritage legislation). It could also have a 'levelling down' effect on provincial laws governing such matters as land rights and indigenous natural resource use because of any inconsistencies between the laws of different provinces which a transnational corporation might expose. The consequence of this is that provincial laws might be 'levelled down' to the least restrictive (in terms of accommodating corporate interests) of the provincial laws. It could also impact on amendments to or provisions of intellectual property laws intended to extend protection to indigenous communal intellectual property rights, or to other measures intended in some way to specifically protect the IPRs of indigenous communities (for example, prior informed consent measures regarding access to indigenous natural resources and traditional knowledge). And finally, because indigenous peoples do not have the status of peoples in international law, international treaties which might be used to protect the rights of peoples against corporate predations perpetrated under an investment treaty such as the proposed MAI will be of no avail (see also Clarke and Barlow, 1997: 158–60).

As Tony Clarke concludes, if implemented, the MAI will amount to a declaration of global corporate rule, designed to enhance the political rights, power and security of transnational corporations on a worldwide scale. It should not be forgotten that the twenty-nine OECD member countries are headquarters to over 95 per cent of the world's largest transnational corporations (based on Global Fortune 500 ratings; Clarke, 1997: 2). The OECD countries are also the headquarters for the top ten corporations in each of the categories of the life industry (mentioned above). These corporations include Monsanto (USA), Novartis (Switzerland), Glaxo-Wellcome (UK), Merck (USA), Hoechst (Germany) and Limagrain (France) (RAFI, undated: 62–70). While many countries consider the TRIPs agreement to be antithetical to the cause of biodiversity conservation and sustainable use (see, for example, Organization of African Unity, 1998a), treaties such as the proposed MAI will serve to further bolster the predatory activities of such corporations in indigenous communities and further undermine both cultural and biological diversity.

Conclusion

Whether they are intended to be or not, instruments such as the Rio Declaration on Environment and Development and the Convention on Biological Diversity are as much human rights treaties as they are treaties for the environment, and to ignore this reality is to severely diminish their usefulness. If the biodiversity on indigenous territories continues to be eroded as a result of inappropriate government policies and practices, aspects of their cultures will die and much valuable TEK will be lost. If indigenous peoples are forced to relinquish their traditional customs and languages through, for example, assimilatory programmes which emphasize the conversion of their traditional economies based on biodiverse agricultural and hunter-gatherer ecosystems to cash economies based on monocultural systems of resource exploitation, then both cultural and biological diversity will suffer.

In the interests of both global environmental justice and justice to indigenous peoples it is fundamental that our rights to our traditional lands and waters be secured, that our rights to be self-governing be respected, that our traditional knowledge systems be respected and protected, and that we should be enabled to live our cultures in peace, free of unwanted outside interference. Our own environmental ethic will then ensure that we respect our Mother and all that she nurtures and sustains, and that in turn she will provide for us. In that way both human and environmental justice will be served.

Notes

1 The convention concerns Indigenous and Tribal Peoples in Independent Countries (Centre for Human Rights, 1994: 475–89). It has only been ratified by a handful of countries – three (see Deer, 1994: 102–105).
2 For an overview of these and other instruments concerning the recognition of indigenous natural resource rights see Craig (1996: 91–104) and Sutherland (1996: 67–85).
3 Adopted by the Sub-Commission on Prevention of Discrimination and Protection of Minorities at its forty-sixth session (Working Group on Indigenous Populations, 1994).
4 Both the foregoing are cited in Grifo and Downes (1996: 302).
5 The study concerned the inclusion of TEK in an Environmental Impact Study concerning a Beaufort Sea Oil and Gas Lease Sale conducted with the Alaska Eskimo Whaling Commission and, among others, the native peoples of Barrow, Nuiqsut and Kaktovik.

References

Agarwal, A. and Narain, S. (1996) 'Pirates in the Garden of India', *New Scientist* 26: 14–15.
Akerele, O. (1995) 'Medicinal Plants and Protected Areas', in J. A. McNeely (ed.) *Expanding Partnerships in Conservation*, Washington, DC: Island Press.
Altieri, M. A. (ed.) (1987) *Agroecology: The Scientific Basis of Alternative Agriculture*, Boulder, CO: Westview Press.
Balick, M. (1990) 'Ethnology and the Identification of Therapeutic Agents from the Rainforest', in D. J. Chadwick and J. Marsh (eds) *Bioactive Compounds from Plants*, Chichester: John Wiley, pp. 145–59.
Barber, C. V. and La Vina, A. (1997) 'Regulating Access to Genetic Resources: Philippine's Access to Genetic Resources Regime', in J. Mugabe, C. V. Barber, G. Henne,

L. Glowka and A. La Vina (eds) *Access to Genetic Resources: Strategies*, Nairobi: African Centre for Technology Studies Press. pp. 115–41.

Berkes, F., Colding, J. and Folke, C. (1997) 'Rediscovery of Traditional Ecological, Knowledge as Adaptive Management', Beijer Discussion Paper Series No. 109, Beijer International Institute of Ecological Economics, Stockholm: Royal Swedish Academy of Sciences.

Bragdon, S. H., Downes, D. with Shand, H. and Halewood, M. (1998) *Recent Policy Trends and Development Related to the Conservation, Use and Development of Genetic Resources* (pre-publication copy), Rome: International Plant Genetic Resources Institute.

Brush, S. (1989) 'Rethinking Crop Genetic Resource Conservation', *Conservation Biology* 3(1): 19–29.

Centre for Human Rights (1994) *Human Rights: A Compilation of International Instruments*, vol. 1: *Universal Instruments*, United Nations: New York and Geneva.

Chandler, M. (1993) 'The Biodiversity Convention: Selected Issues of Interest to the International Lawyer', *Colorado Journal of International Environmental Law and Policy* 4: 139–42.

Clarke, T. (1997) *The Corporate Rule Treaty: The Multilateral Agreement on Investments (MAI) Seeks to Consolidate Global Corporate Rule*, Ottawa: Canadian Centre for Policy Alternatives.

Clarke, T. and Barlow, M. (1997) *MAI: The Multilateral Agreement on Investment and the Threat to Canadian Sovereignty*, Toronto: Stoddart.

Clarkson, L., Morrissette, V. and Regallet, G. (1992) *Our Responsibility to the Seventh Generation: Indigenous Peoples and Sustainable Development*, Winnipeg: International Institute for Sustainable Development.

COBASE (1998) *Resolution of Rome: Guidelines for the Protection of Cultural Diversity*, International Conference on Diversity as a Resource: Relationship between Cultural Diversity and Environmental-Oriented Society, Rome, 2–6 March, Rome: COBASE (Cooperativa Tecnico-Scientifica di Base).

Colchester, M. (1994) *Salvaging Nature: Indigenous Peoples, Protected Areas and Biodiversity Conservation*, Geneva: United Nations Research Institute for Social Development (UNRISD); Penang, Malaysia: World Rainforest Movement; Gland, Switzerland: World Wide Fund for Nature.

Commission on Environmental Law of IUCN in cooperation with International Council of Environmental Law (1995) *Draft International Covenant on Environment and Development*, Gland, Switzerland: IUCN/The World Conservation Union.

Commission on Genetic Resources for Food and Agriculture (CGRFA) (1997) *Revision of the International Undertaking on Plant Genetic Resources: Fourth Negotiating Draft*, Document CGRFA/IUND/4 Rev. 1, Rome: CGRFA.

Craig, D. (1996) 'Aboriginal and Torres Strait Islander Involvement in Bioregional Planning: Requirements and Opportunities under International and National Law and Policy', in *Approaches to Bioregional Planning Part 2: Background Papers to the Conference*, 30 October–1 November 1995, Melbourne, Canberra: Commonwealth Government, Department of the Environment, pp. 79–144.

Cultural Survival Canada (undated) *Biodiversity Stewardship and the Rights of Indigenous Peoples: A Comparison of Article 8(j) of the Convention on Biological Diversity and Common Themes from Indigenous Peoples' Statements and Declarations*, Ottawa: Cultural Survival Canada.

Cunningham, A. B. (1993) *Ethics, Ethnobiological Research, and Biodiversity*, Gland, Switzerland: World Wide Fund for Nature.

Daes, E. A. (1995) 'Note by the Chairperson-Rapporteur of the Working Group on Indigenous Populations on Criteria Which Might Be Applied when Considering the Concept of Indigenous Peoples', Document E/CN.4/Sub.2/AC.4/1995/3,

Geneva: United Nations Commission on Human Rights Sub-Commission on Prevention of Discrimination and Protection of Minorities.

—— (1996) *Principles and Guidelines for the Protection of the Heritage of Indigenous People*, Document E/CN.4/Sub.2/1996/22, Geneva: United Nations Commission on Human Rights Sub-Commission on Prevention of Discrimination and Protection of Minorities.

Deer, K. (1994) 'The Failure of International Law to Assist Indigenous Peoples', in A. P. Morrison (ed.) *Justice for Natives: Searching for Common Ground*, Montreal: Aboriginal Law Association of McGill University, pp. 100–5.

Diversity (1996) 'Scientific and Ethical Issues Surrounding Biotechnology and Biodiversity, Focus of APHIS Symposium', *Diversity* 12 (4): 16–17.

Downes, D. R. (1995) 'Global Trade, Local Economies and the Convention on Biological Diversity', in W. J. Snape (ed.) *Biodiversity and the Law*, Washington, DC: Island Press.

Durning, A. T. (1992) 'Guardians of the Land: Indigenous Peoples and the Health of the Earth', Worldwatch Paper 112, Washington, DC: Worldwatch Institute.

Dutfield, G. (1997) 'Can the TRIPs Agreement Protect Biological and Cultural Diversity?' Biopolicy International Series No. 19, Nairobi: African Centre for Technology Studies Press.

First International Conference on the Cultural and Intellectual Property Rights of Indigenous Peoples (1993) *The Mataatua Declaration on Cultural and Intellectual Property Rights of Indigenous Peoples*, Whakatane, New Zealand, 12–18 June.

Foss, K. and Taylor, P. (1998) 'Volatile Mix Meant Trouble at Hospital: Corporate Cash Weakens Scientific Ethics', *The Globe and Mail*, 22 August: A1 and A4.

Four Directions Council (1996) *Forests, Indigenous Peoples and Biodiversity: Contribution of the Four Directions Council*, Submission to the Secretariat of the Convention on Biological Diversity, Montreal.

Fourmile, H. L. (1998) 'The Convention on Biological Diversity, Intellectual Property Rights and the Protection of Traditional Ecological Knowledge', Master's dissertation, Law School, Macquarie University, Sydney.

Gadgil, M., Berkes, F. and Folke, C. (1993) 'Indigenous Knowledge for Biodiversity Conservation', *Ambio* 22 (2–3): 151–56.

GRAIN (Genetic Resources Action International) (undated). *Biodiversity Farming Produces More*, Background Paper, Barcelona: GRAIN.

Grifo, F. T. and Downes, D. R. (1996) 'Agreements to Collect Biodiversity for Pharmaceutical Research: Major Issues and Proposed Principles', in S. B. Brush and D. Stabinsky (eds) *Valuing Local Knowledge: Indigenous Peoples and Intellectual Property Rights*, Washington, DC: Island Press, pp. 281–304.

Gunderson, L., Holling, C. S. and Light, S. (eds) (1995) *Barriers and Bridges to the Renewal of Ecosystems and Institutions*, New York: Columbia University Press.

Gunderson, L. H., Holling, C. S., Pritchard, L. and Peterson, G. (1997) 'Resilience in Ecosystems, Institutions and Societies', Beijer Discussion Paper 95, Stockholm: Beijer International Institute of Ecological Economics, Royal Swedish Academy of Sciences.

Haugerud, A. and Collinson, M. P. (1991) *Plants, Genes and People: Improving the Relevance of Plant Breeding*. Gatekeeper Series No. 30, London: International Institute for Environment and Development.

Hecht, S. B. and Posey, D. A. (1989) 'Preliminary Results on Soil Management Techniques of the Kayapo Indians', *Advances in Economic Botany* 7: 174–88.

Ho, M. W. (1998) *Genetic Engineering Dream or Nightmare? The Brave New World of Bad Science and Big Business*, Bath: Gateway Books; Penang, Malaysia: Third World Network.

Ho, M. W., Traavik, T., Olsvik, R., Midtvedt, T., Tappeser, B., Howard, V., von Weiz-säcker, C. and McGavin, G. (1998) *Gene Technology and Gene Ecology of Infectious Diseases*, Penang, Malaysia: Third World Network.

Holling, C. S. (1986) 'The Resilience of Terrestrial Ecosystems: Local Surprise and Global Change', in W. C. Clark and R. E. Munn (eds) *Sustainable Development of the Biosphere*, Cambridge: Cambridge University Press, pp. 292–317.

Holling, C. S., Berkes, F. and Folke, C. (1997) 'Science, Sustainability, and Resource Management', in F. Berkes and C. Folke (eds) *Linking Social and Ecological Systems: Management Practices and Social Mechanisms for Building Resilience*, Cambridge: Cambridge University Press.

Indigenous Peoples of the Western Hemisphere (1995) *Phoenix Declaration of Indigenous Peoples of the Western Hemisphere Regarding the Human Genome Diversity Project*, Phoenix, AZ: 19 February.

Indigenous-Tribal Peoples of the Tropical Forests (1992) Charter of the Indigenous-Tribal Peoples of the Tropical Forests, Penang, Malaysia, 15 February.

Inglis, J. T. (ed.) (1993) *Traditional Ecological Knowledge: Concepts and Cases*, Ottawa: International Programme on Traditional Ecological Knowledge and International Development Research Centre.

International Alliance of Indigenous-Tribal Peoples of the Tropical Forests and International Work Group for Indigenous Affairs (undated), *Indigenous Peoples, Forests and Biodiversity*, London: IAITPTF; Copenhagen: IWGIA.

International Society of Ethnobiology (1988) *The Declaration of Belém*, First International Congress of Ethnobiology, Belém, Brazil., in Posey and Dutfield (1996) p. 2.

Inuit Circumpolar Conference (undated), *Agenda 21 from an Inuit Perspective*, Ottawa: ICC.

Johannes, R. E. (ed.) (1989) *Traditional Ecological Knowledge: A Collection of Essays*, Gland, Switzerland: International Conservation Union (IUCN).

Johannes, R. E. and Ruddle, K. (1993) 'Human Interactions in Tropical Coastal and Marine Areas: Lessons from Traditional Resource Use', in A. Price and S. Humphrey (eds) *Applications of the Biosphere Reserve Concept to Coastal Marine Areas*, Gland, Switzerland: IUCN, pp. 19–25.

Johnson, M. (ed.) (1992) *Lore: Capturing Traditional Environmental Knowledge*, Hay River (Canada): Dene Cultural Institute; Ottowa: International Development Research Centre.

Lalonde, A. (1993) *The Federal Environmental Assessment and Review Process and Traditional Ecological Knowledge*, Ottawa: Prepared for the Environmental Assessment Branch, Environment Canada.

Latz, P. (1995) *Bushfires and Bushtucker: Aboriginal Plant Use in Central Australia*, Alice Springs: Institute of Aboriginal Development.

Ludwig, D., Hilborn, R. and Walters, C. (1993) 'Uncertainty, Resource Exploitation and Conservation: Lessons from History', *Science* 260: 1736.

McDonald, M., Arragutainaq, L. and Novalinga, Z. (1997) *Voices from the Bay: Traditional Ecological Knowledge of Inuit and Cree in the Hudson Bay Bioregion*, Ottawa: Canadian Arctic Resources Committee and Municipality of Sanikiluaq.

Mann, H. (1997) *Indigenous Peoples and the Use of Intellectual Property Rights in Canada: Case Studies Relating to Intellectual Property Rights and the Protection of Biodiversity*, Consultant's Report to the Intellectual Property Policy Directorate, Ottawa: Corporate Governance Branch, Industry Canada.

Mittermeier, R. A., Gil, P. R. and Mittermeier, C. G. (1997) *Megadiversity: Earth's Biologically Wealthiest Nations*, Washington, DC: Conservation International.

National Institutes of Health (1992) *Conclusions of the Workshop on Drug Development, Biological Diversity and Economic Growth*, Bethesda, MD: NIH.

New Zealand Conservation Authority (1997) *Maori Customary Use of Native Birds, Plants and Other Traditional Materials: Interim Report and Discussion Paper*. Wellington: NZCA.

Nijar, G. S. (1996a) *In Defence of Local Community Knowledge and Biodiversity: A Conceptual Framework and the Essential Elements of a Rights Regime*, Penang, Malaysia: Third World Network.

—— (1996b) *TRIPs and Biodiversity – The Threat and Responses: A Third World View*, Penang, Malaysia: Third World Network.

North American Indigenous Peoples Summit on Biological Diversity and Biological Ethics (1997) *The 'Heart of the Peoples' Declaration*, Gros Ventre and Assinboine Nations' Territories, Fort Belknap Reservation, Montana, 7 August.

Oldfield, M. L. and Alcorn, J. B. (1991) 'Conservation and Traditional Agroecosystems', in M. L. Oldfield and J. B. Alcorn (eds) *Biodiversity: Culture, Conservation and Ecodevelopment*, Boulder, CO: Westview Press, pp. 37–58.

O'Neill, G. (1995) 'Lab Team Tackles the Toughest Nut', *The Sunday Mail* (Brisbane), 5 February: 12–13.

Organization of African Unity, Scientific, Technical and Research Commission (1998a) *Declaration by the OAU/STRC Task Force on Community Rights and Access to Biological Resources*, Addis Ababa: OAU/STRC.

—— (1998b) *Draft Legislation on Community Rights and Access to Biological Resources*, Addis Ababa.

Pereira, W. and Gupta, A. K. (1993) 'A Dialogue on Indigenous Knowledge', *Honey Bee* 4 (4): 6–10.

Peteru, C. (1995) *Treaty for a Lifeforms Patent-Free Pacific and Related Protocols*, Suva, Fiji, April.

Posey, D. A. (1985) 'Management of Tropical Forest Ecosystems: The Case of the Kayapo Indians of the Brazilian Amazon', *Agroforestry Systems* 3 (2): 139–58.

Posey, D. A. and Dutfield, G. (1996) *Beyond Intellectual Property: Towards Traditional Resource Rights for Indigenous Peoples and Local Communities*, Ottawa: International Development Research Centre.

President's Committee of Advisers on Science and Technology, Panel on Biodiversity and Ecosystems (PCAST) (1998) *Teaming with Life: Investing in Science to Understand and Use America's Living Capital*, Washington, DC: PCAST Secretariat.

Regier, H. A. and Baskerville, G. L. (1986) 'Sustainable Redevelopment of Regional Ecosystems Degraded by Exploitive Development', in W. C. Clark and R. E. Munn (eds) *Sustainable Development of the Biosphere*, Cambridge: Cambridge University Press, pp. 75–101.

Reid, W., Barber, C. and Miller, K. (1992) *Global Biodiversity Strategy: Guidelines for Action to Save, Study and Use Earth's Biotic Wealth Sustainably and Equitably*, New York: World Resources Institute; Gland, Switzerland: The World Conservation Union, and Nairobi: United Nations Environment Programme.

Rosell, M. (1977) 'Access to Genetic Resources: A Critical Approach to Decision 391 "Common Regime on Access to Genetic Resources", of the Commission of the Cartagena Agreement', *RECIEL* 6 (3): 274–83.

Rural Advancement Foundation International – (RAFI) (1993a) 'Endod: A Case Study of the Use of African Indigenous Knowledge to Address Global Health and Environmental Problems', RAFI communiqué, March.

—— (1993b) 'Bio-piracy: The Story of Natural Coloured Cottons of the Americas', RAFI communiqué, November.

—— (1996) '1996 Biopiracy Update: US Patents Claim Exclusive Monopoly Control of Food Crop, Medicinal Plants, Soil Microbes and Traditional Knowledge from the South', RAFI communiqué: RAFI: December.

—— (1998a) 'The Australian PBR Scandal: UPOV Meets a Scandal "Down Under" by Burying Its Head in the Sand', RAFI communiqué, Ottawa: January/February.

—— (1998b) 'Terminating Food Security? The Terminator Technology That Sterilizes Seed also Threatens the Food Security of 1.4 Billion People and Must Be Terminated', press release, Ottawa: 20 March.

—— (undated) *Enclosures of the Mind: Intellectual Monopolies – A Resource Kit on Community Knowledge, Biodiversity and Intellectual Property*, Ottawa: RAFI.

Schmidt, A. (1990) *The Loss of Australia's Aboriginal Language Heritage*, Canberra: Aboriginal Studies Press.

Secretariat for the Convention on Biological Diversity (1996a) Compilation of International Guidelines Concerning Indigenous and Local Communities, Document UNEP/CBD/COP/3/Inf.24, 30 October, Montreal: United Nations Environment Programme.

—— (1996b) 'Knowledge, Innovations and Practices of Indigenous and Local Communities: Implementation of Article 8(j)', Document UNEP/CBD/COP/3/19, 18 September, Montreal: United Nations Environment Programme.

—— (1996c) 'Traditional Related Knowledge and the Convention on Biological Diversity', Contribution by the Executive Secretary to the Preparation of the Report of the Secretary-General for Programme Element 1.3 of the Intergovernmental Panel on Forests, Document UNEP/CBD/SBSTTA/2/Inf.3, 19 July 1996, Montreal: United Nations Environment Programme.

—— (1997a) *Traditional Knowledge and Biological Diversity*, Document UNEP/CBD/TKBD/1/2, 18 October, Montreal: United Nations Environment Programme.

—— (1997b) 'Report of the Workshop on Traditional Knowledge and Biological Diversity', Document UNEP/CBD/TKBD/1/3, 15 December, Montreal: United Nations Environment Programme.

—— (1997c) 'A Compilation of Case Studies Submitted by Governments and Indigenous and Local Communities Organisations', Document UNEP/CBD/TKBD/1/Inf.1, 30 October, Montreal: United Nations Environment Programme.

Seventh Asian Symposium on Medicinal Plants, Spices and Other Natural Products (1992) *The Manila Declaration*, Manila, Philippines.

Shiva, V. (1998a) *Betting on Biodiversity: Why Genetic Engineering Will Not Feed the Hungry*, New Delhi: Research Foundation for Science, Technology and Ecology (RFSTE).

—— (1998b) 'Monocultures, Monopolies, Myths and the Masculinization of Agriculture', Statement presented at the International Conference on 'Women in Agriculture', Washington, DC, 28 June, New Delhi: Research Foundation for Science, Technology and Ecology.

Simpson, T. (1997) *The Cultural and Intellectual Property Rights of Indigenous Peoples*, prepared on behalf of the Forest Peoples' Programme, Copenhagen: International Work Group on Indigenous Affairs (IWGIA).

Singer, T. and Orbuch, P., with Stumberg, R. (1997) 'Multilateral Agreement on Investment: Potential Effects on State and Local Government', Paper prepared for the Western Governors' Association, Denver, CO.

Society for Research and Initiatives for Sustainable Technologies and Institutions (undated) *Getting Creative Individuals and Communities Their Due: Framework for Operationalizing Articles 8(j) and 10(c)*, submission to the Secretariat for the Convention on Biological Diversity, Ahmedabad: SRISTI.

Sporrong, U. (1997) 'Dalecarlia in Central Sweden before 1800: A Society of Social and Ecological Resilience', in F. Berkes and C. Folke (eds) *Linking Social and Ecological Systems: Management Practices and Social Mechanisms for Building Resilience*, Cambridge: Cambridge University Press, pp. 67–94.

Sutherland, J. (1996) 'Legislative Options and Constraints', in D. Smyth and J. Sutherland (eds) *Indigenous Protected Areas: Conservation Partnerships with Indigenous Landholders*, vol. 1, Canberra: Environment Australia, pp. 1–94.

Tobin, B. (1997) *Protecting Collective Property Rights in Peru: The Search for an Interim Solution*, Lima: Asociación para la Defensa de Los Derechos Naturales (ADN).

UNEP (United Nations Environment Programme) (1997) *The Biodiversity Agenda: Decisions from the Third Meeting of the Conference of the Parties to the Convention on Biological Diversity*, 2nd edition, Buenos Aires, 4–15 November 1996, New York and Geneva: United Nations.

UNESCO (1993) *Amendments to the Draft Programme and Budget for 1994–1995 (27 C/5), Item 5 of the Provisional Agenda (27 C/DR.321)*, Paris: UNESCO.

United Nations Meeting of Experts on Human Rights and the Environment (1994) *Draft Declaration of Principles on Human Rights and the Environment*, Geneva, 16–18 May.

Warren, D. M. (1992) *Indigenous Knowledge, Biodiversity Conservation and Development. Keynote Address at the International Conference on Conservation of Biodiversity in Africa: Local Initiatives and Institutional Roles*, National Museums of Kenya, 30 August–3 September 1992, Nairobi.

Warren, D. M., Slikkerveer, L. J. and Brokensha, D. (eds) (1995) *The Cultural Dimensions of Development: Indigenous Knowledge Systems*, London: Intermediate Technology Publications.

Williams, N. M. and Baines, G. (eds) (1993) *Traditional Ecological Knowledge: Wisdom for Sustainable Development*, Centre for Resource and Environmental Studies, Canberra: Australian National University.

Williams, O. (1997) *TRIPs and the Convention on Biological Diversity: Conflicting Objectives towards Biodiversity*, Consultant's report to the Gaia Foundation, London.

Witherell, W. (1996) 'OECD's Multilateral Agreement on Investment and Its Work on Trade and Competition Policy', Paper presented at Overseas Development Council Conference on Shaping the Trading System for Global Growth and Employment, 12 November, Washington, DC.

Working Group on Indigenous Populations (1994) United Nations Draft Declaration on the Rights of Indigenous Peoples, Document E/CN.4/Sub.2/1994/2/Add.1, Geneva: Commission on Human Rights Sub-Commission on the Prevention of Discrimination and Protection of Minorities.

World Conference of Indigenous Peoples on Territory, Environment and Development (1992) *Kari-Oca Declaration and the Indigenous Peoples' Earth Charter*, Kari-Oca, Brazil, 30 May.

World Intellectual Property Organization (WIPO) (1998a) *Fact-Finding Missions on the Traditional Knowledge, Innovations and Practices of Indigenous and Local Communities: Terms of Reference*, Geneva: WIPO.

—— (1998b) *Main Programme 11: Global Intellectual Property Issues*, Document A/32/2,WO/BC/18/2, Geneva: WIPO.

Worthington, L. F. (1996) 'Transferring Research to Farmers' Fields to Eradicate World Hunger, Focus of International Crop Science Congress', *Diversity* 12 (4): 9–12.

14 Fairness matters

The role of equity in international regime formation

Oran R. Young

Introduction: the problem of fairness

Most realists and neo-realists dismiss fairness or equity as a force to be reckoned with in explaining or predicting the course of events in world affairs. Whatever their feelings about the attractions of various principles of fairness or equity in normative terms, they see no need to resort to such considerations in accounting for what actually happens at the international level. Nor are they inclined to alter this general assessment in analysing the role of international regimes or, more broadly, institutions operating in international society. What is more, this view is shared in large measure by an array of analysts who do not regard themselves as realists or neo-realists. David Victor, for example, has recently written that 'for most states most of the time, the decisionmaking process is mainly a selfish one. Consequently, there exists very little evidence that fairness exerts a strong influence on international policy decisions' (Victor, 1996: 3).

Yet interest in the role of fairness at the international level is both strong and growing stronger. Practitioners who devote their time and energy either to creating international regimes or to operating institutional arrangements once they are in place persist in employing concepts of fairness or equity in characterizing the activities in which they are engaged. Of course, some may see this as little more than a form of self-serving behaviour or even self-delusion on the part of those who are actually engaged in hard-nosed political processes. But interest in the role of fairness is now spreading rapidly within the academic community concerned with the establishment of international regimes designed to solve various problems. As Victor puts it in his discussion of the climate change convention, 'The search for "fair" or "equitable" agreements to slow global warming has become a cottage industry' (Victor, 1996: 1). Much the same can be said of efforts to solve other large-scale or global problems, like ozone depletion or the loss of biological diversity.

How can this be? Are those who focus on considerations of fairness or equity in discussing these issues operating exclusively in the realm of normative analysis, engaging in a form of wishful thinking, or simply fooling themselves? Or is there something significant going on here that the realists, neo-realists and other tough-minded analysts are missing in their effort to explain and predict outcomes without reference to fairness or equity? One obvious response to these queries is to question the utilitarian assumptions that these analysts employ – either explicitly or implicitly – in thinking about the behaviour of the actors in international society. Thus, we might construct behavioural models that stress the role of norms and values in contrast to calculations of benefits and costs and compare the explanatory power of such models with conventional models that start from the assumptions that actors behave in a rational and self-interested manner.

In this chapter, however, I adopt a different approach. Rather than scrapping the assumptions of rational and self-interested behaviour, I ask whether there are circumstances under which those driven by utilitarian concerns will experience incentives to pay serious attention to considerations of fairness or equity in conjunction with international regimes. In order to examine this issue in depth, I devote particular attention to the processes involved in forming institutional arrangements at the international level, though the same question arises regarding the processes involved in operating regimes once they are in place. Some readers may regard this line of analysis as an intellectual dodge, saying that it reduces the concern for fairness or equity to a form of enlightened self-interest that accords no independent role to fairness or equity as a driver of behaviour. In my judgement, however, this reaction misses the point. If considerations of fairness or equity are excluded from utilitarian analyses by definition, then there will be no escaping the morass of arguments pitting fundamentally different and often incommensurable behavioural models against one another. But if, by contrast, we examine carefully the incentives of various actors to pay attention to considerations of fairness and equity, I believe we can account for a lot of behaviour relating to international regimes that is difficult or impossible to explain in terms of narrow or shortsighted calculations of self-interest.

The result of this analysis is by no means a blanket endorsement of the claims of those who see considerations of fairness or equity as a powerful driver of behaviour relating to international regimes. Rather, I reach two conclusions that are considerably more modest but by no means uninteresting in this context. Fairness matters in the creation of international regimes but only as one element in a kind of causal soup that typically includes a variety of other factors operative at the same time. In addition, the role of fairness is a variable that assumes substantially different values across issue areas and even across regimes within the same issue area. Drawing on examples relating largely to environmental cases, I consider this variance throughout the chapter and seek to understand the reasons why actors have stronger or weaker incentives to pay attention to fairness depending upon specific characteristics of the problem at stake or the institutions they create to deal with them.

The role of fairness in regime formation

For the most part, international regimes are products of bargaining processes. This is not to deny the role of spontaneous developments that give rise to customary rules, or what some observers have called tacit bargaining in international society (Axelrod, 1984; Downs and Rocke, 1990). But at the heart of almost every lasting regime lies a constitutive contract that is ordinarily articulated in the form of one or more explicit – though not necessarily legally binding – agreements whose provisions are products of negotiations. Like negotiations taking place in other social settings, institutional bargaining at the international level involves conscious efforts on the part of self-interested actors to pursue their own ends by seeking to hammer out agreements with others. Those who for one reason or another have exceptional bargaining leverage in such processes tend to fare particularly well when bargains are ultimately struck, and there is no general principle or precept calling on them to eschew such advantages in the interests of conforming to community standards of fairness or equity.

At the same time, institutional bargaining at the international level has a number of distinctive characteristics that have significant implications for the pursuit of fairness or equity (Young, 1994: chs 4 and 5). For the most part, efforts to form regimes are exercises in problem-solving rather than efforts to reach agreement on some particular point along a contract curve – a set of feasible outcomes that all participants would prefer to an outcome of no agreement – whose locus is fixed and known to the parties at the outset. Much of the effort in processes of regime formation goes into discussions of the pros and cons of different ways of framing the problem in contrast to extracting concessions from other participants. It makes a big difference in terms of the discourse of problem-solving, for instance, whether an issue is discussed in the language of environmental protection or in the language of sustainable development.[1]

Institutional bargaining typically focuses on the development of projects (e.g. the phasing out of chlorofluorocarbons, the establishment of protected natural areas) designed to solve problems rather than on the deployment of tactics intended to get others to agree to settlements at particular places on relevant contract curves. Under these conditions, those engaged in bargaining may never discover the exact locus or shape of the contract curve, and they are apt to engage in integrative as well as distributive moves as they seek to refine projects expected to form the heart of regimes capable of solving significant collective action problems.[2] What is more, institutional bargaining generally features an effort to build maximum coalitions – ideally, the coalition of the whole – in contrast to minimum winning coalitions.[3] This is not to say that no progress can be made in regime formation in the absence of consensus; consider the sulphur and nitrogen protocols to the European regime on transboundary air pollution as cases in which parties have taken institutional initiatives in the absence of consensus. Yet the goal of institutional bargaining is to establish governance systems that are acceptable to as many of the participants as possible.

How do considerations of fairness come into play in a contractarian setting of this sort (Rawls, 1971; Buchanan, 1975)? In this account, I examine three distinct ways in which taking fairness seriously can facilitate the achievement of mutually beneficial outcomes: lowering transaction costs; accommodating the veil of uncertainty; and respecting negotiation norms.

Lowering transaction costs

Institutional bargaining can and often does become a protracted and costly affair. Even in cases that ultimately reach successful conclusions, progress is frequently slow and tortuous. Consider the eight years of intensive negotiations required to produce the 1982 United Nations Convention on the Law of the Sea (UNCLOS), not to mention the twelve years that elapsed between the signing of this convention and its formal coming into force (Sebenius, 1984). And there is no guarantee that the resources expended on regime formation will yield any clear-cut benefits at all. The 1988 Convention on the Regulation of Antarctic Mineral Resource Activities (CRAMRA) took six years to negotiate. Yet it was abandoned shortly after the completion of an agreed text owing to defections on the part of Australia and France followed by a host of other countries (Orrego Vicuna, 1988). None of this alters the fact that institutional bargaining is an interest-based activity in which participants drive hard bargains designed to promote their own causes. But rational participants will be acutely aware of the costs involved in regime formation, and they can be expected to weigh these costs against any future benefits expected to flow from the operation of a regime once it is up and running.

What can cost-conscious participants do about the problem of the high transaction costs involved in institutional bargaining? It goes without saying that they will not set aside their own interests and transform themselves into institutional altruists. Nonetheless, there is often good reason to pay attention to considerations of fairness or equity as a means of avoiding protracted disagreements or even stalemate in institutional bargaining. Put another way, parties who are seeking to build maximum coalitions and to avoid the occurrence of gridlock in bargaining processes will experience a distinct incentive to put themselves into the shoes of other participants and to craft negotiating texts that seem fair or equitable to all the parties or negotiating blocs involved in processes of regime formation. In general, the larger the benefits expected to flow from a regime once it becomes operational and the greater the probability of stalemate occurring in the process of regime formation, the stronger will be the incentives of individual participants to craft formulae that all those involved can accept as meeting basic standards of fairness or equity.

This instrumental approach can and often does yield outcomes that conform to or reflect prevailing standards of fairness or equity, even though it is possible to explain their adoption in terms of calculations that take the form of enlightened self-interest. The acceptance of differential obligations among regime members (e.g. the grace period granted to developing countries with regard to the phasing out of ozone-depleting substances) is properly understood as an acknowledgement of what it is reasonable to expect from member states whose circumstances differ markedly (Parson, 1993). Redefinitions of roles (e.g. the rights and duties of coastal, flag and port states with regard to the law of the sea) often reflect considerations of fairness as well as pragmatic calculations of what is needed to solve specific problems, like the conservation of fish stocks located in coastal waters. The establishment of financial mechanisms, such as the Multilateral Fund of the ozone regime or the Global Environment Facility (GEF), designed to help poorer countries meet their obligations under the provisions of international regimes, involves a recognition of the need to treat these parties fairly in order to gain their acceptance of the institutional bargains embedded in the constitutive provisions of individual regimes. Would it be possible to produce accounts of the adoption of these measures based exclusively on the give-and-take involved in hard-nosed bargaining? Perhaps so. Is there nonetheless a sense in which the actual content of these measures designed to circumvent gridlock in processes of regime formation reflects conceptions of fairness or equity prevalent in international society at the time regime formation takes place? I would argue that it is difficult to make sense of the content of the measures chosen without a careful examination of community standards of fairness or equity.

That said, there is no reason to expect that incentives to adopt equitable arrangements as a means of lowering transaction costs will be equally strong across the entire range of processes of regime formation at the international level. Consider, in this connection, several sets of circumstances in which one or more of the key participants are likely to ignore considerations of fairness or to abandon the search for fair or equitable standards before mutually acceptable formulae are devised. Powerful participants sometimes conclude that the potential gains flowing from a regime are not worth the effort involved in institutional bargaining. As a result, they simply walk away from the process of regime formation without any agreement in hand. The OECD countries led by the United States eventually adopted a posture of this sort towards the effort to create a new international economic order (NIEO) during the 1970s (Hart, 1983). In other cases, subgroups of those participating in processes of regime formation proceed to adopt relatively strong commitments applicable to their own members rather than accepting watered-down provisions on the issues at stake acceptable to all participants in processes of regime formation. The diplomatic procedure which arose in connection with acid rain in Europe and which led to developments like the 1985 agreement among a subset of the members of the overarching regime to cut sulphur emissions by 30 per cent exemplifies this case (Levy, 1993). Still other cases involve relationships in which one or more of the key participants are perfectly happy with situations that feature continuing gridlock. In the case of acid rain in North America, for instance, the United States could argue that it was negotiating with Canada in good faith, while avoiding any serious commitments regarding the reduction of sulphur emissions as a result of the effects of stalemate at the bargaining table. The eventual emergence of the 1991 bilateral Air Quality Agreement (AQA) is widely understood as owing more to domestic political developments within the United States than to the success of international diplomacy (Munton *et al.*, 1999).

These observations should not be read as contradicting the argument articulated in the preceding paragraphs about the role of considerations of fairness or equity in lowering transaction costs. Rather, they are intended to demonstrate that the influence of considerations of fairness or equity is a variable that will take on different values from one situation to another. For better or for worse, there is no reason to expect that the prominence of such considerations will itself conform to some meta-standard regarding the role of fairness. That is, the extent to which participants in processes of regime formation are motivated by considerations of fairness or equity may not conform well to outside judgements concerning what they ought to do. But this in no way undermines the conclusion that fairness matters in some processes of institutional bargaining at the international level.

Accommodating the veil of uncertainty

Some actors, especially individuals who pride themselves on their diplomatic skills, derive pleasure or satisfaction from participation in the process of institutional bargaining itself. For the most part, however, those engaged in institutional bargaining are interested primarily in the benefits and costs to themselves that they believe will flow over the course of time from the operation of the regimes they create. It follows that participants in efforts to form regimes will experience strong incentives to make projections regarding the probable impact of alternative institutional designs on their own welfare. Although they may clothe their arguments in language that emphasizes the common good, those engaged in such bargaining processes exhibit a marked tendency to find virtue in arrangements they believe will work to their own benefit and to find fault with arrangements that seem more attractive to their opposite numbers.

Even so, institutional bargaining differs from ordinary, everyday bargaining in ways that make it difficult for participants to anticipate with any precision how the outcomes will affect their interests over time. When bargaining focuses on the selection of a particular point along a contract curve (e.g. the purchase/sale price of a house), it is readily apparent how the content of the bargain struck will affect the welfare of the parties. Everyone understands that one party's gain is the other party's loss. Where bargaining deals with the design of institutional arrangements expected to operate over an indefinite period of time, on the other hand, it is typically difficult and sometimes impossible to foresee how the outcome will impact the welfare of the participants. To be sure, there are some design features whose distributive implications are obvious to all. The fact that the creation of Exclusive Economic Zones (EEZs) and Fishery Conservation Zones (FCZs) would prove advantageous to domestic fishers in contrast to foreign fishers, for instance, was foreseen by virtually everyone. Similarly, no one is likely to be surprised by the resistance of coal and oil producers to arrangements featuring serious targets and timetables governing reductions in emissions of greenhouse gases.

Beyond these simple cases, however, lie a range of conditions that complicate efforts to make accurate projections of the distributive consequences of social institutions. Many regimes are just too complex or multifaceted to allow observers to be sure of their impacts *ex ante*. To take a relatively simple example, most commentators expected the 1987 Montreal Protocol dealing with the phasing out of ozone-depleting substances to impose significant costs on major chemical companies like DuPont. But this has not turned out to be the case. The leading chemical companies have been able to dominate the search for substitutes for CFCs and other ozone-depleting substances, thereby opening up new markets that are just as profitable as those being phased out by the ozone regime.

In other cases, regimes turn out to harbour unforeseen loopholes that aggressive actors are able to exploit to their own advantage or are implemented in ways that produce sizeable gaps between the arrangements described in constitutive contracts and the regimes that operate in practice. In arrangements featuring bans on the production and consumption of various goods (e.g. CFCs, certain animal parts), to take a well-known example, there is always the possibility that some actors will do extremely well by participating in black markets that arise to meet continuing demands for the goods in question. Similarly, the operation of a regime can lead to profound but largely unanticipated changes in the way the relevant problem is framed. In 1979, for example, no one foresaw how the operation of the Environmental Monitoring and Evaluation Programme (EMEP) would alter conceptions of air pollution and turn the long-range transboundary air pollution (LRTAP) regime into a force to be reckoned with by domestic policymakers in a number of member states. Beyond this, there are cases in which institutional arrangements themselves change or evolve in a manner that is unpredictable at the outset but that has profound consequences for the interests of their founders. A classic case in point is the regime for whales and whaling, which started out as a conservation arrangement designed to promote the common interests of whaling states but developed over the course of several decades – as a result of the influx of a sizeable number of non-whaling states – into an arena dominated by preservationists.

What are the implications of these institutional uncertainties for the role of fairness or equity in processes of regime formation?[4] Broadly speaking, as James Buchanan and his colleagues have argued in describing what they call the veil of uncertainty, these conditions give participants in institutional bargaining a distinct incentive to devise arrangements that conform to well-defined or widely accepted criteria of fairness (Brennan and Buchanan, 1985).[5] The logic underlying this argument is straightforward. If an actor does not know or cannot predict how the operation of a regime will affect its interests over the course of time, it will experience an unmistakable incentive to favour the creation of arrangements that can be counted on to generate outcomes that are fair to everyone regardless of where they stand on specific issues. Otherwise, the actor runs the risk of advocating the creation of arrangements that later on turn out to produce results that are detrimental to its own welfare. If all those engaged in the effort to form regimes experience the same incentives, the result will be a process in which considerations of fairness or equity play a significant role in the crafting of constitutive contracts. This is what accounts for the attention delegates typically devote to considerations of fairness or equity in constitutional conventions in contrast to ordinary legislative arenas at the domestic level. And the forces at work are essentially the same at the inter-national level. The same actors who fight hard for outcomes advantageous to themselves in the day-to-day operations of established regimes often exhibit a genuine interest in questions of fairness or equity in the course of bargaining over the design of these institutional arrangements in the first place.

Not surprisingly, institutional uncertainty or, in other words, the thickness of the veil of uncertainty can vary greatly from one situation to another. In certain cases, differences in the roles of regime members are so clear-cut that there is no mistaking how the provisions of regimes will affect the welfare of various groups of members. In the case of the funding mechanisms associated with the regimes dealing with ozone depletion and climate change, for example, every-one understands that the provisions dealing with financial transfers and the decision-making procedures set up to handle them are intended to make these institutional arrangements palatable to the developing countries. As a number of commentators have observed, however, it is reasonable to suppose in more general terms that the broader the functional scope of a regime and the longer it is expected to remain in place, the thicker the veil of uncertainty will be. This is what makes discussions of fairness or equity particularly prominent in situa-tions involving the framing of comprehensive social contracts expected to define the terms of a society's overall governance system for an indefinite period of time. But other factors are also relevant in determining the thickness of the veil of uncertainty. Provisions that make it relatively easy for new parties to join a regime or for existing members to alter the content of a regime, for instance, will increase institutional uncertainty and make it harder to project the results of a regime's operation over time. Similarly, technology often plays a role in such matters, since the development and diffusion of new technologies can have drastic but largely unforeseen consequences for the results regimes produce. The introduction of high-endurance factory trawlers, to take a well-known case, turned fisheries regimes that once did a tolerable job of conserving stocks into mechanisms that are incapable of preventing severe depletions or even collapses of important fish stocks.[6]

It is worth noting as well that the veil of uncertainty and incentives to devise formulae that are fair or equitable can become mutually reinforcing factors in processes of regime formation. To satisfy standards of fairness or equity, and therefore prove effective as responses to the veil of uncertainty, institutional arrangements must operate in such a way that it is difficult or impossible to tell in advance what their distributive consequences will be. A voting system that parties regard as fair, for example, must not produce outcomes that are easy for all concerned to predict in advance. But this is exactly the condition that gives rise to the veil of uncertainty in the first place. It follows that efforts to cope with the veil of uncertainty by establishing equitable institutions may actually reinforce pre-existing incentives to pay attention to considerations of fairness or equity in designing institutions expected to remain in place over the long run.

Needless to say, there is no guarantee that the results actually flowing from the operation of regimes *ex post* will conform to any well-defined standard of fairness or equity. Even arrangements that strike everyone as fair at the design stage can operate in practice in ways that systematically privilege the interests of some members of the social system over others. It would have been difficult to foresee at the outset, for instance, that the regime for whales and whaling would end up as an arena that is highly advantageous to the interests of preservationist groups. More generally, because regimes are arenas for action rather than actors in their own right, the possibility is ever-present that they will be captured by particular interest groups using stratagems that were not foreseen at the time of their creation. But none of this undermines the argument under consideration here. On the contrary, it actually reinforces the incentives of those engaged in institutional bargaining to incorporate considerations of fairness into the institutions they create. Knowing full well that individual regimes can be captured by particular interests at various stages in their life cycles, those who are present at the creation will have all the more incentive to preserve uncertainty by erecting barriers to the success of efforts to hijack the institutions they create once they are up and running.

Respecting negotiation norms

To this point, I have been addressing the role of fairness or equity in consequentialist terms. That is, I have focused on the outcomes flowing from the operation of regimes and asked why those engaged in institutional bargaining might have incentives to pay attention to fairness and equity in deciding on the provisions of constitutive contracts likely to affect the content of these outcomes. As many thoughtful observers have pointed out, however, it is helpful to distinguish between procedures and outcomes in thinking about fairness or equity. Procedures (e.g. democratic voting systems) that seem fair to most participants can produce outcomes that few would regard as meeting even the most minimal standards of fairness or equity. Conversely, outcomes that are perfectly acceptable from the standpoint of equity can arise from the use of procedures that are hard to justify in terms of fairness. Some writers – Robert Nozick is a prime example – have stressed the importance of procedural justice and argued that a very wide range of substantive outcomes should be regarded as acceptable, so long as they arise from the use of procedures that meet community standards of fairness (Nozick, 1974). But there is no need to adopt a relatively extreme position of this sort in order to reach the conclusion that we should pay serious attention to procedural as well as consequentialist conceptions of fairness or equity in thinking about the processes of institutional bargaining involved in the formation of international regimes.

How does fairness in the procedural sense arise in the context of institutional bargaining, and why would those engaged in bargaining of this type be motivated to pay attention to such concerns? Institutional bargaining at the international level is not an *ad hoc* process that participants reinvent every time they start to negotiate over the terms of a constitutive contract setting out the provisions of a specific regime. On the contrary, bargaining of this sort is a recognized social practice that is well known to the members of international society and that features an array of procedural rules spelling out the proper way to go about negotiating the terms of constitutive contracts. In effect, these rules are normative guidelines that are accepted by participants in the process of regime formation as reasonable codes of conduct for institutional bargaining. Most procedural rules are not legally binding; there are seldom any formal sanctions imposed on those who fail to adhere to them in specific cases. Rather, they owe their influence to a widespread sense among the members of international society that they are legitimate (Franck, 1995). Or to put it another way, they achieve behavioural significance not only because they are expected to guide the bargaining process towards convergence on mutually agreeable outcomes but also because they are seen as reflecting community standards about what constitutes procedural fairness in the context of institutional bargaining.

Why do those engaged in regime formation – including Great Powers as well as smaller powers – abide by negotiation norms? For one thing, it is efficient for them to do so. Redefining procedural rules at the start of each episode of institutional bargaining would be a time-consuming and costly process, especially if some of the participants were to hold out for the adoption of their favourite rules against the sustained opposition of others. Under the circumstances, it makes sense from everyone's point of view to devise a set of negotiation norms that reflect community standards of fairness or equity and then to accept their use as a matter of course rather than wasting time and resources bargaining over such matters before embarking on substantive negotiations dealing with regime formation itself. For another thing, the acceptance of a set of negotiation norms that conform to community standards of fairness provides some assurance that participants in institutional bargaining will have a reasonable chance to pursue their interests in specific cases, regardless of what part they play in the bargaining process. The same actor may assume the role of pusher in some cases of regime formation, for instance, while adopting the posture of laggard in other cases (Sprinz and Vaahtoranta, 1994). Even within the group of those favouring the establishment of new regimes, the same actor may sometimes advocate the use of the framework/protocol approach in the interests of putting some institutional arrangement in place quickly, while preferring to move more slowly in other cases in the interests of incorporating more substance into a regime at the outset. In this connection, the role of negotiation norms is to provide a uniform set of procedural rules that participants can count on regardless of the substantive position they espouse in particular cases. Compared with the prospect of having to renegotiate the rules on a case-by-case basis, this uniformity is apt to seem quite appealing under normal conditions.

What is the content of these negotiation norms in international society? Given the nature of institutional bargaining at the international level, it will come as no surprise that most of the norms prevailing at this time focus on the process through which negotiating texts are framed and gradually moulded into mutually acceptable agreements.[7] They specify, for example, that all states participating in a given process of regime formation have an equal right to contribute proposed language to the negotiating text, that those in charge of the negotiating process be given some leeway in integrating proposals into a unified negotiating text, that negotiations focus increasingly as the process goes forward on efforts to remove brackets surrounding specific language that is controversial, and that parties adopt an agreed timetable in the sense of identifying a target date for the production of a completed and generally acceptable agreement. Needless to say, these procedural rules are not meant to become straitjackets in the sense that they allow no exceptions or modifications. Those participating in processes of regime formation can and do adjust the procedures they follow to fit the circumstances arising in specific cases. Nonetheless, procedural rules of this sort, which are expressions of prevailing community standards, do serve to provide participants in institutional bargaining with common expectations about the nature of the process and legitimate ways to pursue their own interests within it.

None of this is meant to suggest that negotiation norms operative in international society are static; far from it. Recent experience, for instance, reflects a striking trend towards greater openness and a willingness to allow non-state actors to participate in the process, especially in negotiations relating to the creation of environmental regimes.[8] Despite the appeal of the Wilsonian precept calling for open covenants, openly arrived at, institutional bargaining in international society has traditionally been a somewhat secretive process carried out by skilled diplomats operating behind closed doors (Nicholson, 1963). The justification for this practice presumably rests on the proposition that institutional bargaining is a messy process involving the making of deals and the crafting of compromises that the public would find hard to understand and quite likely odious. Even today, this view of the process remains widespread, especially among foreign ministry personnel who are responsible for the conduct of negotiations. In this regard, the growing capacity of non-state actors – ranging from the World Wide Fund for Nature to Greenpeace – to intervene directly in processes of regime formation regarding environmental matters constitutes a remarkable development (Wapner, 1996). It will be interesting to see whether this shift in negotiation norms spreads to other issue areas, including economic problems and even security problems, during the near future.

As is true in the cases of transaction costs and the veil of uncertainty, the role of negotiation norms as a mechanism for introducing a concern for fairness or equity into processes of regime formation is a variable. Several types of variance are readily distinguishable in this regard. Even when negotiation norms appear to be based on community standards, they may reflect the preferences of dominant actors in international society. It does not require an appeal to complex ideas like the concept of Gramscian hegemony to understand this.[9] Dominant actors – the upper class in domestic settings or the Great Powers in international society – exert a disproportionate influence on the content of community standards of fairness or equity in every social setting. Yet it is easy to exaggerate this phenomenon. Once negotiation norms are established, it is not a simple matter – even for a hegemon – to redefine them. As the United States has discovered, even the structural advantages associated with being the sole remaining superpower are not sufficient to allow it simply to change the procedural rules applying to the process of regime formation whenever it dislikes the direction that the process is taking. How else can we explain the development of arrangements, like those embedded in the London Amendments to the Montreal Protocol, the Convention on Biological Diversity, and the Framework Convention on Climate Change, with respect to which the United States has clearly and persistently adopted a laggard posture (von Moltke, 1997)?

Note also that negotiation norms are not always applied uniformly and that parties break the procedural rules arising from these norms from time to time. As is to be expected, Great Powers are in a better position to flout such rules than smaller powers. The United States, for example, was able to renege on prior commitments during the final stages of negotiating the 1982 Convention on the Law of these Sea and to refuse to sign the Convention on Biological Diversity during the June 1992 United Nations Conference on Environment and Development with relative impunity, even though it was widely criticized in both instances for violating procedural rules relating to the process of regime formation (Sebenius, 1984). But breaches of this sort are by no means limited to the actions of dominant powers. The scuttling of the 1988 Convention on the Regulation of Antarctic Mineral Resource Activities (CRAMRA) brought about by the defection first of Australia and then of France during the immediate aftermath of the completion of an agreed text was dramatic, at least in part, because it constituted a sharp break with the norms relating to the process of regime formation (Chaturvedi, 1996: ch. 5).

Such behaviour makes it clear that the role of negotiation norms in regime formation does indeed vary from case to case. Yet it is worth observing that disregard for procedural rules is not cost free, even for major powers. Most of the provisions of the Convention on the Law of the Sea – including some that the United States did not at the time regard with favour – soon passed into customary international law. The Convention on Biological Diversity has entered into force, even in the absence of the United States as a member. In the meantime, the United States has earned a reputation for being a 'bad guy' when it comes to institutional bargaining at the international level, a condition that makes others understandably wary in their dealings with the sole remaining superpower. As for Australia and France, their action was widely hailed as a progressive move by an array of environmental groups and certainly played a key role in sidetracking CRAMRA and replacing it in short order with the 1991 Protocol on Environmental Protection (Stokke and Vidas, 1996). Yet it is probably accurate to say that this breach of the procedural rules provoked the most significant crisis that has occurred in the life of the Antarctic Treaty System, and it has left many with serious reservations about the reliability of Australia and France as negotiating partners when it comes to regime formation. In the final analysis, it is indisputable that the significance of negotiating norms does vary from one case of regime formation to another. But it would be a mistake to suppose that participants in this process, even when they enjoy the prerogatives of Great Powers, can get away with ignoring or flouting these norms without incurring significant costs.

The future of fairness in regime formation

Should all this lead us to expect a flowering of concern for fairness or equity among those engaged in institutional bargaining at the international level? Not necessarily. There is no reason to doubt the sincerity of those who claim that they are motivated by considerations of fairness as participants in these processes. Yet it is often difficult to know how best to interpret the behaviour of these individuals. The same behaviour may strike some observers as reflecting hard-nosed calculations of relative bargaining power yet seem to others to constitute clear-cut evidence of the influence of considerations of fairness or equity in international affairs.

Consider a few examples drawn from recent efforts to solve global problems like ozone depletion, climate change and the loss of biological diversity (Haas *et al.*, 1993). The ozone regime lays down substantially different obligations for those who are heavy users of CFCs and related chemicals in contrast to those who do not have a record of heavy dependence on ozone-depleting substances. Since the addition of the 1990 London Amendments, it also acknowledges the legitimacy of requests for financial assistance on the part of developing countries seeking to reap the rewards of economic growth while avoiding significant increases in the use of ozone-depleting substances. For its part, the climate regime clearly acknowledges the primary responsibility of the advanced industrial countries – those listed in Annex 1 of the Framework Convention on Climate Change – for increases in levels of greenhouse gases resident in the atmosphere over the past century. It calls on these countries both to take the lead in devising targets and timetables covering reductions in greenhouse gas emissions and to provide substantial assistance to other countries desiring to benefit from economic growth while avoiding or at least controlling increases in greenhouse gas emissions. The regime dealing with biological diversity acknowledges that countries that are sources of genetic materials that become the basis for commercially valuable products have a right to benefit from such developments, though it is only fair to note that the apportionment of such benefits remains a particularly contentious issue (Bowman and Redgwell, 1996). More broadly, this arrangement recognizes the essential need to promote mutual respect and mutually rewarding partnerships between developing countries that hold jurisdiction over much of the world's genetic resources and interest groups based in the developed countries that have come to place an increasingly high value on the protection of biological diversity.

What are we to make of these relationships? There is no doubt that a conventional assessment of bargaining strength can illuminate some of these developments (Young 1994: ch. 5). In the absence of the Multilateral Fund established under the 1990 London Amendments to the Montreal Protocol, China and India would simply have refused to become members of the ozone regime. There is little chance of generating, much less sustaining, interest in the climate regime on the part of countries like China in the absence of a serious effort on the part of the advanced industrial countries to make significant reductions in their emissions of greenhouse gases, a fact whose importance is impossible to deny in the aftermath of the 1997 special session of the UN General Assembly called to assess progress regarding environmental issues five years after the Earth Summit held in Rio de Janeiro in June 1992. Anyone who thinks it is possible to deal with threats to biological diversity without striking mutually acceptable bargains with countries like Brazil, Indonesia and Malaysia is simply out of touch with reality.

Yet it would be a mistake to dismiss the role of considerations of fairness or equity in the creation and operation of these regimes. Although some may see the addition of the Multilateral Fund to the ozone regime as a concession to the bargaining strength of China and India, many interpret this development as the least that could be done in the interests of achieving a fair deal for those being asked to forgo the use of CFCs and related chemicals in their quest for economic development. Similar comments are in order regarding the acknowledgement of responsibility for greenhouse gas emissions on the part of the advanced industrial states and the right of countries possessing plant genetic resources to benefit from these resources and to receive the support of others who have come to value the preservation of these resources highly. It is perfectly true that advocacy of these and related institutional developments in the name of fairness or equity is compatible with a broadly utilitarian perspective on human behaviour. There are compelling reasons to conclude that those who are serious about protecting stratospheric ozone, avoiding climate change and slowing the loss of biological diversity should be strong supporters of institutional arrangements that seem fair to the world's developing countries. But in the final analysis, this support is more likely to flow from an appreciation of the sorts of linkages I have been exploring in this chapter than from some simplistic attempt to compute the benefits and costs of options relating to such matters in the quantitative terms envisioned in conventional utilitarian assessments.

Perhaps we are witnessing here the rise of a different form of utilitarianism, one which continues to focus on incentives in interpreting behaviour but which is rendered more realistic by a willingness to devote serious attention to normative factors like considerations of fairness or equity. To the extent that this is true, there is much to be said not only for accepting the proposition that fairness matters in processes of institutional bargaining but also for devoting time and energy to thinking systematically about the application of community standards of fairness to specific instances of regime formation in international society.

Notes

1 For a general discussion of the role of discourses in regime formation see Litfin (1994).
2 While distributive moves focus on the allocation of a fixed pie, integrative moves emphasize efforts to expand the size of the pie available for distribution. See Walton and McKersie (1965) for a prominent account of integrative bargaining and Krasner (1991) for an analysis of what is likely to happen when the players know the locus and shape of the contract curve.
3 For a theory that emphasizes the role of minimum winning coalitions – the smallest number of members of a group needed to arrive at collective choices for the entire group – in legislative settings see Riker (1962).
4 For a general discussion of the idea of institutional uncertainties see Young (1998).
5 The veil of uncertainty differs substantially from the Rawlsian concept of the veil of ignorance (Rawls, 1971) because actors know who they are and what their utility functions look like in Buchanan's world. It follows that the circumstances of those engaged in institutional bargaining differ fundamentally from the circumstances of those operating in Rawls's original position.
6 The story of deepwater trawling in the North Atlantic is told in vivid detail by Warner (1977). For a more general account of these and related problems in the fisheries see Peterson (1993).
7 For illustrative accounts relating to the negotiations dealing with ozone depletion and climate change see Benedick (1991), Bodansky (1993) and Mintzer and Leonard (1994).
8 For accounts exploring the growing importance of non-state actors – especially in environmental affairs – see Princen and Finger (1994) and Wapner (1996).
9 Gramsci developed the idea of cognitive dominance as a counterpoint to the more familiar realist and neo-realist idea of structural or material dominance. For a variety of perspectives on Gramsci's thought and its influence see Gill (1993).

References

Axelrod, R. (1984) *The Evolution of Cooperation*, New York: Basic Books.
Benedick, R. E. (1991) *Ozone Diplomacy: New Directions in Safeguarding the Planet*, Cambridge, MA: Harvard University Press.
Bodansky, D. (1993) 'The United Nations Framework Convention on Climate Change', *Yale Journal of International Law* 18: 453–558.

Bowman, M. and Redgwell, C. (eds) (1996) *International Law and the Conservation of Biological Diversity*, London: Kluwer Law International.

Brennan, G. and Buchanan, J. M. (1985) *The Reason of Rules: Constitutional Political Economy*, Cambridge: Cambridge University Press.

Buchanan, J. M. (1975) *The Limits of Liberty: Between Anarchy and Leviathan*, Chicago: University of Chicago Press.

Chaturvedi, S. (1996) *The Polar Regions: A Political Geography*, Chichester: John Wiley.

Downs, G. W. and Rocke, D. M. (1990) *Tacit Bargaining: Arms Races and Arms Control*, Ann Arbor: University of Michigan Press.

Franck, T. M. (1995) *Fairness in International Law and Institutions*, Oxford: Clarendon Press.

Gill, S. (ed.) (1993) *Gramsci, Historical Materialism, and International Relations*, Cambridge: Cambridge University Press.

Haas, P. M., Keohane, R. O. and Levy, M. A. (eds) (1993) *Institutions for the Earth: Sources of Effective International Environmental Protection*, Cambridge, MA: MIT Press.

Hart, J. A. (1983) *The New International Economic Order: Cooperation and Conflict in North–South Economic Relations*, New York: St Martin's Press.

Krasner, S. D. (1991) 'Global Communications and National Power: Life on the Pareto Frontier', *World Politics* 43: 336–66.

Levy, M. A. (1993) 'European Acid Rain: The Power of Toteboard Diplomacy', in P. M. Haas, R. O. Keohane and M. A. Levy (eds) *Institutions for the Earth: Sources of Effective International Environmental Protection*. Cambridge: MIT Press, pp. 75–132.

Litfin, K. T. (1995) *Ozone Discourses: Science and Politics in Global Environmental Cooperation*, New York: Columbia University Press.

Mintzer, I. M. and Leonard, J. A. (eds) (1994) *Negotiating Climate Change: The Inside Story of the Rio Convention*, Cambridge: Cambridge University Press.

Munton, D., Soroos, M., Nikitina, E. and Levy, M. A. (1999) 'Acid Rain in Europe and North America', in O. R. Young (ed.) *The Effectiveness of International Environmental Regimes: Causal Connections and Behavioral Mechanisms*, Cambridge, MA: MIT Press.

Nicholson, H. (1963) *Diplomacy*, 3rd edition, New York: Harcourt, Brace.

Nozick, R. (1974) *Anarchy, State, and Utopia*, New York: Basic Books.

Orrego Vicuna, F. (1988) *Antarctic Mineral Exploitation: The Emerging Legal Framework*, Cambridge: Cambridge University Press.

Parson, E. A. (1993) 'Protecting the Ozone Layer', in P. M. Haas, R. O. Keohane and M. A. Levy (eds) *Institutions for the Earth: Sources of Effective International Environmental Protection*, Cambridge, MA: MIT Press, pp. 27–74.

Peterson, M. J. (1993) 'International Fisheries Management', in P. M. Haas, R. O. Keohane and M. A. Levy (eds) *Institutions for the Earth: Sources of Effective International Environmental Protection*, Cambridge, MA: MIT Press, pp. 249–305.

Princen, T. and Finger, M. (1994) *Environmental NGOs in World Politics: Linking the Local and the Global*, London: Routledge.

Rawls, J. (1971) *A Theory of Justice*, Cambridge, MA: Harvard University Press.

Riker, W. H. (1962) *The Theory of Political Coalitions*, New Haven, CT: Yale University Press.

Sebenius, J. K. (1984) *Negotiating the Law of the Sea*, Cambridge, MA: Harvard University Press.

Sprinz, D. and Vaahtoranta, T. (1994) 'The Interest Based Explanation of International Environmental Policy', *International Organization* 48: 77–106.

Stokke, O. S. and Vidas, D. (eds) (1996) *Governing the Antarctic: The Effectiveness and Legitimacy of the Antarctic Treaty System*, Cambridge: Cambridge University Press.

Victor, D. G. (1996) 'The Regulation of Greenhouse Gases – Does Fairness Matter?', unpublished paper dated 13 December.

von Moltke, K. (1997) 'Institutional Interactions: The Structure of Regimes for Trade and the Environment', in O. R. Young (ed.) *Global Governance: Drawing Lessons from the Environmental Experience*, Cambridge, MA: MIT Press, pp. 247–71.

Walton, R. E. and McKersie, R. B. (1965) *A Behavioral Theory of Labor Negotiations*, New York: McGraw-Hill.

Wapner, P. (1996) *Environmental Activism and World Civic Politics*, Albany: State University of New York Press.

Warner, W. W. (1977) *Distant Water: The Fate of the North Atlantic Fisherman*, Boston: Little, Brown.

Young, O. R. (1994) *International Governance: Protecting the Environment in a Stateless Society*, Ithaca, NY: Cornell University Press.

—— (1998) 'Institutional Uncertainties in International Fisheries Management', *Fisheries Research* 761: 1.14.

15 Global ecological democracy

John S. Dryzek

Introduction

At first sight the international system does not seem like a good place to look for either environmental justice or democracy. Formal institutions in this system have always been quite weak, and seemingly no match for the power of sovereign states and multinational corporations. Inasmuch as international organizations and regimes are now being strengthened, it is generally with economic concerns in mind – think, most notably, of the recent establishment of the World Trade Organization (WTO) to preside over the expansion of global free trade. As Vandana Shiva argues in her contribution to this volume (Chapter 4), the WTO in many ways represents the latest phase of an exploitative and hierarchical global order in which the world's poor and their environments are further subordinated to the interests of the world's wealthy, now organized into global capitalism. Certainly there are plenty of reasons to be pessimistic about the prospects for ecological rationality, justice and democracy in the contemporary international order.

I will suggest that, appearances to the contrary notwithstanding, the international system should be one of the first places to look for felicitous combinations of democratic structure and ecological concern, 'ecological democracy' for short. But the search will involve some rethinking of what constitutes the substance of democracy, and looking for it in some unusual and novel places. It is important to press this search to the limit, rather than give up in the face of some very large and seemingly overwhelming problems. For it hardly needs saying that some of the larger issues of ecological destruction and environmental injustice arise at the international level. If one believes that ecological irrationality and injustice can only be remedied through democratic means – and shortly I will try to establish that such is indeed the case – then the search for global environmental justice just has to involve a search for enhanced democracy in the global system.

By way of preliminaries I want to establish that the effective resolution of environmental problems, be it in the international system or elsewhere, is indeed best pursued through democratic means, and that this is especially true for issues of environmental justice. I will treat these matters briefly, because much ink has been spilled lately on the subject of democracy and the environment, and I do not wish to get enmeshed in the details of that debate here (for my own contributions see Dryzek, 1990b, 1992, 1996b, c; for further treatment see Lafferty and Meadowcroft, 1996; Mathews, 1996; Doherty and de Geus, 1996; Goodin, 1996). My main concern is with exploring the prospects for ecological democracy in the international system in particular. These prospects prove surprisingly bright, though only if we are prepared to think about democracy in unconventional ways.

On behalf of ecological democracy

One can approach the conclusion that democracy is conducive – perhaps essential – to the effective resolution of environmental problems from a variety of directions. Paehlke (1988, 1996) notes that the rise of ecological issues since the late 1960s has been accompanied by more in the way of democracy, in the form of community consultation, right-to-know legislation, public inquiries and alternative dispute resolution. These expansions are a far cry indeed from the kind of ecological authoritarianism favoured by survivalist thinkers such as Garrett Hardin, Robert Heilbroner and William Ophuls in the 1970s (all three remain unrepentant in the 1990s). Martin Jänicke and his associates have demonstrated in a series of cross-national comparisons that democracy is positively correlated with environmental policy success (Jänicke, 1992). Both Paehlke and Jänicke are mostly concerned with the extension of existing liberal constitutional forms of democracy rather than any radical alternative (though Jänicke is attuned to the possibility of the failure of the liberal democratic state, in the environmental realm and elsewhere). There are plenty of theoretical arguments on behalf of democracy in an ecological context, many of them building on exposés of the irrationality of hierarchical and authoritarian alternatives (Dryzek, 1987: 88–109; Orr and Hill, 1978; Walker, 1988).

More positively, it may be that participation by partisans of a variety of perspectives is the only real way to come to grips with wicked or complex environmental problems (Dryzek, 1990a: 57–76; Fischer, 1993). Certainly it seems that the standard response on the part of governments to especially tricky environmental conflicts is to call in the people, be it through exercises in public consultation, policy dialogue, public inquiries and so forth. Sometimes the participation so achieved is merely symbolic in character, but still it speaks to the felt need to legitimize government action on such issues through participatory exercises (for details on such exercises see Press, 1994; Williams and Matheny, 1995).

But what about issues of environmental justice in particular? Should we expect the prospects for environmental justice to be enhanced by democratic participation? Here the conventional wisdom of political philosophers is no help, for it argues that justice is a substantive value, while democracy is a matter of procedure. Thus there is no guarantee that the procedures of democracy will produce more in the way of social justice, however one defines justice. While there are reasons to dispute this conventional wisdom for social justice in general, especially given that historically struggles for justice and democracy have frequently gone hand in hand (Shapiro, 1996), I would argue that environmental justice in particular stands to benefit from democratization.

Environmental justice is concerned first and foremost with the distribution of environmental risks and harms across individuals and social groups. Normally the issue can be captured in spatial terms; somebody, somewhere is getting dumped on, and somebody, somewhere else is benefiting. That dumping may come in the form of toxic waste disposal, the siting of noxious facilities, destruction of local ecosystems through mining, logging or intensive export-oriented agriculture, the expropriation of local biological knowledge, and so forth. The theory of democracy suggests that if the people being dumped on can achieve political equality with those who benefit from the harms inflicted upon them, then they are more able to take preventive action.

Issues of environmental justice arise through the displacement of the negative environmental effects of an activity on to distant others (Wapner, 1995). Such displacement can take the form of destroying someone else's rain forest ecosystem in order to secure one's timber supplies, exporting one's toxic wastes, building tall smokestacks to send one's air pollution to distant localities, and so forth. To the extent that such displacement can be prevented, it is not just environmental justice that will benefit, but ecological rationality in general, for the producers of environmental hazards will be forced to take into account the full environmental costs of their activities. Along these lines, Lois Gibbs of the Citizens' Clearinghouse for Hazardous Wastes in the United States claims that her goal is to plug the toilet on toxic wastes: to make it clear to toxic waste producers that there is nowhere for them to go (Dowie, 1995: 126). The local actions which her organization coordinates were once derided as selfish 'not in my back yard' or NIMBY actions. But the extension to 'not in anybody's backyard' or NIABY means that something more than limited, defensive and local resistance is at issue. For if the toilet is well and truly plugged, then the operating principles of the entire industrial political economy have to change: rather than being treated as unfortunate by-products for which a dump must be found, toxic wastes are conceptualized as chemicals that should not be generated in the first instance, or if they do appear, then a way should be found of recycling them back into the process of production. In short, democratic equality that extends across the perpetrators and victims of environmental risks makes displacement less likely and so promotes ecological rationality.

When it comes to the positive impact of democratic equality on the reduction of environmental risk, the *practice* of the environmental justice movement, especially in North America, is actually more decisive than any political theory. If we look at the movement's successes, they have not come in the form of benign distant governments spurred by nationally organized environmental lobbying groups financed by private foundations coming to the rescue of sufferers in particular localities. Rather, these successes have come in the form of local actions, generally organized on a participatory basis, and often building up into regional and national networks (Schlosberg, 1999; see also Chapter 3 by Bullard in this volume). This situation for environmental justice is quite different from that on many other environmental issues, notably wilderness protection, where national action has often had to overcome local recalcitrance in environmental protection. This is certainly true in Australia, where all the major conflicts that have been resolved in favour of wilderness protection featured federal government quashing of state and local governments that valued exploitation over environmental protection.

The weight of evidence and argument leads, I think, to a conclusion that the more political arrangements are pushed in a democratic direction, the more can environmental justice in particular and ecological values in general expect to benefit. Such advancement can come in the form of enlargement of either the size of the democratic franchise, the scope of issues falling under democratic control, or the authenticity of that control (Dryzek, 1996a: 5–6). Authenticity here means that control is real rather than symbolic, informed rather than ignorant, and engaged by competent actors. Thus, to advocate ecological democratization we do not actually need any sort of blueprint for what the perfect ecological democracy would look like. Democracy is an open-ended project, such that a key feature of democratic society is experimentation with what democracy can mean. Exactly what it can mean varies radically by context. In short: even if we cannot specify what a true democracy is (let alone a true ecological democracy), we can still recognize when a democratic advance is being made. In this light, what then can ecological democratization mean in the context of the international system?

The international system and its sources of order

The most striking political feature of the international political system is that, for better or for worse, it lacks any kind of central authority analogous to the state. For realist theorists of international relations this has always been for the worse: the absence of a state or state-analogue means anarchy, and operating successfully within an anarchical system means having first and foremost to maximize state survival in an environment of actually or potentially hostile competitors. Thus the international anarchy is perceived in Hobbesian terms, as a dangerous place where self-help is the only prudent strategy. To realists, this is the way things are and will always be. Some environmentalists accepted this basic diagnosis of the international system but saw it in more contingent terms: thus they argued that stronger global-level institutions guided by eco-logical values were necessary to resolve the global environmental crisis. Such analyses were more popular and more plausible in the 1970s than in the 1990s, and some of their more prominent 1970s advocates have now recanted (see, for example, Falk, 1995: 10–14 for one such apology). Today, proposals for stronger systems-level global institutions are more likely to be seen as implausible to begin with and authoritarian to the extent that they could be realized (see, for example, Chapter 16 by Altvater in this volume). In present circumstances it is hard to imagine them being guided by anything other than the market liberal imperatives that currently give globalization such a bad name among those concerned with the global environment – and especially global environmental justice (for example, Shiva in Chapter 4 of this volume).

The most sanguine outlook on the prospects for effective international cooperation has generally been associated with the liberal tradition in international relations theory (which should not be confused with market liberalism or neo-liberalism as currently practised in the international political economy). Liberals such as Oran Young look at the international system and see something very different than do the realists: a system already populated by sources of order such as international regimes, international law and international organizations. Liberal international relations theorists have not generally been especially interested in the prospects for international democracy, though they do have a long-standing concern with social justice. Recent work by David Held (1995; see also Archibugi and Held, 1995) on cosmopolitan democracy could, I think, mesh quite nicely with established liberal concerns. Cosmopolitan democracy would involve strengthening existing international institutions such as the United Nations, various regional groupings, and international law and, in the long run, establishing a global parliament which would preside over a demilitarized global system, and policy interventions to secure social justice in the world polity.

Such benign proposals, though clearly more plausible than any kind of global state, have an air of unreality about them in this era when global institution-building seems devoted to economic affairs as defined by market liberals. To use the terminology of Falk (1997), I believe that the prospects for transnational democracy may currently be advanced best by 'globalization from below' to counter the market liberal 'globalization from above'. Falk and others have stressed the role of transnational civil society in this global democratic project. While consistent with this emphasis on global civil society, my own approach looks to a source of order to which liberals and civil society advocates alike have devoted little if any attention, but which proves especially interesting in the light of the prospects for ecological democratization.

The order I wish to stress arises in connection with the *discourses* present in the system. A discourse is a shared set of assumptions and capabilities embedded in language that enables its adherents to assemble bits of sensory information that come their way into coherent wholes. So any discourse involves a shared set of basic, often unspoken, understandings. The concept was popularized by Michel Foucault, who laid bare the history of discourses on sexuality, criminality, illness and so forth. Discourses are not just (or to Foucault, hardly at all) created by human subjects; they also help to create subjects, by giving them a framework with which to make sense of the world and their lives in it. Because discourses are social as well as personal, they act as a source of order by coordinating the behaviour of the individuals who subscribe to them. Foucault himself believes that this coordination functions in mostly oppressive fashion, such that the only proper thing to do once one becomes aware of a discourse to which one is subject is to rebel against it. I would argue in contrast that some discourses are better than others, and that it is possible to reflect upon discourses and even reconstruct them at the margins, rather than simply give them a wholesale rejection (support for this position can itself be found in some of Foucault's later writings).

Discourses are intertwined with institutions; if formal rules constitute institutional hardware, then discourses constitute institutional software. In the international system, the hardware is not well developed, which means that the software becomes more important still. Many observers of the international system lament the absence of institutional hardware, but it may turn out that its absence can be turned to good democratic use, especially if institutional software is less resistant to political change than hardware is.

To demonstrate the importance of discourses, their role in coordinating action, and the possibilities they allow for reasoned reflection and reconstruction, let me introduce some discourses present in the international system.

Anarchy

Anarchy is treated by realist theorists of international relations as a given and immutable feature of the international system, but what theorists, state leaders and other actors *make* of anarchy is in large measure a social construction (Wendt, 1992). As constructed by realist actors and analysts, anarchy is Hobbesian: the only entities in the international system that really matter are states. States maximize power and security, and their natural relationship is rivalry or conflict in which violence is always possible. Non-state actors are treated as unimportant by assumption, and interactions transcending state boundaries involving such actors are ignored.

Market liberalism

Market liberalism has emerged as a very different discourse to that of anarchy. Market liberals stress relationships not between states, but between economic actors such as corporations, banks and consumers. States enter in a secondary role, to facilitate the conditions for markets to operate. Natural relationships involve competition rather than conflict, with the potential for positive-sum outcomes in which everyone gains from trade (realists deal only in zero-sum relationships). From market liberalism's microeconomic origins comes a metaphorical structure that is essentially mechanistic.

Other examples of such discourses in the international system include human rights and the rules of war. But for present purposes, discourses with direct and obvious implications for environmental affairs are more interesting. Foremost among these in the contemporary international system is the discourse of sustainable development, which, as we shall see, turns out to be quite interesting from the perspective of democratization.

Sustainable development

The first thing to be said about sustainable development is that it is not a concept that can or should be defined with any precision. Rather, it is indeed a discourse. One of the essential aspects of this discourse is dispute over what sustainable development can actually mean. Yet this contest over the meaning of sustainable development takes place within boundaries. The first boundary is set by an older discourse which sustainable development largely displaced in the 1980s: the discourse of limits and survival which reached its zenith with the efforts of the Club of Rome in the early 1970s (Meadows *et al.*, 1972), echoed a decade later in the *Global 2000 Report to the President* (for details of the global transition from a discourse of limits to one of sustainable development see Torgerson, 1995). Sustainable development denies the existence of global ecological limits; it cannot demonstrate (or even really argue) that such limits do not exist, but instead it assumes them away. The ground is thus cleared for commitments to continued economic development and environmental conservation proceeding in tandem – with social justice both within and across generations thrown in for good measure. A natural relationship of harmony between these various values is assumed.

Sustainable development differs from market liberalism in having an organic rather than a mechanical metaphorical structure, recognizing the existence of nested social and ecological systems. The actors and agents highlighted in the discourse are not realism's states or market liberalism's economic actors, but rather political bodies above and below the state, international organizations and citizens' groups of various kinds. Thus sustainable development is a discourse of and for international civil society (Lafferty, 1996: 203; Dryzek, 1997: 131). Sustainable development is further delimited by its rejection of a green radicalism which looks to a future beyond capitalism; for the moment at least, sustainable development is accommodated to the capitalist economic system (though there are those who wish to stretch it as far as possible within these confines).

Sustainable development's discursive role in the international system is to provide a conceptual meeting place for many actors, and a shared set of assumptions for their communication and joint action. It is these shared assumptions and capabilities which define the discourse. Nowhere is sustainable development an accomplished fact, or even a demonstrable possibility, or even a concept that can be defined with any precision. The concept is essentially contestable. Just as one can nowhere see or even envisage a 'true democracy', one cannot see or envisage true sustainable development.

Sustainable development exists in a constellation with a number of other discourses: survivalism (the discourse of limits), market liberalism and green radicalism. The interplay of these discourses is quite crucial in setting the agenda for global international affairs. But what does this interplay have to do with the prospects for global ecological democracy? Some of these discourses are indeed more conducive to democracy that others: for example, green radicalism is far more hospitable to citizen participation than is the grim authoritarianism of survivalism. Sustainable development seems reasonably conducive to democracy inasmuch as it emphasizes the role of transnational civil society (I will return to this point in due course). More importantly, though, the balance of these discourses is quite crucial, more crucial than in domestic society, because discourses play a much greater role in coordinating action in international politics than they do in domestic society. So it does matter a great deal which of these discourses holds sway, or, more likely, what balance of them obtains, and how each of them changes over time. This balance or interplay can be brought under conscious, collective and ultimately democratic control – and, I will now argue, this is more possible today than ever before.

Reflexive modernization in international politics

Ours is an age when self-conscious and self-critical political reshaping is more plausible than ever before. The idea that societies can chart their own course into the future, rather than adapt to the flow of events, is encapsulated in recent work by theorists such as Ulrich Beck, Anthony Giddens and Scott Lash (1994) on the concept of reflexive modernization. To Beck (1992), the semi-modernity of industrial society is being supplanted by an emerging risk society. In industrial society, politics was mostly organized around distributive issues relating to social class. Deindustrialization now signals the fading of class politics. Industrial society is or was only semi-modern because large areas of life, notably those connected to the trajectories of technological change and economic development, were placed off-limits to democratic control. In risk society, in contrast, politics is organized increasingly around issues of risk relating to chemical, nuclear, biotechnological and other environmental hazards. Such hazards existed in industrial society, but in risk society they can no longer be treated as side effects, to be solved piecemeal. Instead, they call into question the whole legitimacy of the political economy. One solution to this legitimation problem is an extension of democratic participation in the selection, allocation, distribution and amelioration of risks. The basic trajectories of economic development and technological change, once treated as matters to which we just had to adjust, become instead the targets of collective control. Beck gives several scenarios about how this control might be exercised, and how we might chart our social course very self-consciously (i.e. reflexively) into the future. One of them is a new kind of participatory politics, under which citizens who refuse to accept the authority of states and their hired risk apologists increasingly demand an effective and decisive say on key economic and technological issues.

In passing, I should note that Beck's own analysis of risk politics is on the face of it incapable of dealing with issues of environmental justice. Indeed, he

seems to deny the existence of such issues in arguing at length that environmental risks fall equally upon rich and poor alike; 'smog is democratic', as he
puts it (Beck, 1992: 36). A few risks may have this characteristic (ozone layer
depletion, the fallout from Chernobyl); but many do not. Beck believes that
the equal incidence of risk across social classes is conducive to cross-class political mobilization, one of the ingredients of his risk society. Still, the notion of a
reflexive modernity can be maintained while dropping this denial of environmental justice issues.

Theorists of reflexive modernization have said little explicitly about the structure of the international system, but it can be analysed in quite similar terms.
Of course, issues of environmental risk arise in the international system too,
and so Beck's argument concerning democratization in the presence of risk can
be applied here as well. But just as in domestic political systems old verities
about the need to adjust to economic development and technological change
have lost plausibility, so in the international system the fixed and stable context
provided by the Cold War has gone. Just what is to replace it remains
uncertain. But the basic parameters of international action are fluid as they have
not been for a long time. Acting intelligently can no longer mean acting within
a fixed context: it is increasingly true that actions create and help constitute
contexts. And a key part of the context so constituted is the prevailing constellation of discourses. Thus rational action on the part of any actor means not
just choosing the best means to a given end, but also contemplation of how
that action helps constitute the emerging international order – and, crucially,
the balance of discourses in that order.

This order and its discourses will never take any final and fixed form, so there
is no point in trying to specify any sort of institutional blueprint for democracy – in the international system or, for that matter, anywhere else. Democracy
is a quintessentially open-ended project, in which experimentation with what
democracy can mean is an essential part of democracy itself. Reflexive action in
large measure means development of this project.

Some discursive contexts

If the preceding discussion of reflexive modernization in the international
system sounds somewhat abstract, let me give some examples of how discourses
make a difference, and how their content and relative weight can be affected by
the actions of international actors, for better or for worse.

The law of the sea

The agreements negotiated in the long and painstaking processes of the United
Nations Conference on the Law of the Sea in the 1970s represented an attempt
at liberal institution-building that eventually secured the near-unanimous
approval of the world's states. In 1981 the newly elected Reagan administration
announced that the United States was withdrawing its support from the agree-

ments, thus vitiating them. Equally important, this action at a stroke established that the deep ocean environment would henceforth be subject to a discourse of Hobbesian anarchy in which no authority has any right to restrict access to the deep ocean. The implications were felt beyond the specific issue of the law of the sea. The reinstated discourse of anarchy meant that no significant global environmental agreements were reached in the 1980s (with the important exception of the 1987 Montreal Protocol for the protection of the ozone layer and its subsequent elaborations).

Ozone discourses

In her study of the ozone issue, Karen Litfin (1994) argues that the breakthrough which moved international negotiations on the ozone issue from impasse to agreement was a shift in the discourse which dominated the proceedings. The breakthrough came with the victory of what she calls a 'discourse of precautionary action', which framed existing – and far from decisive – scientific knowledge in a new way. But this shift did not just *happen*; it came with the rhetorical force of the idea of an 'ozone hole' over the Antarctic, which was a way of simplifying seasonal and variable depletions of ozone levels in the southern hemisphere. By the time of the 1987 Montreal Protocol there was still no proof that CFCs actually damaged stratospheric ozone; but a precautionary discourse means that action does not have to await proof. A key part in shifting the terms of discourse was played by a group of 'knowledge brokers' in the US Environmental Protection Agency, and their like-minded contacts in the environmental agencies in other countries (Litfin, 1994: 81).

Brent Spar

In 1995 the Shell Corporation announced that it planned to dispose of the Brent Spar, a redundant oil storage platform in the British sector of the North Sea, by sinking it in the deep waters of the North Atlantic. The British government approved this action. As the platform was about to be towed to its planned dumping ground, it was occupied by several Greenpeace protestors. In conjunction with the occupation, a boycott of Shell was organized in several European countries. Shell soon backed down, to the great annoyance of the British government, which was preparing to use force to dislodge the protestors. Alternative plans were developed to tow the platform to Norway, dismantle it and dispose of it on land. The Brent Spar case has already become a staple in the literature on international environmental politics: most commentary treats it as illustrative of the power of non-government organizations (NGOs) and international civil society, but worries about the fact that disposal on land may have worse environmental consequences than disposal in the North Atlantic.

Here I want to stress the discourse aspects of this incident. In this light, the actions of Shell are not especially interesting because they can be explained in straightforward commercial terms: seeing the threat to its image and its profits, Shell changed course. The anger of the British government can be explained in large measure because the Greenpeace action constituted a threat to established discourses of national sovereignty and the sanctity of property – both of which the British government was keen to preserve for all kinds of reasons, related to environmental matters and otherwise. The Greenpeace actions, in contrast, were instrumental in strengthening very different kinds of discourse: where obsolete industrial plant cannot be dumped in the nearest available hole, where the governance of environmental affairs need not be under the exclusive control of governments. So even if the environmental costs of onshore disposal do prove to exceed those of sinking the platform in the North Atlantic, the Greenpeace action should not be condemned as ill-conceived or counter-productive. For this action has helped, if only marginally, to shift international discourse on environmental issues in a direction which emphasizes the need to recycle rather than dispose, where one cannot search for distant dumping grounds into which to displace one's problems, and where states cannot do as they please in an international anarchy. (For a detailed study of how Greenpeace actions in general are aimed at changing ecological sensibilities or, in my terms, discourses see Wapner, 1996: 41–71.)

Whaling

A moratorium on commercial whaling was imposed by the International Whaling Commission (IWC) beginning in 1986; exceptions were allowed for Iceland, Japan and Norway to engage in 'scientific whaling'. Scientific whaling is actually a cover for a continued (but sharply reduced) commercial catch; the whale products are still used as before, and precious little science comes from the dead whales. These three countries wish to resume full-scale commercial whaling, arguing that stocks of species such as the minke whale have recovered sufficiently to allow a sustainable harvest. While some might dispute this point, the real issue here is the discursive construction of the whale. What is a whale? On the one hand, it is a natural resource, providing useful products such as meat and baleen. On the other hand, a whale is a magnificent sentient creature whose intelligence perhaps equals or surpasses that of we humans, with an intrinsic right to exist and flourish. The IWC moratorium reflects not a sober consideration of the depletion of whale stocks and what might be done to restore them; evidence of exhaustion verging on extinction of many species had been available to the IWC for decades, with precious little response. Rather, the moratorium is the outcome of a discursive shift, a reconceptualization of what a whale is, and what ocean ecosystems are. Eventually this newer discourse captured a large majority of the national delegations to the IWC. Iceland, Norway and Japan have an interest in a shift away from this newer discourse. But rather than just return to the older discourse of resource harvesting, they frequently invoke the idea of science: the whaling they want to engage in is now 'scientific' though it still means killing whales to eat them.

Sustainable development versus market liberalism

I have already noted that sustainable development has over the past decade or so largely (but not totally) displaced the competing global discourse of limits and survival. However, that decade has also witnessed the rise of a competing global discourse: market liberalism (whose roots of course go back a long way). In terms of impact on the shape of the transnational political economy, market liberalism has been much more successful than sustainable development in that period. So far the clash of the two discourses has taken the form of a few border skirmishes. For example, to a number of business interests (represented, for example, in the World Business Council on Sustainable Development, and Global Climate Coalition), sustainable development is a good thing because 'sustainable' means 'continued' and 'development' means 'growth'. Thus can market liberalism press its claims on sustainable development, and try to push the discourse in a particular pro-business direction. Other border skirmishes have taken place in connection with international trade. For example, under World Trade Organization (WTO) rules it is clear that environmental restraints on trade must give way; the WTO's environmental committee deals almost exclusively with such restraints. Such restraints include government subsidy of pollution control equipment and limits on the kinds of pesticides that can be used on imported food.

Prior to the establishment of the WTO, in 1991 a committee of the General Agreement on Tariffs and Trade (GATT) ruled that the United States ban on tuna imports caught in ways that involved the deaths of large numbers of dolphins was a violation of the GATT. The GATT also ruled against Indonesia's ban on raw log exports, thus diminishing the prospects for sustainable forestry. When it comes to national developmental paths, it is clear market liberalism precludes sustainable development conceptualized as anything other than unrestricted economic growth. For example, countries such as Mexico which commit themselves to export-oriented development along lines approved by the International Monetary Fund and World Bank find that in so doing their locally grown produce cannot compete with cheap imports. Peasants are forced from the land, giving way to export-oriented agribusiness, which generally farms in ecologically unsound fashion. The displaced peasants either exacerbate congestion and pollution in Mexico City or move across the border to the United States.

Contesting sustainable development

At the Rio Earth Summit in 1992 there were two sets of proceedings: the official meeting of heads of government and other state representatives, and the unofficial Global Forum made up mostly of NGO activists. In both fora the discourse of sustainable development held sway. But there was still a contest between the sustainable development being negotiated in the official UNCED proceedings and the kinds of sustainable development on display in the Global

Forum. The former were more closely linked to conventional industrialist notions of economic growth and its benefits, the latter to more radical redistribution of resources from rich to poor, and more radical redistribution of power from government and business to community and citizens.

The contest remains; there is quite a gulf between the vision of the World Business Council on Sustainable Development and the various environmental groups which have also endorsed the concept. It is unlikely that any of the sides will score a clear discursive victory; but clearly the relative weight of these competing conceptions matters enormously for the future of sustainable development and its global impact. Rhetorical interventions by particular actors continually change this balance – both locally and globally. So, for example, when Kai Lee (1993) declares that sustainable development has been successfully launched in the Columbia River Basin in the Pacific Northwest of the United States, he advances a notion of sustainable development as centrally managed ecosystem restoration. When the World Business Council on Sustainable Development collects success stories from its member corporations, it advances the notion of sustainable development as an environmental gloss on corporate business as usual (Schmidheiny, 1992).

In the examples I have given, it is probably fair to say that the actors involved did not have any conception of a theory of reflexive modernity and how their own actions took effect within it. Nevertheless, I would argue that the more intelligent among them may well have had a good tacit grasp of the constitutive effects of their actions, and how they affected the prevailing content and balance of discourses – though again, they may not have framed the question in this sort of language.

Democratic theory and global practice

Democratic action in the international system is rooted in reflexive control of the prevailing balance of discourses. I will now try to be a bit more explicit about how and by whom such control is exercised. Here some recent developments in the theory and practice of democracy may be deployed usefully.

Political theorists used to think about democracy solely in terms of the ideal of a self-governing community within precise territorial boundaries (many of them still do). Extending this idea of democracy to the international system has always been quite hard. But this extension is made easier by the degree to which recent thinking argues that the essence of democratic legitimacy is to be found not in voting or representation of persons or interests, but rather in deliberation (e.g. Cohen, 1989; Benhabib, 1994; Habermas, 1995). In this light, an outcome is legitimate to the extent its production has involved authentic deliberation on the part of the people subject to it. Thus deliberation or communication is the central feature of democracy. Such a discursive or communicative model of democracy is particularly conducive to international society, because unlike older models of democracy, it can dispense with the problem of boundaries. These older models always saw the first task in their application as specification of the boundaries of a political community. Deliberation and communication, in contrast, can cope with fluid boundaries, and the production of outcomes across boundaries. For we can now look for democracy in the character of political interaction, without worrying about whether or not it is confined to particular territorial entities.

The idea that we can have democracy without boundaries means that the intimate link between democracy and the state can be severed. Democracy need no longer be confined to the processes of the state. As I have already emphasized, there is no state-analogue for the international system. Moreover, it would be a mistake to try to create one – for example, through a strengthened United Nations. Such a state analogue is both infeasible and undesirable. It is undesirable because states in today's confidently capitalist world face increasing constraints on their operations which severely restrict the degree of democratic participation which they can allow (Lindblom, 1982; Dryzek, 1996a: 10–11). I would argue that state-like systems at the global level would be guided by all the economic imperatives that currently constrain states organized at the national level. The first concern of all such states is to prevent disinvestment and its concomitant economic downturn, which translates into maintaining the confidence of actual and potential investors. A global state would not be constrained by the threat of capital flight (there are as yet no *maquiladoras* on Mars), but even states in relatively closed economies are governed first and foremost by this economic imperative, as the history of the (relatively closed) economy of the United States illustrates.

In contemplating democracy beyond the state, some recent thinking on democracy can again be put to good use, inasmuch as this thinking now emphasizes civil society as well as the state (Cohen and Arato, 1992). Civil society is in many ways a realm of freedom in political innovation that stands in marked contrast to the realm of necessity occupied by the state (Dryzek, 1996a, d). Civil society is another contested concept, but for the moment let us define it in terms of political association and public action not encompassed by the state on the one hand or the economy on the other. A particularly useful definition of civil society is provided by Martin Jänicke, who defines it in functional terms as public action in response to failure in either the state or the economy. For domestic civil society such action can take the form of protest against and pressure upon the state; or it can involve a 'paragovernmental' activity that involves dealing directly with economic actors, without reference to government. The Brent Spar case introduced earlier involves such paragovernmental activity.

Civil society also exists in the international system, and nowhere is it more in evidence than in the environmental realm. Jänicke's functional definition is less easy to apply here because there is no system-level state to fail; but there are formally constituted political bodies such as the United Nations and its agencies, the World Bank, and the World Trade Organization, which can fairly frequently and sometimes spectacularly. So while there is not state failure, there is plenty of political failure to which civil society action can respond. The role of transnational civil society in environmental affairs has been explored at length in works by Paul Wapner (1996) and Ronnie Lipschutz (1995). The most prominent actors here are NGOs and what Wapner calls TEAGS, 'transnational environmental activist organizations'. These groups work within, across and (often) against states. Politics in transnational civil society is also a matter of promoting the ecological sensibilities of citizens and of exerting pressure on multinational corporations to behave in more environmentally responsible ways. Environmental civic politics of this sort was in particular evidence at the 1992 Rio Earth Summit, where such groups and activists constituted the unofficial Global Forum as a counterpart to the official proceedings.

Observers and (especially) critics of transnational civil society often ask about the extent and sources of its power. To those schooled in conventional analyses of power, that possessed by civil society actors looks very weak and indirect, something like that of interest groups or pressure groups in domestic politics, only less influential because there are few well-developed channels for pressure to be exerted. But the kind of discursive analysis which I have developed highlights the real power of transnational civil society, which is communicative power (in Habermas's, 1995, long-awaited explicit statement of his democratic theory, he speaks of the need to subordinate 'administrative power' to 'communicative power', but he only has politics within state boundaries in mind).

How does this communicative power work? The politics of transnational civil society is largely about networking with others across national boundaries, questioning, criticizing and publicizing. Again,, a cynic might see this as capable of generating only a few pinpricks in the hides of established powers such as states, intergovernmental organizations (IGOs) and multinational corporations. But – crucially – these actions can also change the terms of discourse and the balance of different components in the international constellation of discourses which I have described. Compared to the realm of states and their interactions, civil society is a realm of relatively unconstrained communication; its actors are not bound by reasons of state, by the conventions of diplomatic niceties, by the fear of upsetting allied or rival states, or (more important) by fear of upsetting actual and potential investors in one's country, or the financial markets. For these reasons, one should expect civil society and its actors to take the lead when it comes to interventions affecting and helping to constitute and reconstitute the discursive aspect of international environmental politics. Indeed, the main cumulative effect of the past thirty years of environmental activism may be precisely such reconstruction. Before the 1960s there was no area of politics and policy styled 'environmental'.

This capacity to affect the terms of discourse and change the balance of competing discourses is widely distributed. One does not need an army, control over governmental bureaucracies, massive wealth or even large numbers of activists to be effective. One does need a certain minimum of conventional political resources: money, personnel, access to the media, and credibility. More importantly, actors need to be astute enough to recognize the importance of the discursive realm and its increasing importance with reflexive modernization, and to figure out how to act effectively within this realm. Effective action here means reasoning constitutively: instead of asking the instrumental question 'will action X promote goal Y?' one asks the constitutive question 'will action X help bring into being the kind of world I find attractive?' – especially the terms of communication and discourse in that world. For example, in instrumental terms the Greenpeace action against Shell which I discussed earlier was not especially productive: disposal of the platform on land may have greater environmental costs than if it had been sunk in the deep ocean. But the constitutive effects of that action were profound and positive, undermining a discourse of absolute state sovereignty in international environmental affairs and advancing a discourse of environmental responsibility.

Conclusion

Reflexive modernization is especially at home in international environmental issues because, first, these issues are largely about risk (global warming, ozone layer depletion, etc.), and so Beck's risk society argument can be applied directly; and second, the disappearance of the Cold War context means that there is a world to create, as well as accommodate, through a politics that has its own reshaping perpetually in mind. Such reshaping can come in the form of thinking constitutively about discourses. When one adds to this the relative weakness of formal institutions providing order in the international system (given the absence of any state-analogue and effective legal system at the system level), we can see that the prospects for a democratic international environmental politics rooted in transnational civil society may be more propitious than ever before.

Despite these positive prospects, the struggle for this democracy also faces some fairly major obstacles – as does the struggle for environmental values in general in the international system. Today we are indeed witnessing the sort of transnationalization of civil society which I have emphasized. But we are also witnessing powerful economic globalization. International trade is not, of course, new, but the discourses and institutions associated with free trade and economic development imperatives are stronger and more confident than ever before. The constraints they impose are both discursive and material. I have already noted that market liberalism is powerful as a discourse. But even if people resist the discourse, they may find themselves unable to resist material economic imperatives. States in particular are heavily constrained: the first concern of any state operating in the international economic system is to maintain and cultivate the confidence of actual and potential investors. Correspondingly, the main fear of any state is disinvestment and capital flight. So even if governments want to (say) introduce a progressive environmental policy, they may find these impersonal economic constraints militating against it. On the one hand, these constraints on the state further justify democratization in civil society rather than the state. One the other hand, the same constraints enable states to resist pressure from transnational civil society.

These constraints dissolve to the extent that it can be shown that environmental protection really is good for business. Recent developments in the theory and practice of ecological modernization, especially in Germany and the Netherlands, point to such possibilities: a clean, low-pollution environment indicates efficient use of materials, and is conducive to a happy and healthy workforce. A full discussion of ecological modernization is beyond my scope here (see Dryzek, 1997: 137–52 for my own analysis). On the positive side, it is possible to tie a strong version of ecological modernization to the idea of a reflexive modernity (see Christoff, 1996). On the negative side, nobody has yet given any idea how ecological modernization might apply beyond the boundaries of highly developed states.

Until the possibilities associated with ecological modernization are extended to the global arena, what we now witness in the international system is a conflict between material forces and discursive ones. There are echoes here of the clash between Marxist materialism and Hegelian idealism. The discursive realm itself is not one of unalloyed freedom to construct any discourses we wish; as Foucault and his followers have emphasized, discourses themselves can constrain and repress. But as I have tried to show, discourses can also be amenable to conscious reconstruction. The prospects for ecological democracy are positive to the extent that discursive processes involving transnational civil society can make themselves felt in reflexive reconstruction of the international political economy. This is a tall order; the prospects are in many ways more positive than ever before, but a major struggle with market liberalism as both discourse and material force looms.

Parts of this chapter were previously published in Dryzek, J. S. (1999) 'A Transnational Democracy', *Journal of Political Philosophy* 7 (1): 30–51.

References

Archibugi, D. and Held, D. (1995) *Cosmopolitan Democracy: An Agenda for a New World Order*, Cambridge: Polity.

Beck, U. (1992) *Risk Society: Towards a New Modernity*, London: Sage.

Beck, U., Giddens, A. and Lash, S. (1994) *Reflexive Modernization: Politics, Tradition and Aesthetics in the Modern Social Order*, Cambridge: Polity.

Benhabib, S. (1994) 'Deliberative Rationality and Models of Democratic Legitimacy', *Constellations* 1: 26–52.

Christoff, P. (1994) 'Ecological Modernization, Ecological Modernities', *Environmental Politics* 5: 476–500.

Cohen, J. (1989) 'Deliberation and Democratic Legitimacy', in A. Hamlin and P. Pettit (eds) *The Good Polity: Normative Analysis of the State*, Oxford: Blackwell, pp. 17–34.

Cohen, J. L. and Arato, A. (1992) *Civil Society and Political Theory*, Cambridge, MA: MIT Press.

Doherty, B. and de Geus, M. (1996) *Democracy and Green Political Thought: Sustainability, Rights and Citizenship*, London: Routledge.

Dowie, M. (1995) *Losing Ground: American Environmentalism at the Close of the Twentieth Century*, Cambridge, MA: MIT Press.

Dryzek, J. S. (1987) *Rational Ecology: Environment and Political Economy*, New York: Blackwell.

—— (1990a) *Discursive Democracy: Politics, Policy, and Political Science*, New York: Cambridge University Press.

—— (1990b) 'Green Reason: Communicative Ethics for the Biosphere', *Environmental Ethics* 12: 195–210.

—— (1992) 'Ecology and Discursive Democracy: Beyond Liberal Capitalism and the Administrative State', *Capitalism, Nature, Socialism* 3 (2): 18–42.

—— (1996a) *Democracy in Capitalist Times: Ideals, Limits, and Struggles*, New York: Oxford University Press.

—— (1996b) 'Political and Ecological Communication', in F. Mathews (ed.) *Ecology and Democracy*, London: Frank Cass, pp. 13–30.

—— (1996c) 'Strategies of Ecological Democratization', in W. M. Lafferty and J. Meadowcroft (eds) *Democracy and the Environment: Problems and Prospects*, Cheltenham: Edward Elgar, pp. 108–23.

—— (1996d) 'Political Inclusion and the Dynamics of Democratization', *American Political Science Review* 90: 475–87.

—— (1997) *The Politics of the Earth: Environmental Discourses*, Oxford: Oxford University Press.

Falk, R. (1995) 'Environmental Protection in an Era of Globalization', in G. Handl (ed.) *Yearbook of International Law* 6, Oxford: Clarendon Press, pp. 1–25.

—— (1997) 'Resisting "Globalisation-from-Above" through "Globalisation-from-Below"', *New Political Economy* 2: 17–24.

Fischer, F. (1993) 'Citizen Participation and the Democratization of Policy Expertise: From Theoretical Inquiry to Practical Cases', *Policy Sciences* 26: 165–87.

Goodin, R. E. (1996) 'Enfranchising the Earth, and Its Alternatives', *Political Studies* 44: 835–49.

Habermas, J. (1995) *Between Facts and Norms: Contributions to a Discourse Theory of Law and Democracy*, Cambridge: Polity.

Held, D. (1995) *Democracy and the Global Order: From the Nation State to Cosmopolitan Governance*, Cambridge: Polity Press.

Jänicke, Martin (1992) 'Conditions for Environmental Policy Success: An International Comparison', *The Environmentalist* 12: 47–58.

Lafferty, W. M. (1996) 'The Politics of Sustainable Development: Global Norms for National Implementation', *Environmental Politics* 5: 185–208.

Lafferty, W. M. and Meadowcroft, J. (eds) (1996) *Democracy and the Environment: Problems and Prospects*, Cheltenham: Edward Elgar.

Lee, K. N. (1993) *Compass and Gyroscope: Integrating Science and Politics for the Environment*, Washington, DC: Island Press.

Lindblom, C. E. (1982) 'The Market as Prison', *Journal of Politics* 44: 324–36.

Lipschutz, R. (1995) *Global Civil Society and Global Environmental Governance: The Politics of Nature from Place to Planet*, Albany: State University of New York Press.

Litfin, K. T. (1994) *Ozone Discourses: Science and Politics in Global Environmental Cooperation*, New York: Columbia University Press.

Mathews, F. (ed.) (1996) *Ecology and Democracy*, London: Frank Cass.

Meadows, D. H., Meadows, D. L., Randers, J. and Behrens, W. H. III (1972) *The Limits to Growth*, New York: Universe Books.

Orr, D. W. and Hill, S. (1978) 'Leviathan, the Open Society, and the Crisis of Ecology', *Western Political Quarterly* 31. 457–69.

Paehlke, R. (1988) 'Democracy, Bureaucracy, and Environmentalism', *Environmental Ethics* 10: 291–308.

282 John S. Dryzek

—— (1996) 'Environmental Challenges to Democratic Practice', in W. M. Lafferty and J. Meadowcroft (eds) *Democracy and the Environment: Problems and Prospects*, Cheltenham: Edward Elgar, pp. 18–38.

Press, D. (1994) *Democratic Dilemmas in the Age of Ecology: Trees and Toxics in the American West*, Durham, NC: Duke University Press.

Schlosberg, D. (1999) *Environmental Justice and the New Pluralism: The Challenge of Difference for Environmentalism*, Oxford: Oxford University Press.

Schmidheiny, S. (1992) *Changing Course: A Global Business Perspective on Development and the Environment*, Cambridge, MA: MIT Press.

Shapiro, I. (1996) 'Elements of Democratic Justice', *Political Theory* 24: 579–619.

Torgerson, D. (1995) 'The Uncertain Quest for Sustainability: Public Discourse and the Politics of Environmentalism', in F. Fischer and M. Black (eds) *Greening Environmental Policy: The Politics of a Sustainable Future*, Liverpool: Paul Chapman, pp. 3–20.

Walker, K. J. (1988) 'The Environmental Crisis: A Critique of Neo-Hobbesian Approaches', *Polity* 21: 67–81.

Wapner, P. (1995) 'Environmental Ethics and World Politics', Paper presented to the Annual Meeting of the International Studies Association, Chicago.

—— (1996) *Environmental Activism and World Civic Politics*, Albany: State University of New York Press.

Wendt, A. (1992) 'Anarchy Is What States Make of It: The Social Construction of Power Politics', *International Organization* 46: 391–425.

Williams, B. A. and Matheny, A. R. (1995) *Democracy, Dialogue, and Environmental Dispute*, New Haven, CT: Yale University Press.

16 Restructuring the space of democracy

The effects of capitalist globalization and the ecological crisis on the form and substance of democracy

Elmar Altvater

Introduction: paradigmatic changes

Globalization means change to the system of coordinates within which theoretical political projects are performed or within which people organize their work and daily life. The conditions of economic globalization cause diffusion of the political space within which democratic deliberations and procedures can be carried out and governability secured; that is, effective regulation of economic processes by government institutions on different levels or by central banks. This is the main reason for the crisis of the Keynesian paradigm and for the predominance of neo-liberal thinking in scientific as well as in political discourses.

Neo-liberals are theoretical and political promoters of deregulation, of limiting the regulative power of the nation state. The dissolution of state borders is the consequence of globalization. Moreover, sensitive observers have noticed for some time the appearance of new ecological limits on the far horizon. Until recently this observation had not touched popular consciousness. Now a new term for these limits of globalization has been coined: the limit of the 'carrying capacity' or 'environmental space' of planet Earth (e.g. Opschoor, 1992; BUND/Misereor, 1996). As long as the 'limits to growth' – to employ the discourse of the Club of Rome (Meadows *et al.*, 1973) – are far distant, the constraints of nature can simply be neglected. Approaching the limit, however, is coming to be recognized as a global environmental crisis. Resources are limited in an objective way by the bounds of planet Earth, even if the limits are politically constituted by people's discursive practices.

In the following it is not possible to address the whole range of problems associated with the paradigmatic crises of economic globalization and political and ecological limits. In the first part of this chapter I discuss the challenges of globalization and the consequences of dissolution of national sovereignty for democracy and theories of democracy. In the second part I touch upon the question of the constitution of a new place for democracy in the context of new ecological limits. The conclusion discusses the necessary paradigm shifts with regard to questions of democracy and human rights in the coming century.

Capitalist globalization and democratic discourse

Dilemmas of democracy

For the determination both of its range of application and the duration of its procedures, democracy requires coordinates in space and time to secure 'governability', since, of course, governability cannot refer to the global system as a whole. *Governability* has to be distinguished from *governance*. The latter is a project of a global reach, as is shown by the establishment of the 'World Commission on Global Governance' (Commission on Global Governance, 1993).

The change in the discourse from 'governability' to 'governance' is interesting. In the 1970s the Trilateral Commission tried to spell out the prerequisites for overcoming non-governability and for re-establishing the conditions of governability of Western *nation states*. The Commission on Global Governance twenty years later aims to establish 'soft' global rules for *new forms of institutionalized global cooperation between states, private economic actors, international organizations and non-governmental organizations (NGOs)* (for the Trilateral Commission see Sklar, 1980; for global governance see Commission on Global Governance, 1993; Falk, 1995).

One major reason for this shift is that the state has itself been 'globalized' or 'internationalized':

> the policy orientation of the state has been pulled away from its territorial constituencies and shifted outwards, with state action characteristically operating as an instrumental agent on behalf of non-territorial regional and global market forces, as manipulated by transnational corporations and banks, and increasingly also by financial traders.
>
> (Falk, 1997: 129; similarly Broadhead, 1996)

'Governance without governments' (Rosenau and Czempiel, 1992) is the challenge to which responses have to be developed, responses to *la nébuleuse*, as Robert Cox (1996) called the unclear constellation of political and economic factors under conditions of globalization.

National borders are already necessary to secure the formal working of democratic procedures. In addition, they are the framework within which *substantial* rights of both individuals (human rights) and peoples (peoples' rights) can be asserted and maintained: 'States won because their institutional logic gave them an advantage in mobilizing their societies' resources' (Spruyt, 1994: 185). The rationality of participation is established not only formally in a spaceless world, but substantially in the coordinates of historical time and space. The participants in the 'democratic game', with all its paradoxes and dilemmas, are therefore equipped with, first, an integrity as persons, and thus, second, not just with formal but specific and substantial rights (human rights of the 'second' and 'third' generation). These rights are historical 'achievements' which have become social and political standards of citizenship. Therefore Peter Christoff is right to speak about even 'ecological citizenship' which is 'centrally defined by its attempt to extend social welfare discourse to recognise "universal" principles relating to environmental rights and centrally incorporate these in law, culture and politics' (Christoff, 1996: 161).

Although globalization exerts a unifying pressure on historical standards these rights, therefore, are essentially different in different cultures. In the postwar period, human rights of the first and second generation have become prerequisites of the modern democratic discourse and in some cases they claim the dignity of constitutional principles. But the recently emerging globalization processes, including the dissolution of political sovereignty on the one hand and the ecological crisis on the other, have undermined claims to substantial rights. The democratic order therefore faces a number of new dilemmas, some of which we shall briefly discuss.

Democracy and the market

The theoretical equality of citizens in politics is thwarted by the inequality in practice between bourgeoisie and proletariat. The welfare state's minimal standards of social equality, and to a certain extent the claims of economic and social human rights in international politics, function as countervailing principles to this inequality. In part this may be because the formal democratic order necessitates the establishment of minimal substantial standards. This question will be discussed later.

The contradiction between the national borders of politics and the globalized economy's boundlessness, however, is more important. The importance of the procedural rationality of democracy can be traced back to the emergence of modernity after the Renaissance, the great expeditions and the triumph of the European rationality of world domination (Max Weber). In the historical process of the 'Great Transformation' (Polanyi, 1978 [1944]) since the seventeenth century, the economy broke out of social control and subjugated society to the laws of capitalist accumulation and to the inherent rationality of acquisition. As Beetham (1993) convincingly showed, this economic rationality is not fully compatible with political rationality, even in the formal Schumpeterian sense of the democratic process (Schumpeter, 1976).

Economic decision-makers either deny political territoriality or take it as an opportunity for arbitrage speculation: to reduce it to an economic calculus by exploiting differentials in time and between spaces. Thus their instrumental, formal economic rationality surpasses political deliberations and the 'bed' of social relations, disembedding them, as Polanyi (1978 [1944]) suggests.

This contradiction between formality and materiality of social (and political) decisions, well known since Max Weber wrote about it, is apparently intensifying at the end of the twentieth century and involves an increasing tension between globalization and nation states, 'systemic constraints', and political deregulation. These factors indicate the political-administrative system's loss of control over essential economic variables.

Democracy as an ingredient of structural adjustment programmes

Democracy, at least in principle, has no enemies in the 'new world order'. 'Global democratization' is one of the most striking and unchallenged characteristics of globalization (Group of Lisbon, 1997). The reason for this state of affairs is that the democratic question has been depoliticized. Market forces exert those constraints on social systems which oblige them to follow their necessities. This is the rule even in global *policy* enacted by global institutions such as the International Monetary Fund (IMF) or the World Bank. These institutions impose 'structural adjustment programmes' on nation states; that is, rules of the world market to be implemented at a national level. It is not the state that is withering away, but rather sovereignty in economic and social policy. The space of civic disputes on alternatives under the pressure of structural adjustment, therefore, shrinks together with democracy's place. Therefore Stephen Gill (1996) mentions the paradox that globalization on the one hand is a constraint upon democracy and, on the other hand, promotes formal democratization.

Procedural democracy and the congruence of decision-making

The contrast between politics and economics, which has been aggravated dramatically by globalization, is woven into the capitalist world system's long history. Nation states are defined by borders, which they set and defend, both domestically by the exclusion of those who are considered as not belonging to the citizenry, and externally, against other nation states and their citizens. Thus, there arises the question of citizenship on the one hand and, on the other, the organization of the 'pluriverse' of sovereign nation states (i.e. the constitution of an international order). This order, however, only works so long as a certain congruence between the political, social and economic system exists. Therefore borders are of the utmost importance. State borders, defining the spatial and temporal reach for the set of formal rules and procedures, are a prerequisite of the territorial *congruence* of decision-making. This is the only way a nation can be a 'community of fate' (Held, 1991: *passim*). The procedural set of rules attached to this principle – in the minimal sense of Bobbio – in practice defines a democratic system of checks and balances as a minimal prerequisite of a democratic order (Bobbio, 1987: 193: 'What is a democracy if not a set of rules [the so-called rules of the game] for the solution of conflict without bloodshed?').

As Collier and Levitsky (1997: 433) observe, '[it] deliberately focuses on the smallest possible number of attributes that are still seen as producing a viable standard for democracy; not surprisingly, there is disagreement about which attributes are needed for the definition to be viable'. Therefore, 'Democracy, even formal democracy, is a matter of power and power sharing' (Huber *et al.*, 1997: 325). Democracy is thus dependent on the balance of class power, the structure of the state and state–society relations, and on 'transnational structures of power . . . grounded in the international economy and the system of states' (*ibid.*). Thus the transformation of the global system exerts an influence on procedural (Schumpeterian) democracy in nation states since it 'strongly affects the structure and capacity of the state, the constraints faced by state policymakers, state–society relations, and even the balance of class power within society' (*ibid.*: 326). Participation in decision making procedures only makes sense so long as there is room for decisions on alternatives. In the event that substantial alternatives do not exist, democratic procedures become hollow and empty processes, not only substantially but also in a formal sense.

Unconstitutional powers

The existence must be noted of unconstitutional powers which shore up market societies. Politically, globalization and concomitant deregulation also means that privatized decision-making is depoliticized: it no longer needs citizens' legitimation. The 'unconstitutional powers' in the economy or the world of media need only ensure the existence of an attractive market supply to the *customers*, yield a profit to the *shareholder* and achieve a high *audience* rating. They have only to obey the rules of the economic (and media) sphere. The 'unconstitutional powers' are not tied to political decisions. The citizens who form a political community are primarily interesting as economic subjects, particularly as consumers. Hence, globalization raises completely new questions which were not on the agenda as long as the systemic constraint of the world market was not seriously in question and the sovereignty of the state over a certain territory was a natural and self-understanding assumption.

The emergence of an economic constitution of economic agents, i.e. of transnational corporations instead of a political constitution of free citizens, is reflected in the attempts of the Organisation for Economic Co-operation and Development (OECD) and the World Trade Organization (WTO) to institutionalize a Multinational Agreement on Investment (MAI) which gives far-reaching political rights to economic agents at the same time as dismantling token rights of political citizens to participate in decisions concerning their life and working conditions. 'Unconstitutional powers' therefore, on the one hand, tend to become partly transformed into constitutional powers. On the other hand, the economic constitution partly takes the place of the political constitution.

The nation state as a competition state

The consequence of the above historical changes is the tendential substitution of the 'binary' political logic of nation states for the multiple economic principle of competition, in so far as the economic sphere is characterized by competitors, not by (political) enemies. Thus, with the exception of a bilateral monopoly, political binary logic is not applicable in the economy. The nation state in this process does not 'wither away' or vanish. However, it changes its character. The political nation (and welfare) state

> has been replaced by the 'competition state'. This shift is leading to a potential crisis of liberal democracy as we have known it – as the international and transnational constraints limit the things that states can do (and therefore the things people can expect from even the best-run government, democratic or authoritarian) – and is creating a new role for the state as the 'enforcer' of decisions which emerge from world markets, transnational 'private interest governments', and international quango-like regimes [a quango is a quasi-autonomous non-governmental organizations] . . . the competition state pursues increased marketization to make economic activities located within the national territory . . . more competitive in international and transnational terms.
>
> (Czerny, 1996: 633)

Politics of the state in the 'geo-economy' differ from those of the sovereign national state in the 'Westphalian order'. The state makes sure that *competitiveness* of the *national* economy is maintained in *global competition* and, if possible, improved. The state therefore matters. The institutional as well as the regulatory system remain important for economic performance of 'national' or 'indigenous' capital (Panitch, 1996: 88 with reference to Poulantzas, 1974: 73). In the global currency competition, at least, nation states are competing to attract highly mobile and volatile financial capital. The borders of a 'currency-space' today seem to be more important than the territorial borders of the political unit.

This transformation of the nation state and the growing importance of markets are leading to changes in citizenship. Falk continues:

> Appropriately these days, it is business elites that are declaring themselves most ardently to be citizens of Europe, or even global citizens, and thereby apparently most willing to forgo the specific identies of the nation-state . . . this new type of global citizenship is pragmatic, and has grown up without accompanying feelings of regional or global solidarity of the sort associated with a sense of community.
>
> (Falk, 1997: 129)

Even more important than the emergence of a global citizenship of business elites is the development of transnational migration. Owing to the dissolution of the unity between state territory and citizenry, the congruence of political, social and economic spaces is increasingly disappearing. The previously clear territorial outlines of state territory, national power and national people are being eroded. The unequivocal allocations of rights and duties as well as the rules of participation in decisions and the mechanisms of legitimation and *accountability* are no longer clear. In times of transnational migration, questions of how civic rights emerge or disappear find no easy or definite answers. Migration makes the modern invention of (national) citizens seem even more artificial, since fewer and fewer people living in a territory are united in language, origin, religion, ethnic origin, etc. Therefore citizenry and citizenship on the one hand are 'fragmented' (Wiener, 1997: 488–96) owing to the loss of congruence between the political and economic order. On the other hand, the endowment of men and women with unrenounceable human rights is becoming a human property of growing importance (see also Christoff, 1996).

Power and political interferences

Not just political decisions, but also economic decisions made within one nation state can turn out to have effects on other nation states. This problem has been known for some time and has been thoroughly discussed, for example with respect to the influence of transnational corporations on governmental decisions in developing countries. German Bundesbank decisions on the prime interest rate affect employment and exchange rates from Portugal to Poland, which has been interpreted as a sign of increasing global interdependence as well as of the extraordinary power of certain central banks.

The power of central banks and governments, however, is not autonomous and politically constituted. They hardly have any political option other than to follow the external course determined by the capital markets. Economic constraints restrict the political space of participatory decision-making. Mexico suffered an especially drastic experience in this respect in December 1994: the reduction of short-term capital due to decisions made by the US Federal Reserve System halved the value of the Mexican currency within a mere two weeks. Since the middle of 1997, Asian countries, formerly called 'emerging markets', have been suffering a similar experience of currency devaluation and externally enforced hard internal adjustment measures which affect not only the economy but all segments of society and social life. The idea of sovereignty having a territorial character is rendered ridiculous in times of globalization. In shaping the governing institutions of Mexico, the World Bank, the IMF and the US Treasury assumed primary importance; the Asian crisis a few years later displayed the same constellation of power and political interference.

Ecological globalization and the time–space coordinates of democratic decisions

The ecological crisis also has consequences for the form and substance of democracy. The radioactive fallout in Chernobyl not only affected citizens in Ukraine, but *citizens nearly everywhere on the globe*: from the Scandinavian countries to Poland, Germany and even the United States (see Chapter 5 by Shrader-Frechette). Even the global citizens of the transnational business elite were concerned. Their health was more or less strongly affected, but they cannot react to this as *national citizens* in any substantially and procedurally *decisive* way.

The idea of the sovereign nation state as a 'national community of fate' has become an anachronism not only in view of economic globalization and the global media world, but also as a consequence of the global ecological crisis. Democratic procedures in the age of global social and ecological problems are rendered questionable merely because the time frame (nuclear material has half-life periods of up to tens of thousands of years) and the expansion in space (across the whole planet Earth) have become far too big for the human dimension of rational decision-making. The congruence of decision, concern and control has gone. It is impossible to decide on the effects of the radioactive fallout from Chernobyl or the construction and use of the atomic bomb democratically in 'democracy's place' (Shapiro, 1996: 1–15).

The political authority of the world market and the revival of formal democracy

The connection between *globalization, deregulation* and *depoliticization* not only creates dilemmas for democracy, but also presents a paradox for the 'democratic question'. Authoritarian political systems lose their 'sense' in view of the economic *authority of the world market*. They simply become dysfunctional and make room for formal democratic systems. Political power counts less and less compared to economic power. The transition from the 'bureaucratic authoritarian state' (O'Donnell *et al.*, 1986) to democratic political systems in Latin America during the 1980s (the *abertura*) and in Eastern Europe about one decade later (transformation or transition) are a politically adequate reaction to globalization. Despite many differences, these two sets of events are therefore comparable – although the comparative approach should not be overstretched as in Munck and Skalnik Leff (1997). In all cases, the transition – unlike such events in previous history – took place in a surprisingly orderly fashion, nearly without violence, and without the representatives of the authoritarian regimes clinging to their power and defending it violently against the democratic participation requested by the popular masses. They accommodated readily to democratic regimes, and vice versa. This is evident in the amnesty laws in Argentina, Chile and Brazil as well as in the continuity of political elites in many former 'socialist' countries.

The direct repression of authoritarian political systems (Latin American dictatorships of the developing state, and 'socialist' one-party political systems and planning economies in Central and Eastern Europe) has been replaced by the 'systemic constraint' imposed by the world market, no less effective and harsh than the previous authoritarian political regimes. However, it is not arbitrary political power, but the objective constraints of the market which rule social and individual lives. The politics of democratic governments nowadays often consist of more or less intelligent 'structural adjustment' to the challenges of the world market, and such adjustments are often enough demanded by the world market's institutions – the IMF, the World Bank, G7, etc. – and associated interested political fora. In Latin America, the nation state's central role

> changed from one of encouraging development and providing public services to overseeing the foreign debt and implementing IMF-inspired structural adjustments. . . . The 'highly transnationalized and weak' state acts as a 'liquidator of its own bankruptcy', and the process depoliticizes, demobilizes, privatizes, and insures that any democratic opening will be limited. . . . Latin America today faces the reality that the basic terms not just of its broadly construed political economy but of the details of its state budget and social trade-offs are often determined outside its borders.
>
> (Volk, 1997: 10)

With limited sovereignty over (global) economic procedures, the nation state cannot enforce political aims – not even with authoritarian measures – or carry out macroeconomic (national) plans (as in Brazil during the 1970s) against microeconomic powers. The rationality of economic deregulation supersedes the rationality of political regulation. Therefore, capitalism and democracy are actually compatible. Economic opening and political opening are conditioning each other. Under these new conditions, however, there is no room for the 'domestic social contract between the state and society, which had become integral to the programme of welfare capitalism and social democracy' (Falk, 1997: 130). To achieve simultaneously competitiveness on the world market, democratic participation on the national level, and (welfare state) systems for social security is a political project as difficult to realize as 'squaring the circle' (Dahrendorf, 1995), and this is even more so for less developed countries.

Meanwhile, in all international institutions in which the mighty Western societies are involved, there is consent on the fact that countries seeking membership in international institutions or development assistance from the 'international community' or association with the EU, NATO, OECD, etc. need to pass the 'democracy test'. They have to secure human rights, the rule of law, a market-friendly political environment, i.e. deregulation as far as possible, ecological minimum standards in order to achieve a state of sustainability, and, last but not least, democratic principles, such as freedom of speech and organization, and the secret ballot (this is the catalogue issued by the German minister of development cooperation in 1991). These principles are understood as the unquestionable conditions of 'good governance', and therefore they secure entrance into the regulatory order of the IMF, the World Bank, or the WTO. This again sheds light on the above-mentioned correspondence between the functioning mode of the economic and the political order respectively.

The consequences of new borders for democratic discourse

Democracy and the limits of environmental space

Is formal democracy a viable project? Dankwart Rustow's distinction between 'factors that keep a democracy stable . . . [and] the ones that brought it into existence' (Rustow, 1970: 346) is a key concept with which to analyse democracy during periods of transition from an authoritarian to a politically free order. In the transition process a formal democratic order may perform excellently, but it is questionable whether the factors of democratic stability during the transition process also work under conditions of systemic normality. Then, a formal democratic order requires embeddedness into the social system; that is, substantial rights for those participating in formal procedures. In the 1950s Seymour M. Lipset referred to the substantial preconditions of formal democracy: 'the more well-to-do a nation, the greater the chances that it will sustain democracy' (Lipset, 1959: 75).

Adam Przeworski holds that an empirical examination shows that no democratic order has failed since the Second World War in a country with a per capita income of more than US$4,335 (Przeworski, 1994). Under the assumption of equal distribution, a per capita income of about US$4,400 for a world population of about 6 billion requires a global GNP of about US$26,500 billion. This is less than the present gross global product and less than half the value of financial derivatives on global financial markets. This comparison demonstrates that the substantial democratic question is more about distribution than about production and productivity (at the given level of resource consumption and emission production). Inequality of wealth and income (in each national society as well as in the world as a whole) is not a good precondition for the development of democratic institutions. In a substantially unequal world, where 20 per cent of humanity has access to 80 per cent of the resources and 80 per cent of humanity can use only 20 per cent for themselves (UNDP, 1994), no formal democratic procedure can exert a compensatory effect. Therefore, even the establishment of global structures of governance is a difficult undertaking.

This statement might be considered within another theoretical discourse. The substantial ingredients of wealth and income (a rich bundle of goods and services) become 'oligarchic' (Harrod, 1958) or 'positional goods' (Hirsch, 1980) the higher the level of income. Positional or oligarchic goods cannot be increased at the same pace as the economy is expanding. The 'Western' standard of life cannot be generalized on Earth without destroying the planet's nature to the extent that human life on Earth is jeopardized. Western formal (procedural) democracy could only be globalized if the 'Western way of life' (substantially, and not only as an abstract model) could be globalized. Or, in categories of thermodynamic economics: far from the entropic equilibrium, all possibilities of development and democratic participation are open (see Georgescu-Roegen, 1971). At the limits of environmental space, however, the environmental goods needed for production and consumption become 'oligarchic goods'; that is, reserved for a money oligarchy which secures its access to resources with monetary measures. Those who do not have monetary wealth are excluded from the consumption of goods and services produced. Therefore, economic and financial globalization is a mighty tendency; but it is impossible that the *situation of globality* (i.e. a world society based on equality and reciprocity, if not on solidarity) can be achieved in the *process of capitalist globalization* (see Altvater and Mahnkopf, 1996). David Harvey (1996: 139 and *passim*), however, rejects this argument because of its Malthusian bias. He is correct in interpreting Malthusian reasoning as deriving, from the assumption of scarce ecological resources, the necessity and God-givenness of social inequality:

> Thus is the Enlightenment project reserved for a small elite while everyone else is condemned to live by natural law. This is an appalling instance of that awful habit of denying one section of our species the right to be considered human.
>
> (Harvey, 1996: 145)

He is also correct in alluding to 'even the short history of capitalism', which shows 'that resources are not fixed. . . . What exists "in nature" is in a constant state of transformation' (Harvey, 1996: 147). Certainly, but the problem of capitalism is that since the Industrial Revolution it has relied on non-renewable (fossil) and not on renewable (e.g. solar) energy. The capitalist mode of production is based on energetic stocks (respectively on flows which need some millions of years) and not on energetic flows. Stocks can be used only once, unless the time and space frame can be extended to a dimension which transforms stocks into flows. This is the kernel of the thermodynamic argument which Harvey harshly criticizes:

> It is one thing to argue that the second law of thermodynamics and the laws of ecological dynamics are necessary conditions within which all human societies have their being, but quite another to treat them as sufficient conditions for the understanding of human history.
>
> (Harvey, 1996: 140)

Of course, it is not sufficient to explain human history and the societal relation of human beings to nature by referring only to the second law. But it makes a difference whether we make this statement far away from the limits of the 'environmental space' or as we approach them. Serious studies on the carrying capacity of global ecosystems or of the 'environmental space' doubtless have demonstrated that these limits objectively matter (BUND/Misereor, 1996; Worldwatch Institute Annual Reports) and that they therefore must be included into discourses on ecological sustainability as well as into those on social justice and the democratic order. It is common sense that not only are fossil resources limited, but their combustion is responsible for the greenhouse effect and other ecological harm.

This is the *factum brutum* and the tragedy, so to speak, of the democratic process: the formal rules of the game are not matched by the stakes in the game. This discrepancy is decisive: here, the rules of the game, i.e. the rules of *formal* democracy, are bent for *substantial* reasons. In the social sciences the dilemma has been described by Garrett Hardin as the 'tragedy of the commons' (Hardin, 1968), and can also be applied to the rationality of democratic procedures. It is not necessary to accept either the conclusion which Hardin has drawn from his statement (he proposed the privatization of the commons) or his point of departure, which Harvey calls a 'fusion of contemporary Darwinian thinking, the mathematical logic of diminishing returns, and a political economy of an individualized, utility-maximizing, property-owning democracy' (Harvey, 1996: 372).

Harvey's critique does not touch upon necessary transformations of the social and ecological discourse in the event that the carrying capacity of resources and sinks tends to be overloaded. Two different reactions are possible: either not all human beings can use natural resources on the level attained by 'Westerners' (the oligarchic solution), or all human beings, including 'Westerners', have to reduce their levels of consumption in order to reduce the overload (the democratic-egalitarian solution). Consequently, the rules of the 'democratic game' cannot be sufficiently discussed without considering *historical (and therefore political) space and time and the (ecological) carrying capacity of the (global) commons.* Far from their boundaries, substance does not matter for formal rules of decision-making; near the 'limits of growth' (or of the 'environmental space') they are of decisive importance and must be taken into account. This substantial change leads to a number of consequences.

Substantial wealth and high real income levels depend on the level and growth rate of productivity. Productivity increase is one of the main features of industrial capitalism in general and of the Fordist system especially. Increase of productivity is only possible by using more fixed capital and consuming growing quantities of matter and energy – efficiency increases are taken for granted. Also, the 'spatial fix' as a prerequisite of productivity increase has to be produced, for example material infrastructure in order to eliminate all spatial barriers for the acceleration of the circulation of capital (Harvey, 1996: 295, 412). Under these assumptions it is hardly sensible to determine the central political concept of 'power', as Franz Neumann and others did half a century ago:

The concept of power comprises two constituent facts: domination of nature and domination of human beings. Domination of nature is intellectual domination which results from the recognition of the lawfulness of external nature. This knowledge is the foundation of the productivity of society. This domination has no power. As such, it does not include domination of other human beings.

(Neumann 1978 [1950]: 385)

This quotation shows that a paradigmatic discourse can lose its validity in a very short time – in less than half a century. The assumption that domination of nature is powerless and thus practically unpolitical cannot be maintained if the ecological and feminist discourse and the discourse of critical Marxism are considered in the light of an understanding of the recognized limits of the environmental space. The domination of knowledge over nature helps increase productivity only as long as the consequences of energy and material throughput within the environmental space can be benignly neglected. Limits to increasing productivity are inevitable as resources are exhausted and the spheres of planet Earth are being polluted and contaminated. It may even be that the productivity of labour, its increase being the fundamental principle of the capitalist market economy and modern society since Adam Smith, has to be reduced for ecological reasons (radical reduction in the use of fossil energy, apart from an 'efficiency revolution' in energy which so many hope for).

James O'Connor in a seminal article (O'Connor, 1989, and the subsequent debate in *Capitalism, Nature, Socialism*) has drawn the conclusion that ecological limits are responsible for a 'second contradiction' in modern capitalism (see also the critique of Altvater, 1993: 218, *passim*). David Harvey, however, rejects this approach (Harvey, 1996: 139, *passim*) because of its Malthusian bias. He understands limits only as discursively construed social limits, never as objective natural limits. But this is not the argument, either of thermodynamic economics or of ecological Marxism. Discourses are not free and ideal imaginations but bound to a transforming environment. The environment matters, for example as additional costs in microeconomic calculations or with regard to the validity of GNP figures, or as a factor influential in the emergence of new social movements. So the objective factor of ecological scarcity is integrated into the intersubjective, social discourses on society's relation to nature, social justice and ecological democracy.

Therefore, in contrast to the economic tendencies of dissolving (political) borders, new ecological borders are emerging. They have an objective character, but their relevance is constituted only in the course of a globalized discourse on ecological sustainability. Society's relation to nature does actually allow different forms of dealing with the borders of the environmental space; the premises under which they are constituted are contested, but nevertheless the natural carrying capacity or the environmental space exhibits new borders. They are discursively *constructed* after the limits of the state territory have been *deconstructed.* Consequently, the democratic question is radicalized from two antagonistic sides. On the one side is the globalization of the economy and the information media which perforate *traditional political borders.* On the other, the ecological crisis creates *new borders* which in the long term cannot be ignored. This leads to the *traditional* question of how to reconcile a boundless (in the global *space*) expanding and deregulated market with the limited *place* of politics.

The new question is directed to the effects of the limits of the environmental space on the possibilities of participation, the legitimation of institutions, the representation of interests, and finally, on the governability of the limited environmental space under the auspices of the politically borderless, 'disembedded', and deregulated economic processes. The answers to the 'democratic question' must therefore comprise the effects of globalization on the economy and the media (*dissolution of borders in politics*) as well as the effects of the ecological crisis (*new borders of politics*). The democratic question, therefore, cannot be discussed without taking the question of ecological sustainability into consideration.

Ecological limits are not congruent with national borders, and therefore political subjects do not gain their political identity in disputes concerning the establishment of territorial borders and social limits; citizens are 'citizens as ecological trustees' (Christoff, 1996: 158). Traditional citizens wear a 'national uniform' which equips them with rights and duties within the state territory. As we already have seen, the dissolution of territorial state borders exerts a substantial impact on the concept of citizenship and citizenry. Christoff explains:

> Ecological citizenship is centrally defined by its attempt to extend social welfare discourse to recognise 'universal' principles relating to environmental rights and centrally incorporate these in law, culture and politics. In part, it seeks to do so by pressing for recognition of the need actively to include human 'non-citizens' (in a territorial and legal sense) in decision making.
>
> (Christoff, 1996: 161)

Moreover, there is an additional factor influencing the development of political identities of citizens. As actors in the environmental space (e.g. in environmentalist movements), citizens face ecological borders: the restrictions on the use of natural resources in production and consumption. This is not a question of different political party affiliations, but of lifestyle and mode of production. Politics in the nation state and politics in the environmental space thus differ in principle, especially under the aspect of economic globalization. On the one hand, there is a strong and even overpowering tendency towards deregulation, and on the other hand, there is the unconditional necessity of regulating society's relation to nature.

Democratic equity and justice in the capitalist world

At this point in our reasoning four conclusions are possible: (a) a disconnection between substantial and formal democracy; (b) reliance on the ingenious capacity of humanity to find a solution to the limitations of globalization; (c) the expectation of a process of global redistribution of resources; and (d) the expectation of an ecologically sustainable reorganization of the mode of production. Let us briefly discuss these four possible conclusions, of which the third and fourth, are of course, the most radical.

Disconnecting formal and substantial democracy

The *first* conclusion suggests disconnecting the conceptual links between substantial and formal democracy. It is possible then to apply the rules of rational decision-making in a democratic order without any reference to substantial preconditions or constraints. The quoted statements about substantial minimum standards of a formal democratic order are perhaps of practical relevance, but theoretically they are not decisive. Democracy and the discourse on the democratic order are reduced to a minimal version without any reference to substantial standards. This is also the case with regard to human rights. They do not contain social and environmental rights – human rights of the 'second' and 'third' generation.

Neo-liberal optimism about ingenious problem solutions

The *second* conclusion follows the optimistic position that in the long run humanity will find solutions to any shortage or scarcity of resources (for the distinction between scarcity and shortage see Altvater, 1993: 70 and *passim*) and therefore will always be able to provide a better standard of life to ever more people in the world (Simon, 1995; critics of this position include Buell and De Luca, 1996: 11). The limits of the environmental space do not matter; productivity and welfare increases are not only desirable but in the long run also possible. Humanity is not approaching the 'limits of growth' or of the environmental space, because each border poses a challenge to overcome it. This Napoleonic view is a very common assumption of neo-liberal reasoning. It is based on the assumption of the objective limitlessness of the environmental space. This position is partly shared by Harvey (1996), with the decisive difference that limits are discursively construed and therefore can be overcome by political action.

Global redistribution and the divide between extraction and production

Third, there is enough evidence about the limits of the environmental space to indicate that the substantial preconditions of formal democracy cannot be established for all societies on the globe at the level of a 'Western lifestyle'. In addition, formal parliamentary democracy in most parts of the world is a fragile order because of its naturally conditioned substantial deficiencies. The unequal distribution of resources, and the unequal location of industrial sites for transforming resources into those goods and services which define the Fordist model of production and consumption, have an impact on the democratic order.

The use of fossil energies needs a worldwide logistical system which requires high technological and organizational competence, finances, economic knowhow, transport facilities and political relationships, which for the foreseeable future can be procured only by highly developed industrial countries. The tendency to unequal globalization not only is due to the functioning of the financial system, but also results from the logic of the energy system of the capitalist mode of production. The possibilities for applying democratic procedures are better in complex *production economies* than in simple *extraction economies*.

The economic and political trajectory is different for production and extraction economies. The Ricardian theorem of comparative cost advantages promises trade gains for all participants in the trading system so long as they concentrate their production and trade on those products for which they have comparative cost advantages. Although this principle has been elevated above all economic principles and laws, it has been harshly criticized because of its conservative character. Those countries endowed with raw agricultural, mineral and energy-producing materials remain extraction economies in most cases, whereas those countries already industrialized with an abundance of capital and high endowment of human capital are supposed to specialize in industrial products (List, 1982 [1841]). These countries are capable of acquiring competitive advantages and, as a result, can follow the development trajectory of production economies. The traditional reading of trade advantages would not find any problem with such a specialization in the world economy. But the debate on the 'Dutch Disease', or on underdevelopment of rich extraction economies, has shown that this kind of specialization has a negative impact on the extraction economies. The reason is that the exploitation of raw materials is more profitable than investment in industrial production sites. Wages are higher in the extraction sector than in manufacturing; the state is interested in the extraction of primaries because of the royalties; and, last but not least, the exchange rate of the currency is overvalued because of flourishing exports of raw materials. It exerts a negative impact on exports of manufactured goods and services and instead favours their imports. 'A Dutch economist moaned, when remembering the strong revaluation of the Dutch guilder after the discovery of minerals, "Never find a raw material!"' (for a comprehensive debate of the 'Dutch disease' see Tilton, 1992).

In modern economics the question of distribution over time (among different generations) and space (among peoples in the contemporary world) is formally resolved by applying a rate of discount. Thus it seems to be possible to weigh costs and benefits over time and space in terms of a common monetary unit. The premises of a positive discount rate are of course that benefits and costs can be compared in monetary terms. But if costs (and benefits) do not have monetary value, how is it then possible to construct a discount rate? And what happens when costs and benefits over time and over space are not commensurable? In that case a rational and clear answer on the question of distribution of costs and benefits of environmental degradation is excluded. In *The Social Costs of Private Enterprise* William Kapp stated that

> Social costs and social benefits have to be considered as extra-market phenomena; they are borne and accrue to society as a whole; they are heterogeneous and cannot be compared quantitatively among themselves and with each other, not even in principle.

> (Kapp, 1950)

Against the mainstream of neo-liberal thinking, the market mechanism is not capable of offering any solution for dealing rationally with the problem of extra-market phenomena and with consequences of effects in time and space beyond the horizon of individual market agents. The neo-classical attempts to overcome these problems show only poor results in reconciling extra-market and market processes; that is, by 'internalizing' social costs into private, micro-economic cost calculations. Social (ecological) movements of civil society, even with all their exaggerations and irrationalities, provide more appropriate means of identifying costs and benefits of industrial production and of finding answers to the challenges of ecological degradation and industrial hazards than scientific counsel based on mainstream economics.

With regard to distribution there is a second problem to be taken into consideration. Present rational choices demand estimations of future incomes and knowledge of the present discount rate. Given this information, alternatives to the rational decision-making process can be compared. But the rationality of allegedly 'rational' choices obviously depends on the assumption of future incomes. Of course, such incomes are not known, and therefore for reasons of relief they might be calculated under the assumption that present incomes will be based on normal remunerations in future times. Obviously present incomes vary between classes and socio-economic strata within a country, and even more between countries, especially when calculating North–South income differentials.

Average income levels reach from less than a hundred US$100 per capita per year in some states to more than US$20,000 per capita per year in highly industrialized countries (data from the World Bank and UNDP). With comparatively low incomes in Third World countries (for the purpose of calculating environmental costs in the future at a given discount rate) it is quite rational to give the advice to translocate polluting industries and waste disposals to those countries with the lowest incomes. It could even be possible to draw the conclusion that low per capita income is an indicator of countries of being 'underpolluted'. Therefore, it would make sense for 'underpolluted countries' to accept pollution in return for monetary compensation. It was US Treasury Secretary Lawrence Summers, then a World Bank official, who argued, 'I think the economic logic behind dumping a load of toxic waste in the lowest wage country is impeccable and we should face up to that'. (quoted in Buell and De Luca, 1996: 44; for a good discussion of this argument see Harvey, 1996: 366 and *passim*). Since compensation for environmental disasters depends on income levels, hazards are 'less expensive' and in monetary terms 'less damaging' in poor countries than to people in rich countries. This is the lesson given by the responses of economic and political agents to the environmental disaster of Bhopal.

The slippery ground of industrial democracy: the limits of productivity increases

We now reach the *fourth* conclusion. For decades, in the Western capitalist industrial societies, *industrial democracy* has made increasing income the expected norm. Even where income increases have been lacking for several years, as in the United States for a majority of workers during the Reagan era, and as in most Western European countries in the 1990s, growth of income is still considered the rule and real wage cuts the exception. Income increases are easier to achieve if the monetary (and physical) surplus increases through higher productivity. Growth of labour productivity has a positive effect on capital productivity, the rate of profit and economic growth.

International comparison shows that unit labour costs can be reduced by an increase in labour productivity. The competitive position on the world market is, *ceteris paribus*, improved, and therefore the 'competition state' pushes attempts to promote productivity growth. The increase in productivity is more than just a historical mission of the capitalist mode of production, as Marx thought. It forms the common reformist production interest of all actors in capitalist society: trade unions, entrepreneurs and governments (Sinzheimer, 1976). Productivity increase is the starting and finishing point of the social democratic reform policy which has made history in the twentieth century against conservative resistance, on the one hand, and attempts to transcend the system in socialist societies, on the other. The 'pact for productivity' is the basis of the common production interest between wage-labour and capital, between trade unions and entrepreneurs, and of governments, parties and parliaments. The welfare state is practically speaking the substantial materialization of formal democracy. This is rightly and emphatically pointed out by Eric Hobsbawm (1995).

Fordism, therefore, is not only a technical and social innovation. It also includes a new relation to external nature compared with 'pre-Fordist' modes of production and regulation because the system of production and consumption and the mode of social regulation are heavily based on the use of fossil energy (Altvater, 1992). This 'fossilist' aspect is mostly ignored in studies of Fordism which concentrate either on the organization of wage and labour-relations or elaborate upon macroeconomic conditions of market supply and demand, of money and economic policy (this neglect is a critical failing of the regulation school in its approach to the analysis of social development). A crucial variable in these deliberations is unit labour costs and therefore the relationship between wage and productivity increases. The high input of energy, of mineral and agrarian resources, as well as the technical and social system of transforming energy and materials, are the vehicles for a considerable increase in labour productivity, and thus wealth. Wealth status, as Lipset said, forms the material foundation of the formal democratic procedures.

New subjects of the democratic process

But can we identify attempts to create an ecological democracy within the limits of the environmental space (limits of resources and sinks) mentioned at the beginning? We can, and they can be identified with regard to procedures, subjects and forms. The procedures of democracy are partially transnationalized when the territory of the nation state is no longer the space for which the democratic procedure is designed. In the world of nation states, the place of democracy was identical with the territory of nation states. In the course of globalization and the withering away of territorial places of democratic deliberations, democracy is becoming – as mentioned before – a 'placeless' procedure. The territorial placelessness of democracy, however, is substituted by the formation of new communities (and therefore the emergence of 'communitarianism' in times of globalization) and of new *communication nets*. They are not only of a virtual nature in the Internet; democratic deliberations find new places, taking account of the consequences of new limits of new functional spaces.

New political subjects: non-governmental organizations

New subjects enter this new place because both the concept and the reality of the people's sovereignty change with the deterritorialization of democracy. In traditional democracies, political parties are the main mediators between society and state institutions. Multi-party democracy, however, supplies insufficient opportunities for participation when dealing with questions for which the political institutional system of the nation state cannot find sufficient answers. Why participate in democratic procedures of legitimation and representation in the party democracy if economic processes have already pre-decided contested issues which can only be confirmed politically *post factum* and *festum*, and, moreover, can only be marginally changed?

 If decisions have been made according to the criteria of the market, there is no more space for the application of criteria of political justice. The coordinates in space and time of politics are thus by no means congruent with those of the market or with those of material and energy transformations. Therefore civic interest groups are established which articulate locally restricted and temporary issues which in multi-party democracy cannot be sufficiently resolved because of changes of the state *vis-à-vis* globalization and the environmental challenge. There have always been groups fighting against the destruction of 'their' environment, very often (and very radically) in a NIMBY (Not In My Backyard) manner of 'militant particularism' (Harvey, 1996: 19 and *passim*).

The nearly complete capitalization of modern societies and the consequent dramatic devastation of the environment in connection with the so-called 'colonialization of the living worlds' has several far-reaching effects. The once locally limited, temporary and singular protest has expanded to a ubiquitous and permanent feature of modern (and post-modern) societies. The formerly localized, 'single-issue' interest groups are then transformed into permanent social movements and also into parties, on the one hand, and NGOs on the other hand. Parties are still (even in unifying Europe) tied to a (national) territory; new social movements and, even more clearly, NGOs are not. They represent certain social interests (especially the conservation of nature) much more directly and flexibly than parties. They become important actors in the arena of environmental policy. NGOs, as far as they are lobby organizations, and operate in the foreground of (nation) state bureaucracies, have often been co-established by the latter to improve governance in the policy domains of environment or development. Today, NGOs are broadly accepted in international regime formation (Nelson, 1996; Jasanoff, 1997; Altvater *et al.*, 1997).

Form and institution

The forms of democratic processes change when organizations of 'civil society' speak directly without relying on parties, which usually filter the intervention as they carry it into the state institutions. Between market and state, 'civil society' speaks. This concept has a long tradition. It characterizes the space of social life which is not constituted by the state (unlike the *'società politica'* or the 'ideological and repressive state apparatus'). Even this concept of civil society, whether in Hegel's or in Gramsci's tradition, has a territorial dimension which coincides with the nation state. With sovereignty being undermined and *'società politica'* subsequently changing its meaning, the horizon of the *società civile* is expanding beyond the respective national borders as a consequence of the globalization processes already mentioned, particularly those of communication, transport and migration.

The procedures are mostly discursive. Only one thing is indisputable: the 'insight into the necessity' of politics subjected to harsh ecological restrictions facing the limits of the (global) environmental space. The knowledge of the necessity of ecological sustainability thus transforms objective limits into subjectively disputed issues. Moreover, it determines the discourse of the appropriateness of the market, as well as the discourse on the procedures for democracy: what would be the implications of the necessary reduction of the consumption of natural resources by up to 90 per cent (in highly industrialized, rich countries like Germany, according to calculations by the Wuppertal Institute (BUND/Misereor, 1996), for the organization of substantial (and industrial) democracy, for a regime based on the 'pact for production and productivity' between capital and labour (Sinzheimer, 1976)? When ecological limits challenge the substance of participation, which rationality is able to guide formal democratic procedures?

All these questions indicate contested issues. Therefore, it is naive to assume that an international civil society could emerge without heavy social and political conflicts that would alter the relations between market actors, governments of nation states, NGOs, international organizations, etc. Global governance is not an institutional setting for the global regulation of global problems (such as drug traffic, debts, AIDS, migration, etc.), but an arena of conflicts waged by social, economic and political actors. The outcome is uncertain and not predictable, since it is the result of chaotic social, political and economic processes.

Conclusion: globalization, environmental limits, sustainability and human rights

The consequences of globalization and ecological crisis for the democratic question can be sketched as shown in Figure 16.1. The substantial preconditions of modern democracy, such as economic growth, social welfare, institutional modernization and national sovereignty, can no longer claim global validity. The crisis of Western democracy is a challenge for the democratic discourse, but not in such a way that the right of political participation – the most crucial element of democracy – could become debatable. Ecological sustainability needs participation, and globalization can be regulated only through the establishment of bodies of 'global governance'. It has already been stressed that the institutionalization of citizen participation and networks of global governance is a conflict-ridden, rather than harmonious, process.

Although in the past few decades a body of international environmental law has emerged, its applicability still remains restricted:

Compte tenu de l'inventaire sans cesse croissant des problèmes écologiques globaux et des risques technologiques majeurs, il faut malheureusement constater que le droit de l'environnement n'a guère fait ses preuves. Au contraire, la coexistence d'un droit prolifique de l'environnement et d'une crise écologique de plus en plus aigue, montre à quel point le premier baigne dans l'équivoque et l'ambiguïté. [Considering the endlessly growing list of global ecological problems and major technological risks, it must unfortunately be said that the right to the environment has scarcely proved itself. On the contrary, the coexistence of a widespread right to the environment and the increasingly acute ecological crisis demonstrates to what degree the former is bathed in equivocation and ambiguity.]

(Gutwirth, 1993: 75; translation by ed.)

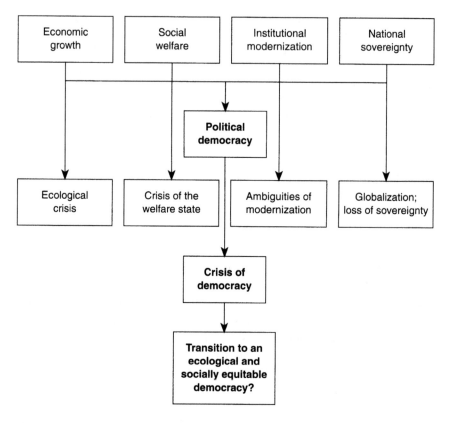

Figure 16.1 Conditions and consequences of globalization, sustainability and democracy

Global governance is 'soft law' in the double sense that the rules are less binding than in other areas of international law, for example in political, military and economic relations between states, and that the jurisdiction of the International Court is limited with regard to admissible cases and plaintiffs. An Environmental World Court, as has been proposed by NGOs today, is nothing but a future ideal even though environmental rules are slowly emerging since the Brundtland Report (1987), the Rio 'Earth Summit' (1992) and the follow-up conferences and institutions. In many nation states at the national level environmental law has been improved. However, the foremost problem today is the global expansion of environmental destruction and the transborder character of territorially limited environmental incidents, such as the pollution of rivers, acid rain, forest fires and so forth.

The human rights system includes environmental rights and can be used as a conceptual framework for environmental law (Pathak, 1992: 205–43; Bierman, 1997). This is because in many parts of the world individuals and people have had to learn that 'structural violence' (in Johan Galtung's terms) today very often consists in ecological degradation and destruction, thus worsening or even destroying the concrete living conditions of human beings. Therefore, environmental degradation, which more and more is a threat to the ecological security of humanity (Timoshenko, 1992: 413–56), requires political and legal regulation in international law as well as in the systems of human rights and of peoples' rights. The debate on sustainability is of the highest importance.

The Brundlandt Report (World Commission on Environment and Development, 1987) came to the conclusion that in the future not only ecological sustainability but also economic development are desirable and possible. In the analyses of the Brundlandt Report no contradiction exists between economic development and environmental sustainability. The Report holds it to be certain that human rights of the second generation (economic and social development) and human rights of the third generation (ecological sustainability and integrity) are reconcilable. In the aftermath of the Rio Conference this assumption has been broadly criticized, especially by ecologists (see Altvater and Mahnkopf, 1996, chs 14–16). In fact, development on the predominant capitalist development path exemplified by the highly developed countries is far from socially equitable and ecologically sustainable. The development model of the industrialized countries is a 'positional good' in the sense discussed earlier. It is good (in some respects) for some, and in the long run bad for all.

Consequently, the question of the development model comes in. This is a challenge for the social sciences. It has already triggered a paradigmatic crisis, and therefore a new discourse is emerging: on the consequences of globalization, i.e. economic boundlessness, and on new boundaries of the environmental space for the democratic question, on contradictions between human rights of the second and the third generation so long as the 'Western' development model is taken for granted. Neo-liberal reforms and the processes of economic globalization have dismantled part of the public regulative capacities necessary to control environmental degradation, as well as those designed to deal with social problems, such as unemployment, poverty or all those problems arising from 'wild urbanization'. Here organizations of so-called civil society, environmental and social movements and NGOs are gaining momentum. They are increasingly becoming part of new structures of *political governance* on a regional and national and even on a global level.

The emerging governance structures are acting as substitutes for traditional state functions, at least in the areas of environmental policy. But the reach of this kind of governance is limited (Broadhead, 1996). It does not extend to 'hard' issues such as the loan conditions of the World Bank, the rules of the World Trade Organization or of military alliances. Nor is it extended to the decisions undertaken in the G7 meetings. The power of NGOs is also limited with regard to private corporations. The reconciliation of human rights of the second and the third generations with the economic world order requires more political regulation guided by the principles of *social justice, political participation and environmental sustainability*. The functioning of the market is good for *economic efficiency* only in a narrow sense; that is, so long as there are no market failures, and so long as there are profitable areas of capital investment. The governments of nation states have to follow the rules of competition and not the necessities of sustainability, unless they are forced to change their attitudes by social subjects. This is a strong reason for the rules of democratic participation and for the necessity to take substantial interests into deliberations on formal procedures. The integration of social subjects into a system of global governance could be interpreted as a first step to some future kind of international regulation which overcomes the fallacies of globalization on the one hand and respects the hard natural constraints of the environmental space on the other hand.

References and further reading

Altvater, E. (1987) *Sachzwang Weltmarkt: Verschuldungkrise, blockierte Industrialisierung, ökologische Gefährdung – der Fall Brasilien*, Hamburg: VSA.
—— (1992) *Der Preis des Wohlstands*, Münster: Westfälisches Dampfboot.
—— (1993) *The Future of the Market*, London: Verso.
—— (1994) 'Die Ordnung rationaler Weltbeherrschung oder: Ein Wettbewerb von Zauberlehrlingen', *PROKLA 95 – Zeitschrift für kritische Sozialwissenschaft*, Jg. 24, No. 1, pp. 186–225.
Altvater, E. and Mahnkopf, B. (1996) *Grenzen der Globalisierung: Ökonomie, Politik, Ökologie in der Weltgesellschaft*, Münster: Westfälisches Dampfboot.
Altvater, E., Brunnengräber, A., Haake, M. and Walk, H. (1997) *Vernetzt und verstrickt: Nicht-Regierungsorganisationen als gesellschaftliche Produktivkraft*, Münster: Westfälisches Dampfboot.
Bealey, F. (1993) 'Capitalism and Democracy', *European Journal of Political Research* 23: 203–23.
Beetham, D. (1993) 'Four Theorems about the Market and Democracy', *European Journal of Political Research* 23: 187–201.
Benvenuti, P., Gargiulo, P. and Lattanzi, F. (eds) (1996) *Nazioni unite e diritti dell'uomo a trent'anni dall'adozione dei patti*, Teramo: Università degli Studi di Teramo.
Bierman, F. (1997) 'Umweltvölkerrecht: Eine Einführung in den Wandel völkerrechtlicher Konzeptionen zur Weltumweltpolitik', Wissenschaftszentrum Berlin, Papers FS II 97–402.
Bobbio, N. (1987) *The Future of Democracy*, Cambridge: Polity Press.
Braudel, F. (1986a) *Sozialgeschichte des 15.–18. Jahrhunderts: Der Handel*, Munich: Kindler.
—— (1986b) *Sozialgeschichte des 15.–18. Jahrhunderts: Aufbruch zur Weltwirtschaft*, Munich: Kindler.
Broadhead, L. A. (1996) 'Commissioning Consent: Globalization and Global Government', *International Journal* 51 (4): 651–68.
Brown, L. *et al.* (1984 etc.) *State of the World: Worldwatch Institute Reports*, New York: W. W. Norton.
Buell, J. and De Luca, T. (1996) *Sustainable Democracy: Individuality and the Politics of the Environment*, London and New Delhi: Thousand Oaks.
BUND/Misereor (ed.) (1996) Wuppertal Institut für Klima, Umwelt, Energie (1996) *Zukunftsfähiges Deutschland: Ein Beitrag zu einer global nachhaltigen Entwicklung*, Basel, Boston and Berlin: Birkhäuser.
Bunker, S. (1985) *Underdeveloping the Amazon*, Urbana and Chicago: University of Illinois Press.

Christoff, P. (1996) 'Ecological Citizens and Ecologically Guided Democracy', in B. Doherty and M. de Geus (eds) *Democracy and Green Political Thought*, London: Routledge, pp. 151–69.

Collier, D. and Levitsky, S. (1997) 'Democracy with Adjectives: Conceptual Innovation in Comparative Research', *World Politics* 49: 430–51.

Commission on Global Governance (1995) *Our Global Neighbourhood*, Oxford: Oxford University Press.

Cox, R. W. (1996) 'A Perspective on Globalization', in J. H. Mittelman (ed.) *Globalization: Critical Reflections*, Boulder, CO: Lynne Rienner, pp. 21–30.

Czerny, P. G. (1996) 'Globalization and Other Stories: The Search for a New Paradigm for International Relations', *International Journal* 51 (4): 617–37.

Dahrendorf, R. (1995) *Quadrare il cerchio: benessere economico, coesione sociale e libertà politiche*, Rome and Bari: Laterza.

Falk, R. (1995) 'Liberalism at the Global Level: The Last of the Independent Commissions?', *Millennium: Journal of International Studies* 24 (3): 563–76.

—— (1997) 'State of Siege: Will Globalization Win Out?', *International Affairs* 73 (1): 123–36.

Georgescu-Roegen, N. (1971) *The Entropy Law and the Economic Process*, Cambridge, MA: Harvard University Press.

Gill, S. (1996) 'Globalization, Democratization and the Politics of Indifference', in J. H. Mittelman (ed.) *Globalization: Critical Reflections*, Boulder, CO: Lynne Rienner, pp. 205–27.

Group of Lisbon [Gruppe von Lissabon] (1997): *Grenzen des Wettbewerbs: Die Globalisierung der Wirtschaft und die Zukunft der Menschheit*, Neuwied: Luchterhand-Verlag.

Gutwirth, S. (1993) 'Autour du contrat naturel', in P. Gérard, F. Ost and M. van de Kerchove (eds) *Images et usages de la nature en droit*, Brussels: Publications des Facultés Universitaires Saint-Louis, pp. 75–131.

Hardin, G. (1968) 'The Tragedy of the Commons', *Science* 162: 1243–48.

Harrod, R. (1958) 'The Possibility of Economic Satiety: Use of Economic Growth for Improving the Quality of Education and Leisure', in *Problems of United States Economic Development*, vol. 1, New York: Committee for Economic Development pp. 207–13.

Harvey, D. (1996) *Justice, Nature and the Geography of Difference*, Oxford: Blackwell.

Held, D. (1991) 'Democracy, the Nation-State and the Global System', *Economy and Society* 20 (2): 138–72.

Hirsch, F. (1980) *Die sozialen Grenzen des Wachstums* (The social limits to growth), Reinbek: Rowohlt.

Hobsbawm, E. (1995) *Das Zeitalter der Extreme: Weltgeschichte des 20. Jahrhunderts* (The age of extremes: world history of the twentieth century), Vienna and Munich: Hanser.

Huber, E., Rueschemeyer, D. and Stephens, J. D. (1997) 'The Paradoxes of Contemporary Democracy: Formal, Participatory, and Social Dimensions', *Comparative Politics*, April: 323–42.

Jasanoff, S. (1997) 'NGOs and the Environment: From Knowledge to Action', *Third World Quarterly* 18 (3): 579–94.

Kapp, K. W. (1950) *The Social Costs of Private Enterprise*, Cambridge, MA: Harvard University Press.

—— (1983) *Social Costs, Economic Development, and Environmental Disruption*, ed. and with an introduction by J.E. Ullmann, Lanham, MD: University Press of America.

Knieper, R. (1991) *Nationale Souveränität: Versuch über Ende und Anfang einer Weltordnung*, Frankfurt: Fischer.

308   *Elmar Altvater*

Kooiman, J. (1993) 'Social Political Governance': Introduction to J. Kooiman (ed.) *Modern Governance*, London: Sage, pp. 1–6.

Lipset, S. M. (1959) 'Some Social Requisites of Democracy: Economic Development and Political Legitimacy', *American Political Science Review*, 53 (1): 69–105.

List, F. (1982 [1841]) *Das nationale System der Politischen Ökonomie*, Stuttgart and Tübingen: Nachdruck Berlin: Akademie-Verlag.

Martinez-Alier, J. and Guha, R. (1997) *Varieties of Environmentalism: Essays North and South*, London: Earthscan.

Meadows, D., Meadows, D., Zahn, E. and Millinger, P. (1973) *Die Grenzen des Wachstums: Bericht des Club of Rome zur Lage der Menschheit*, (The limits to growth: report of the Club of Rome on the situation of humanity), Reinbeck bei Hamburg: Rowohlt.

Messner, D. (1995) *Die Netzwerkgesellschaft: Wirtschaftliche Entwicklung und internationale Wettbewerbsfähigkeit als Probleme gesellschaftlicher Steuerung*, Cologne: Weltforum Verlag.

Munck, G. and Skalnik Leff, C. (1997) 'Modes of Transition and Democratization: South America and Eastern Europe in Comparative Perspective', *Comparative Politics* 29: 343–62.

Narr, W.-D. and Schubert, A. (1994) *Weltökonomie: Die Misere der Politik*, Frankfurt am Main: Suhrkamp.

Nelson, P. J. (1996) 'Internationalising Economic and Environmental Policy: Transnational NGO Networks and the World Bank's Expanding Influence', *Millennium: Journal of International Studies* 25 (3): 605–33.

Neumann, F. (1978 [1950]) 'Die Wissenschaft der Politik in der Demokratie', in A. Söllner (ed.) *Wirtschaft, Staat, Demokratie: Aufsätze 1930–1954*, Frankfurt am Main: Suhrkamp.

O'Connor, J. (1989) 'Political Economy of Ecology of Socialism and Capitalism', *Capitalism, Nature, Socialism* 3: 93–107.

O'Donnell, G., Schmitter, P. C. and Whitehead, L. (1986) *Transitions from Authoritarian Rule*, 3 vols, Baltimore: Johns Hopkins University Press.

Opschoor, J. B. (1992) *Environment, Economics and Sustainable Development*, Groningen: Wolters Noordhoff Publishers.

Panitch, L. (1996) 'Rethinking the Role of the State', in J. H. Mittelman (ed.) *Globalization: Critical Reflections*, Boulder, CO: Lynne Rienner, pp. 83–113.

Pathak (1992) 'The Human Rights System as a Conceptual Framework for Environmental Law', in B. Weiss (ed.) *Environmental Change and International Law*, Tokyo: United Nations University Press, pp. 205–43.

Polanyi, K. (1978 [1944]) *The Great Transformation*, Frankfurt am Main: Suhrkamp.

Poulantzas, N. (1974) *Classes in Contemporary Capitalism*, London: New Left Books.

Przeworski, A. (1994) Paper presented at the International Congress of Political Science, Berlin August 1994; German version: 'Ökonomische und politische Transformationen in Osteuropa: Der aktuelle Stand', in *PROKLA: Zeitschrift für kritische Sozialwissenschaft*, 25. Jg., March 1995, pp. 130–51.

Rhodes, R. A. W. (1996) 'The New Governance: Governing without Government', *Political Studies* 44 (4): 652–67.

Rosenau, J. and Czempiel, E. O. (eds) (1992) *Governance without Government: Order and Change in World Politics*, Cambridge: Cambridge University Press.

Rustow, D. A. (1970) 'Transitions to Democracy: Toward a Dynamic Model', *Comparative Politics* 2 (3): 337–63.

Sassen, S. (1996) 'The Spatial Organization of Information Industries: Implications for the Role of the State', in J. H. Mittelman (ed.) *Globalization: Critical Reflections*, Boulder, CO: Lynne Rienner, pp. 33–52.

Schmitt, C. (1963) *Der Begriff des Politischen: Text von 1932 mit einem Vorwort und drei Corollarien*, Berlin: Duncker & Humboldt.

Schumpeter, J. A. (1976) *Capitalism, Socialism and Democracy*, London: Allen & Unwin.

Shapiro, I. (1996) *Democracy's Place*, Ithaca, NY: Cornell University Press.

Simon, J. A. (1995) 'More People, Greater Wealth, More Resources, Healthier Environment', in J. T. Rourke (ed.) *Taking Sides, Clashing Views on Controversial Issues in World Politics*, New York: Guilford-Dushkin, pp. 297–306.

Sinzheimer, H. (1976): *Arbeitsrecht und Rechtssoziologie: Gesammelte Aufsätze und Reden*, ed. O. Kahn-Freund and T. Ramm, with an introduction by O. Kahn-Freund, 2 vols, Frankfurt and Cologne: Suhrkamp.

Sjölander, C. T. (1996) The Rhetoric of Globalization: What's in a Wor(l)d?', *International Journal*, 51: 603–16.

Sklar, H. (1980*) Trilateralism: The Trilateral Commission and Elite Planning for World Management*, Boston: South End Press.

Spruyt, H. (1994) *The Sovereign State and Its Competitors*, Princeton, NJ: Princeton University Press.

Tilton, J. (ed.) (1992) *Mineral Wealth and Economic Development*, Washington, DC: Resources for the Future.

Timoshenko X. (1992) 'Ecological Security: Response to Global Challenges', in W. Brown (ed.) *Environmental Change and International Law*, Tokyo: United Nations University Press, pp. 413–56.

United Nations Development Programme (UNDP) (1994) *Human Development Report 1994: New Dimensions of Human Security*, New York: Oxford University Press.

Volk, S. (1997) '"Democracy" versus "Democracy"', in *NACLA: Report on the Americas*, 30 (4): 6–12.

Wiener, A. (1996) Editorial, 'Fragmentierte Staatsbürgerschaft', *Zeitschrift für kritische Sozialwissenschaft*, PROKLA 105, 26. Jahrg., No 4, December: 488–96.

World Commission on Environment and Development (1987) *Our Common Future* (the Brundtland Report), Oxford: Oxford University Press.

Index